Trends in Set Theory

CONTEMPORARY MATHEMATICS

752

Trends in Set Theory

Simon Fest
Conference in Honor of Simon Thomas's 60th Birthday
September 15–17, 2017
Rutgers University, Piscataway, New Jersey

Samuel Coskey
Grigor Sargsyan
Editors

AMERICAN MATHEMATICAL SOCIETY

2010 *Mathematics Subject Classification.* Primary 03E15, 03E35, 03E45, 03E60, 03E75, 03C45, 28D05, 05C45 .

Library of Congress Cataloging-in-Publication Data

Names: Coskey, Samuel Gregory, 1982– editor. | Sargsyan, Grigor, 1980– editor.
Title: Trends in set theory : conference in Honor of Simon Thomas's 60th Birthday, Simon Fest, September 15–17, 2017, Rutgers University, Piscataway, New Jersey / Samuel Coskey, Grigor Sargsyan, editors.
Description: Providence, Rhode Island : American Mathematical Society, 2020. | Series: Contemporary mathematics, 0271-4132 ; volume 752 | Includes bibliographical references. |
Identifiers: LCCN 2019051312 | ISBN 9781470443320 (paperback) | ISBN 9781470456092 (ebook)
Subjects: LCSH: Set theory–Congresses. | AMS: Mathematical logic and foundations – Set theory – Descriptive set theory. | Mathematical logic and foundations – Set theory – Consistency and independence results. | Mathematical logic and foundations – Set theory – Inner models, including constructibility, ordinal definability, and core models. | Mathematical logic and foundations – Set theory – Determinacy principles. | Mathematical logic and foundations – Set theory – Applications of set theory. | Mathematical logic and foundations – Model theory – Classification theory, stability and related concepts. | Measure and integration – Measure-theoretic ergodic theory – Measure-preserving transformation | Combinatorics – Graph theory – Eulerian and Hamiltonian graphs.
Classification: LCC QA248 .T67 2020 | DDC 511.3/22–dc23
LC record available at https://lccn.loc.gov/2019051312

Contemporary Mathematics ISSN: 0271-4132 (print); ISSN: 1098-3627 (online)

DOI: https://doi.org/10.1090/conm/752

∞ The paper used in this book is acid-free and falls within the guidelines
established to ensure permanence and durability.
Visit the AMS home page at `https://www.ams.org/`

10 9 8 7 6 5 4 3 2 1 25 24 23 22 21 20

Contents

Preface

The volume is based on the Simon Fest conference, which was held in 2017 in honor of Simon Thomas's contributions to set theory. Thomas's own recent work is in classical descriptive set theory and its connections with other areas of mathematics, such as ergodic theory, C*-algebras and group theory. But Thomas also has great enthusiasm for *good* mathematics in any field. He often teaches courses outside his own research area, and often has lively discussions about many areas of mathematics that are far removed from his own. Therefore we have included articles expositing recent advances in a variety of areas of set theory. We hope that the diversity of the articles that we have selected reflects Thomas's personality.

Thomas was born in 1957 on March 8th in the coastal town of Milford Haven in Wales. As a teenage Welshman he outran all of his fellow countrymen to win an 800m race, but even that didn't distract him from his curiosity in mathematics. He soon left running completely behind and enrolled in Bedford College of the University of London where he studied mathematics between 1975–1978. After completing his degree, he spent the next two years of his life in Denu, Ghana where he taught at St. Paul's Senior High School. He then returned to Bedford College to begin his PhD in mathematics under Wilfrid Hodges, and he completed it in 1983. His dissertation work was on the classification problem for simple locally finite groups. The main theorem of his thesis showed that uncountably categorical simple locally finite groups are exactly those groups that are Chevalley groups over $\overline{\mathbb{F}}_p$.

Following his PhD, Thomas spent his next three years at the University of Freiburg, Yale University and the University of Notre Dame. In 1986, he accepted an assistant professor position at Rutgers, where he has since had an illustrious career. Thomas is now a distinguished professor of mathematics and served as the chair of the mathematics department during 2013–2016. He was an invited speaker at the International Congress of Mathematics in 2006 and was a Fellow of the American Mathematical Society in 2013.

In addition to carrying out his own research programs, Thomas has devoted much of his time at Rutgers to supervising PhD students and to working closely with undergraduate honors students. Many of his undergraduate students have since pursued a PhD in mathematics at outstanding universities around the world.

Thomas often entertains his colleagues and students with his witty British humor. He has been known for masquerading as a certain famous Welsh football player or a famous Welsh poet. During moments when his typically cheerful personality suffers from sadness, Thomas finds refuge in the Statue of Sorrow.Thomas is an avid jazz enthusiast, lover of all things Manhattan (below 23rd Street), and a formidable opponent to those who seek to abuse their political power.

To us, he has been a friend, a teacher, a mentor and a colleague, and we are looking forward to many more years of the same. Thank you, Simon!

Samuel Coskey, Boise State University, Boise, ID
Grigor Sargsyan, Rutgers University, Piscataway, NJ

List of Speakers

Clinton Conley
Carnegie Mellon University

Ilijas Farah
York University

Matt Foreman
University of California, Irvine

Alexander S. Kechris
California Institute of Technology

Andrew Marks
University of California, Los Angeles

Justin Moore
Cornell University

Itay Neeman
University of California, Los Angeles

Saharon Shelah
Rutgers University and Hebrew University of Jerusalem

Robin Tucker-Drob
Texas A&M University

Anush Tserunyan
University of Illinois at Urbana–Champaign

Francis Urquhart
Rutgers University

Contemporary Mathematics
Volume **752**, 2020
https://doi.org/10.1090/conm/752/15127

The open dihypergraph dichotomy and the second level of the Borel hierarchy

Raphaël Carroy, Benjamin D. Miller, and Dániel T. Soukup

Abstract. We show that several dichotomy theorems concerning the second level of the Borel hierarchy are special cases of the $\aleph_0$-dimensional generalization of the open graph dichotomy, which itself follows from the usual proof(s) of the perfect set theorem. Under the axiom of determinacy, we obtain the generalizations of these results from analytic to separable metric spaces. We also consider connections between cardinal invariants and the chromatic numbers of the corresponding dihypergraphs.

Introduction

A topological space is *Polish* if it is second countable and completely metrizable. A subset of a topological space is *Borel* if it is in the smallest σ-algebra containing the open sets, F_σ if it is a countable union of closed sets, G_δ if it is a countable intersection of open sets, and $\mathbf{\Delta}^0_2$ if it is both F_σ and G_δ. A *D-ary relation* on X is a subset of X^D. A *homomorphism* from a D-ary relation R on X to a D-ary relation S on Y is a function $\pi\colon X \to Y$ such that $\pi^D(R) \subseteq S$, a *reduction* of R to S is a homomorphism from R to S that is also a homomorphism from $\sim R$ to $\sim S$, and an *embedding* of R into S is an injective reduction of R to S. The existence of a continuous-embeddability-wise-minimal Borel subset of a Polish space that is not F_σ was established in [**Hur28**].

A set Z *separates* a set X from a set Y if $X \subseteq Z$ and $Y \cap Z = \emptyset$. A subset of a product $\prod_{d \in D} X_d$ is a *hyperrectangle* if it is of the form $\prod_{d \in D} Y_d$, where $Y_d \subseteq X_d$ for all $d \in D$. When $D = 2$, we say that such a set is a *rectangle*. A *D-dimensional dihypergraph* on X is a D-ary relation H on X disjoint from the set $\triangle^D(X)$ of constant sequences. When $D = 2$, we say that a symmetric such set is a *graph*, and use $\triangle(X)$ to denote $\triangle^D(X)$. The *complete D-dimensional hypergraph* on X is the complement of $\triangle^D(X)$, and a set $Y \subseteq X$ is *H-independent* if $H \upharpoonright Y = \emptyset$. A *$\kappa$-coloring* of a D-dimensional dihypergraph H on X is a homomorphism from H to the complete D-dimensional hypergraph on a set Y of cardinality κ, or equivalently, a function $\pi\colon X \to Y$ such that $\pi^{-1}(\{y\})$ is H-independent for all $y \in Y$. The *chromatic number* of H is the least cardinal $\chi(H)$ for which there is a $\chi(H)$-coloring of H. Given a pointclass Γ of subsets of topological spaces, a function $\pi\colon X \to Y$

2010 *Mathematics Subject Classification.* Primary 03E15, 26A21, 28A05, 54H05.
Key words and phrases. Coloring, dichotomy, separation, sigma-continuous.
The authors were supported in part by FWF Grants P28153, P29999, and I1921.

is Γ-*measurable* if $\pi^{-1}(V) \in \Gamma$ for all open sets $V \subseteq Y$. A characterization of the circumstances under which a Borel subset of a product of two Polish spaces can be separated from another subset by a countable union of closed rectangles was given in [**LZ14**], as was a characterization of the circumstances under which a Borel graph on a Polish space admits a $\mathbf{\Delta}^0_2$-measurable $\aleph_0$-coloring. These results were obtained by adapting a classical proof of Hurewicz's original dichotomy theorem. Despite their simplicity, the underlying arguments rely upon finite injury, and do not yield the theorem in its natural generality. Moreover, the resulting minimal objects in some sense fail to be canonical, in that they depend on an arbitrary parameter. Analogous characterizations were obtained at the next level of the Borel hierarchy, although the underlying arguments were quite intricate, as were the corresponding minimal objects. The question as to whether such results hold at still higher levels of the Borel hierarchy remains open.

A topological space is *analytic* if it is a continuous image of a closed subset of $\mathbb{N}^{\mathbb{N}}$. A function is *Borel* if it is Borel-measurable, and *Baire class one* if it is F_σ-measurable. We say that a function $\pi \colon X \to Y$ is *σ-continuous with Γ witnesses* if X is a union of countably-many sets in Γ on which π is continuous. We write $\forall^\infty n \in N\ \phi(n)$ to indicate that $\phi(n)$ holds for all but finitely many $n \in N$. The *eventual domination quasi-order* on $\mathbb{N}^{\mathbb{N}}$ is given by $\mathbf{c} \leq^* \mathbf{d} \leftrightarrow \forall^\infty n \in \mathbb{N}\ \mathbf{c}(n) \leq \mathbf{d}(n)$, and the *dominating number* is the least cardinal $\mathfrak{d}$ for which there is a set $\mathcal{F} \subseteq \mathbb{N}^{\mathbb{N}}$ of cardinality $\mathfrak{d}$ such that $\forall \mathbf{c} \in \mathbb{N}^{\mathbb{N}} \exists \mathbf{d} \in \mathcal{F}\ \mathbf{c} \leq^* \mathbf{d}$. In [**JR82**], it was shown that a function from an analytic metric space to a separable metric space is $\mathbf{\Delta}^0_2$-measurable if and only if it is σ-continuous with closed witnesses. In [**Sol98**], this was derived from the fact that there is a two-element basis, consisting of non-$\mathbf{\Delta}^0_2$-measurable functions, for the family of Baire-class-one functions that are not σ-continuous with closed witnesses. As a corollary, it was also shown that if a Baire-class-one function $\pi \colon X \to Y$ is not σ-continuous with closed witnesses, then $\mathfrak{d}$ is the least cardinal κ for which X is the union of κ-many closed sets on which π is continuous. These results were established using ad-hoc recursive constructions reminiscent of those underlying the Lecomte–Zeleny results. Despite having received quite a bit of attention, the question as to whether these results generalize in their most natural form to higher levels of the Borel hierarchy remains open.

A natural reaction to the aforementioned shortcomings is to seek simpler proofs which more readily adapt to Borel sets of higher complexity. Here we use a straighforward generalization of the perfect set theorem to provide a simple unified explanation of the Kechris–Louveau–Woodin generalization of Hurewicz's original dichotomy theorem (see [**KLW87**]), the level-two results of Lecomte–Zeleny, the Jayne–Rogers theorem, the generalizations of these results from analytic to separable metric spaces under the axiom of determinacy, and the generalization of Solecki's result concerning $\mathfrak{d}$ to Borel functions. Unfortunately, the question as to whether such arguments do indeed generalize to higher levels of the Borel hierarchy remains open.

The underlying dichotomy. We say that a subset of a topological space has a property P of topological spaces if it has property P when equipped with the subspace topology. The *perfect set theorem* for Γ is the statement that for all $X \in \Gamma$, either $|X| \leq \aleph_0$ or there is a continuous injection of $2^{\mathbb{N}}$ into X. In [**Sou17**], this simplest of descriptive set-theoretic dichotomy theorems was established for the pointclass of analytic subsets of Hausdorff spaces. In [**Dav64**], it was generalized to the pointclass of all subsets of analytic Hausdorff spaces under the axiom of determinacy.

The *open graph dichotomy* for Γ is the statement that for all $X \in \Gamma$ and open graphs G on X, either $\chi(G) \leq \aleph_0$ or there is a continuous homomorphism from the *complete graph* $\mathbb{K}_{2^{\mathbb{N}}} = \sim\Delta(2^{\mathbb{N}})$ on $2^{\mathbb{N}}$ to G. In [**Fen93**], this generalization of the perfect set theorem was established for the pointclass of analytic subsets of Hausdorff spaces, as was its generalization to the pointclass of all subsets of analytic Hausdorff spaces under the axiom of determinacy. This was achieved by showing that the usual proofs of the perfect set theorem also yield the open graph dichotomy, giving a sense in which the latter is also among the simplest descriptive set-theoretic dichotomy theorems.

The *box topology* on a product $\prod_{d \in D} X_d$ of topological spaces is the topology generated by the sets of the form $\prod_{d \in D} U_d$, where $U_d \subseteq X_d$ is open for all $d \in D$. Given partial functions $s, t \colon \mathbb{N} \rightharpoonup D$, we write $s \sqsubseteq t$ to indicate that $s = t \upharpoonright \mathrm{dom}(s)$, and we define $\mathcal{N}_t = \{\mathbf{d} \in D^{\mathbb{N}} \mid t \sqsubseteq \mathbf{d}\}$. We use (d) to denote the sequence of length one with value d, and we use $s \frown t$ to denote concatenation of s and t. The *box-open D-dimensional dihypergraph dichotomy* for Γ, or $\mathtt{OGD}^D(\Gamma)$, is the statement that for all $X \in \Gamma$ and box-open D-dimensional dihypergraphs H on X, either $\chi(H) \leq \aleph_0$ or there is a continuous homomorphism from the D-dimensional dihypergraph $\mathbb{H}_{D^{\mathbb{N}}} = \bigcup_{t \in D^{<\mathbb{N}}} \prod_{d \in D} \mathcal{N}_{t \frown (d)}$ on $D^{\mathbb{N}}$ to H.

Organization. In §1, we establish the box-open $\aleph_0$-dimensional dihypergraph dichotomy for the pointclass of analytic subsets of Hausdorff spaces, as well as its generalization to the pointclass of all subsets of analytic Hausdorff spaces under the axiom of determinacy. As in [**Fen93**], we achieve this by showing that the usual proofs of the perfect set theorem easily adapt, giving a sense in which even the latter is among the simplest descriptive set-theoretic dichotomy theorems.

A subset of a topological space is K_σ if it is a union of countably-many compact sets. In §2, we give a first glimpse into how topological properties can be codified into dihypergraphs by showing that the Kechris–Saint Raymond generalizations of Hurewicz's characterization of the circumstances under which a Polish space is K_σ (see [**Kec77, SR75**]) are special cases of the box-open $\aleph_0$-dimensional dihypergraph dichotomy.

In §3, we establish the basic properties of partial compactifications of $\mathbb{N}^{\mathbb{N}}$ arising from the box-open $\aleph_0$-dimensional dihypergraph dichotomy.

In §4, we show that a special case of the box-open $\aleph_0$-dimensional dihypergraph dichotomy yields a characterization of the circumstances under which an analytic subset of a D-fold product of metric spaces can be separated from another subset by a countable union of closed hyperrectangles. As a corollary, we obtain a characterization of the circumstances under which a D-dimensional dihypergraph on an analytic metric space has a $\mathbf{\Delta}^0_2$-measurable $\aleph_0$-coloring. We also obtain the generalizations in which analyticity is weakened to separability under the axiom of determinacy. The Kechris–Louveau–Woodin and level-two Lecomte–Zeleny theorems follow from the special cases of these results where $D = 1$ and $D = 2$.

In §5, we show that a special case of the box-open $\aleph_0$-dimensional dihypergraph dichotomy yields the Jayne–Rogers characterization of the circumstances under which a function from an analytic metric space to a separable metric space is σ-continuous with closed witnesses. We also obtain the generalization in which analyticity is weakened to separability under the axiom of determinacy.

We use $\mathbb{H}'_{\mathbb{N}^{\mathbb{N}}}$ to denote the $\mathbb{N}$-dimensional dihypergraph on $\mathbb{N}^{\mathbb{N}}$ given by $\mathbb{H}'_{\mathbb{N}^{\mathbb{N}}} = \bigcup_{(\mathbf{d},t)\in(\mathbb{N})^{\mathbb{N}}\times\mathbb{N}^{<\mathbb{N}}} \prod_{n\in\mathbb{N}} \mathcal{N}_{t\frown(\mathbf{d}(n))}$, where $(M)^N$ denotes the set of injective elements of M^N. The *covering number* of an ideal $\mathcal{I}$ on a set X is the least cardinal $\mathrm{cov}(\mathcal{I})$ for which X is the union of $\mathrm{cov}(\mathcal{I})$-many sets in $\mathcal{I}$. The *cofinality* of an ideal $\mathcal{I}$ on a set X is the least cardinal $\mathrm{cof}(\mathcal{I})$ for which there is a set $\mathcal{J} \subseteq \mathcal{I}$ of cardinality $\mathrm{cof}(\mathcal{I})$ with the property that $\forall I \in \mathcal{I} \exists J \in \mathcal{J}\ I \subseteq J$. We use $\mathcal{M}$ to denote the σ-ideal on $\mathbb{R}$ consisting of the meager sets, and $\mathcal{N}$ to denote the σ-ideal on $[0,1]$ consisting of the Lebesgue null sets. The *bounding number* is the least cardinal $\mathfrak{b}$ for which there is a set $\mathcal{F} \subseteq \mathbb{N}^{\mathbb{N}}$ of cardinality $\mathfrak{b}$ such that $\forall \mathbf{c} \in \mathbb{N}^{\mathbb{N}} \exists \mathbf{d} \in \mathcal{F}\ \mathbf{d} \not\leq^* \mathbf{c}$. In §6, we note that $\chi(\mathbb{H}_{\mathbb{N}^{\mathbb{N}}}) = \mathrm{cov}(\mathcal{M})$ and $\chi(\mathbb{H}'_{\mathbb{N}^{\mathbb{N}}}) = \mathfrak{d}$. In conjunction with the box-open $\aleph_0$-dimensional dihypergraph dichotomy, the latter fact yields the promised generalization of Solecki's theorem. We also show that if $2 < D < \aleph_0$, then $\chi(\mathbb{H}_{D^{\mathbb{N}}})$ is at least $\mathfrak{b} \cdot \mathrm{cov}(\mathcal{N})$, consistently strictly below $\mathfrak{d}$, and consistently strictly above $\mathrm{cof}(\mathcal{N})$.

Assumptions. We work in the base theory $\mathtt{ZF} + \mathtt{DC}$ throughout, with the exception of the final section, where we work in $\mathtt{ZFC}$ so as to keep our language as transparent as possible.

1. The box-open D-dimensional dihypergraph dichotomy

In this section, we establish two instances of the box-open D-dimensional dihypergraph dichotomy.

THEOREM 1.1. *Suppose that D is a discrete space of cardinality at least two, X is an analytic Hausdorff space, and H is a box-open D-dimensional dihypergraph on X. Then exactly one of the following holds:*

(1) *There is an $\aleph_0$-coloring of H.*
(2) *There is a continuous homomorphism from $\mathbb{H}_{D^{\mathbb{N}}}$ to H.*

PROOF. To see that the two conditions are mutually exclusive, it is sufficient to show that there is no $\aleph_0$-coloring of $\mathbb{H}_{D^{\mathbb{N}}}$. Towards this end, suppose that $X \subseteq D^{\mathbb{N}}$ and $c\colon X \to \mathbb{N}$ is an $\aleph_0$-coloring of $\mathbb{H}_{D^{\mathbb{N}}} \upharpoonright X$, recursively find $d_n \in D$ such that $n \notin c(\mathcal{N}_{(d_m)_{m\leq n}})$ for all $n \in \mathbb{N}$, and observe that $(d_n)_{n\in\mathbb{N}} \notin c^{-1}(\mathbb{N})$, thus $X \neq D^{\mathbb{N}}$.

To see that at least one of the conditions hold, we can assume that $X \neq \emptyset$, in which case there is a continuous surjection $\pi\colon \mathbb{N}^{\mathbb{N}} \to X$. By replacing H with its pullback through π, we can assume that $X = \mathbb{N}^{\mathbb{N}}$. Set $S = \{s \in \mathbb{N}^{<\mathbb{N}} \mid H \upharpoonright \mathcal{N}_s \text{ has an } \aleph_0\text{-coloring}\}$ and $Y = {\sim}\bigcup_{s\in S} \mathcal{N}_s$. Note that if $s \in {\sim}S$, then there is no $\aleph_0$-coloring of $H \upharpoonright (\mathcal{N}_s \cap Y)$, so there exists $(y_d)_{d\in D} \in H \upharpoonright (\mathcal{N}_s \cap Y)$, thus the fact that H is box open yields a sequence $(s_d)_{d\in D} \in ({\sim}S)^D$ of proper extensions of s such that $\prod_{d\in D} \mathcal{N}_{s_d} \subseteq H$. It follows that if there is no $\aleph_0$-coloring of H, then there is a function $f\colon D^{<\mathbb{N}} \to {\sim}S$ such that:

(a) $\forall d \in D \forall t \in D^{<\mathbb{N}}\ f(t) \sqsubset f(t \frown (d))$.
(b) $\forall t \in D^{<\mathbb{N}}\ \prod_{d\in D} \mathcal{N}_{f(t\frown(d))} \subseteq H$.

Condition (a) ensures that we obtain a continuous function $\phi\colon D^{\mathbb{N}} \to Y$ by setting $\phi(\mathbf{d}) = \bigcup_{n\in\mathbb{N}} f(\mathbf{d} \upharpoonright n)$, and condition (b) implies that ϕ is a homomorphism from $\mathbb{H}_{D^{\mathbb{N}}}$ to H. ☐

As in [**Fen93**], the same argument yields the natural generalization to κ-Souslin Hausdorff spaces, and a Cantor–Bendixson-style derivative can be used to avoid the need for copious amounts of choice. However, in order to establish the box-open D-dimensional dihypergraph dichotomy for all subsets of analytic metric spaces under the axiom of determinacy, the natural analog of the game considered in [**Fen93**, §3] is insufficient, as it requires the first player to play $|D|$-many natural numbers in each round. We next show that a slowed-down version of this game can be used instead.

Given a box-open $\mathbb{N}$-dimensional dihypergraph H on $\mathbb{N}^{\mathbb{N}}$ and a set $X \subseteq \mathbb{N}^{\mathbb{N}}$, consider the ω-length two-player game $G(X, H)$ whose n^{th} round consists of the first player playing a sequence $s_n \in \mathbb{N}^{<\mathbb{N}}$ with the property that $i_m = 1 \implies s_m \sqsubset s_n$ for all $m < n$, and then the second playing a natural number $i_n < 2$. There are two types of runs of $G(X, H)$, depending on whether the set $N = \{n \in \mathbb{N} \mid i_n = 1\}$ is finite or infinite. In the former case, the first player wins if and only if $\prod_{n \in \mathbb{N}} \mathcal{N}_{s_{\max(N)+1+n}} \subseteq H$. In the latter, the first player wins if and only if $\bigcup_{n \in N} s_n \in X$.

PROPOSITION 1.2. *Suppose that H is a box-open $\mathbb{N}$-dimensional dihypergraph on $\mathbb{N}^{\mathbb{N}}$ and $X \subseteq \mathbb{N}^{\mathbb{N}}$.*

(1) *The first player has a winning strategy in $G(X, H)$ if and only if there is a continuous homomorphism from $\mathbb{H}_{\mathbb{N}^{\mathbb{N}}}$ to $H \upharpoonright X$.*
(2) *The second player has a winning strategy in $G(X, H)$ if and only if there is an $\aleph_0$-coloring of $H \upharpoonright X$.*

PROOF. Let S denote the set of sequences $s \in 2^{<\mathbb{N}}$ that do not end in zero. Let $[\mathbb{N}]^{\mathbb{N}}$ denote the set of strictly increasing elements of $\mathbb{N}^{\mathbb{N}}$, define $\pi_{\mathbb{N}^{\mathbb{N}}, [\mathbb{N}]^{\mathbb{N}}} \colon \mathbb{N}^{\mathbb{N}} \to [\mathbb{N}]^{\mathbb{N}}$ by $\pi_{\mathbb{N}^{\mathbb{N}}, [\mathbb{N}]^{\mathbb{N}}}(b)(n) = n + \sum_{m \leq n} b(m)$, and define $\pi_{[\mathbb{N}]^{\mathbb{N}}, 2^{\mathbb{N}}} \colon [\mathbb{N}]^{\mathbb{N}} \to 2^{\mathbb{N}}$ by $\pi_{[\mathbb{N}]^{\mathbb{N}}, 2^{\mathbb{N}}}(b) = \mathbf{1}_{b(\mathbb{N})}$. Then the function $\pi_{\mathbb{N}^{\mathbb{N}}, 2^{\mathbb{N}}} = \pi_{[\mathbb{N}]^{\mathbb{N}}, 2^{\mathbb{N}}} \circ \pi_{\mathbb{N}^{\mathbb{N}}, [\mathbb{N}]^{\mathbb{N}}}$ is a homeomorphism from $\mathbb{N}^{\mathbb{N}}$ to the space $C = \{c \in 2^{\mathbb{N}} \mid |\text{supp } c| = \aleph_0\}$, in addition to being an isomorphism of $\mathbb{H}_{\mathbb{N}^{\mathbb{N}}}$ with the restriction of the box-open $\mathbb{N}$-dimensional dihypergraph $H_{2^{\mathbb{N}}} = \bigcup_{s \in S} \prod_{n \in \mathbb{N}} \mathcal{N}_{s \frown (0)^n \frown (1)}$ to C.

Suppose that τ is a winning strategy for the first player in $G(X, H)$, and define $\phi \colon C \to \mathbb{N}^{\mathbb{N}}$ by $\phi(c) = \bigcup_{n \in \text{supp } c} \tau(c \upharpoonright n)$. To see that ϕ is a homomorphism from $H_{2^{\mathbb{N}}} \upharpoonright C$ to H, note that if $(c_n)_{n \in \mathbb{N}} \in H_{2^{\mathbb{N}}} \upharpoonright C$, then there exists $s \in S$ for which $(c_n)_{n \in \mathbb{N}} \in \prod_{n \in \mathbb{N}} \mathcal{N}_{s \frown (0)^n \frown (1)}$, in which case $(\phi(c_n))_{n \in \mathbb{N}} \in \prod_{n \in \mathbb{N}} \mathcal{N}_{\tau(s \frown (0)^n)}$, so the fact that the first player wins runs of $G(X, H)$ of the first type ensures that $(\phi(c_n))_{n \in \mathbb{N}} \in H$. The fact that the first player wins runs of $G(X, H)$ of the second type implies that $\phi(C) \subseteq X$.

Conversely, suppose that $\phi \colon C \to X$ is a continuous homomorphism from $H_{2^{\mathbb{N}}} \upharpoonright C$ to H. For each non-empty sequence $s \in S$, let s^- denote the immediate predecessor of s. We will recursively construct functions $\sigma \colon 2^{<\mathbb{N}} \to S$ and $\tau \colon 2^{<\mathbb{N}} \to \mathbb{N}^{<\mathbb{N}}$ such that:

(a) $\forall s \in S \ \prod_{n \in \mathbb{N}} \mathcal{N}_{\tau(s \frown (0)^n)} \subseteq H$.
(b) $\forall s \in 2^{<\mathbb{N}} \ \phi(\mathcal{N}_{\sigma(s)}) \subseteq \mathcal{N}_{\tau(s)}$.
(c) $\forall n \in \mathbb{N} \forall s \in S \setminus \{\emptyset\} \ \tau(s^-) \sqsubset \tau(s \frown (0)^n)$.
(d) $\forall n \in \mathbb{N} \forall s \in S \setminus \{\emptyset\} \ \sigma(s^-) \frown (0)^n \frown (1) \sqsubseteq \sigma(s \frown (0)^n)$.

Suppose that $s \in S$ and we have already found $\sigma(r)$ and $\tau(r)$ for all $r \sqsubset s$. If $s = \emptyset$, then set $u_s = v_s = \emptyset$, and otherwise define $u_s = \sigma(s^-)$ and $v_s = \tau(s^-)$. As $(u_s \frown (0)^n \frown (1)^\infty)_{n \in \mathbb{N}} \in H_{2^{\mathbb{N}}} \upharpoonright C$, there are strict extensions $\tau(s \frown (0)^n) \sqsubset \phi(u_s \frown (0)^n \frown (1)^\infty)$ of v_s such that $\prod_{n \in \mathbb{N}} \mathcal{N}_{\tau(s \frown (0)^n)} \subseteq H$, as well as positive integers $k_{n,s}$ such that $\phi(\mathcal{N}_{u_s \frown (0)^n \frown (1)^{k_{n,s}}}) \subseteq \mathcal{N}_{\tau(s \frown (0)^n)}$ for all $n \in \mathbb{N}$. We complete the construction by setting $\sigma(s \frown (0)^n) = u_s \frown (0)^n \frown (1)^{k_{n,s}}$ for all $n \in \mathbb{N}$. To see that τ is a winning strategy for the first player in $G(X, H)$, note that the first player wins runs of $G(X, H)$ of the first type by condition (a), whereas the other conditions ensure that if $c \in C$, then $\phi(\bigcup_{n \in \mathrm{supp}\ c} \sigma(c \upharpoonright n)) = \bigcup_{n \in \mathrm{supp}\, c} \tau(c \upharpoonright n)$, thus the first player wins runs of $G(X, H)$ of the second type.

Suppose now that τ is a winning strategy for the second player in $G(X, H)$. Let T denote the set of partial runs of $G(X, H)$ against τ for which $(\tau(t \upharpoonright \{0, \dots, n\}))_{n<|t|} \in S$, and associate with each $t \in T$ the set $X_t = \{x \in X \mid t \neq \emptyset \implies t(|t|-1) \sqsubseteq x\}$. If $(x_n)_{n \in \mathbb{N}} \in H \upharpoonright X_t$, then there are sequences $t_n \sqsubset x_n$ with the property that $\prod_{n \in \mathbb{N}} \mathcal{N}_{t_n} \subseteq H$ and $t \neq \emptyset \implies \forall n \in \mathbb{N}\ t(|t|-1) \sqsubset t_n$, so the fact that the second player wins runs of $G(X, H)$ of the first type therefore yields $n \in \mathbb{N}$ for which $t \frown (t_m)_{m \leq n} \in T$ and $x_n \in X_{t \frown (t_m)_{m \leq n}}$. In particular, it follows that the sets of the form $X_t \setminus \bigcup_{t \sqsubset u, u \in T} X_u$ are H-independent. As the fact that the second player wins runs of $G(X, H)$ of the second type ensures that every $x \in X$ appears in a set of the latter form, it follows that there is an $\aleph_0$-coloring of $H \upharpoonright X$.

Conversely, suppose that $c\colon X \to \mathbb{N}$ is an $\aleph_0$-coloring of $H \upharpoonright X$, and let τ be the strategy for the second player in $G(X, H)$ in which 0 is played in the nth round of the game if and only if $c^{-1}(\{k_n\}) \cap \mathcal{N}_{s_n} \neq \emptyset$, where $k_n = |\{m < n \mid i_m = 1\}|$. The fact that the sets of the form $c^{-1}(\{k\})$ are H-independent ensures that the second player wins runs of $G(X, H)$ of the first type, while the fact that $X \subseteq c^{-1}(\mathbb{N})$ implies that the second player wins runs of $G(X, H)$ of the second type. □

In particular, we obtain the following.

THEOREM 1.3 (AD). *Suppose that Y is an analytic Hausdorff space, H is a box-open $\mathbb{N}$-dimensional dihypergraph on Y, and $X \subseteq Y$. Then exactly one of the following holds:*

1. *There is an $\aleph_0$-coloring of $H \upharpoonright X$.*
2. *There is a continuous homomorphism from $\mathbb{H}_{\mathbb{N}^{\mathbb{N}}}$ to $H \upharpoonright X$.*

PROOF. As noted in the proof of Theorem 1.1, conditions (1) and (2) are mutually exclusive, so it is sufficient to show that at least one of them holds. We can also assume that $Y \neq \emptyset$, in which case there is a continuous surjection $\phi\colon \mathbb{N}^{\mathbb{N}} \to Y$. By replacing H and X with their pullbacks through π, we can assume that $Y = \mathbb{N}^{\mathbb{N}}$. But AD and Proposition 1.2 ensure that one of the two conditions holds. □

2. K_σ sets

Given a topological space X, let H_X denote the $\mathbb{N}$-dimensional dihypergraph on X consisting of all injective sequences $(x_n)_{n \in \mathbb{N}}$ of elements of X with no convergent subsequence.

PROPOSITION 2.1. *Suppose that X is a metric space.*

1. *The dihypergraph H_X is box open.*
2. *There is an $\aleph_0$-coloring of the restriction of H_X to a set $Y \subseteq X$ if and only if Y is contained in a K_σ subset of X.*
3. *A continuous function $\phi\colon \mathbb{N}^{\mathbb{N}} \to X$ is a homomorphism from $\mathbb{H}_{\mathbb{N}^{\mathbb{N}}}$ to H_X if and only if it is an injective closed map.*

PROOF. To see (1), note that if $(x_n)_{n \in \mathbb{N}} \in H_X$, then there exist positive real numbers $\epsilon_n \to 0$ such that $\rho_X(x_m, x_n) \geq 2\epsilon_n$ for all natural numbers $m \neq n$, in which case $\prod_{n \in \mathbb{N}} \mathcal{B}_X(x_n, \epsilon_n) \subseteq H_X$.

To see (2), it is sufficient to observe that a set $Y \subseteq X$ is H_X-independent if and only if its closure is compact.

To see (3), note first that if ϕ is an injective closed map, then the fact that each sequence $(\mathbf{d}_n)_{n \in \mathbb{N}} \in \mathbb{H}_{\mathbb{N}^\mathbb{N}}$ is an injective enumeration of a closed discrete set ensures that the same holds of $(\phi(\mathbf{d}_n))_{n \in \mathbb{N}}$. Conversely, suppose that ϕ is a homomorphism from $\mathbb{H}_{\mathbb{N}^\mathbb{N}}$ to H_X. The fact that H_X consists solely of injective sequences easily implies that ϕ is injective. To see that ϕ is a closed map, it is sufficient to show that every sequence $(\mathbf{d}_n)_{n \in \mathbb{N}}$ of elements of $\mathbb{N}^\mathbb{N}$ for which $(\phi(\mathbf{d}_n))_{n \in \mathbb{N}}$ converges has a convergent subsequence. If there exists $\mathbf{d} \in \mathbb{N}^\mathbb{N}$ such that $\mathbf{d}_n(i) < \mathbf{d}(i)$ for all $i, n \in \mathbb{N}$, then the compactness of $\prod_{i \in \mathbb{N}} \mathbf{d}(i)$ yields the desired subsequence. So suppose, towards a contradiction, that there does not exist such a $\mathbf{d}$. Then there is a least $k \in \mathbb{N}$ for which $\{\mathbf{d}_n(k) \mid n \in \mathbb{N}\}$ is infinite. By passing to a subsequence, we can assume that for all distinct $m, n \in \mathbb{N}$, the sequences $\mathbf{d}_m$ and $\mathbf{d}_n$ differ from one another for the first time on their k^{th} coordinates. By passing to a further subsequence, we can assume that $(\mathbf{d}_n)_{n \in \mathbb{N}}$ is a subsequence of an element of $\mathbb{H}_{\mathbb{N}^\mathbb{N}}$, so $(\phi(\mathbf{d}_n))_{n \in \mathbb{N}}$ is a subsequence of an element of H_X, contradicting the fact that it converges. □

In particular, we obtain the following.

THEOREM 2.2 ($\mathtt{OGD}^\mathbb{N}(\Gamma)$). *Suppose that X is a metric space and $Y \subseteq X$ is in Γ. Then exactly one of the following holds:*

1. *The set Y is contained in a K_σ subset of X.*
2. *There is a closed continuous injection $\phi \colon \mathbb{N}^\mathbb{N} \to X$ with the property that $\phi(\mathbb{N}^\mathbb{N}) \subseteq Y$.*

PROOF. This follows from $\mathtt{OGD}^\mathbb{N}(\Gamma)$ and Proposition 2.1. □

Theorems 1.1 and 1.3 together with the special cases of Theorem 2.2 where Γ is either the pointclass of analytic subsets of metric spaces or the pointclass of all subsets of analytic metric spaces yield the Kechris–Saint Raymond generalizations of Hurewicz's characterization of the circumstances under which a Polish space is K_σ.

3. Partial compactifications

For each topological space X and discrete set $D \subseteq X$, endow the set $\mathrm{Ext}_X(D^\mathbb{N}) = D^\mathbb{N} \cup \{t \frown (x) \mid t \in D^{<\mathbb{N}} \text{ and } x \in {\sim}D\}$ with the topology generated by the sets $\mathcal{N}_{t,U} = \{\mathbf{d} \in \mathrm{Ext}_X(D^\mathbb{N}) \mid t \sqsubset \mathbf{d} \text{ and } \mathbf{d}(|t|) \in U\}$, where $t \in D^{<\mathbb{N}}$ and $U \subseteq X$ is open. As we shall later see, such spaces arise naturally in applications of the box-open D-dimensional dihypergraph dichotomy.

PROPOSITION 3.1. *Suppose that X is a compact space and $D \subseteq X$ is discrete and open. Then $\mathrm{Ext}_X(D^\mathbb{N})$ is compact.*

PROOF. Set $\mathcal{U} = \{\mathcal{N}_{t,U} \mid t \in D^{<\mathbb{N}} \text{ and } U \subseteq X \text{ is open}\}$, and for each sequence $t \in D^{<\mathbb{N}}$ and family $\mathcal{V} \subseteq \mathcal{U}$, let $\mathcal{V}_t$ denote the family of open sets $V \subseteq X$ for which $\mathcal{N}_{t,V} \in \mathcal{V}$. The main observation is as follows:

LEMMA 3.2. *Suppose that* $t \in D^{<\mathbb{N}}$ *and* $\mathcal{V} \subseteq \mathcal{U}$ *covers* $\mathcal{N}_{t,X}$. *If for all* $d \in D$ *there is a finite set* $\mathcal{F}_d \subseteq \mathcal{V}$ *covering* $\mathcal{N}_{t \frown (d),X}$, *then there is a finite set* $\mathcal{F} \subseteq \mathcal{V}$ *covering* $\mathcal{N}_{t,X}$.

PROOF. We can assume that there do not exist $s \sqsubset t$ and an open set $U \subseteq X$ for which $t(|s|) \in U$ and $\mathcal{N}_{s,U} \in \mathcal{V}$. Then there is a finite set $\mathcal{F} \subseteq \mathcal{V}_t$ covering $\sim D$, in which case the set $F = \sim \bigcup \mathcal{F}$ is compact and contained in D, thus finite. But $\mathcal{F} \cup \bigcup_{d \in F} \mathcal{F}_d$ covers $\mathcal{N}_{t,X}$. □

Now observe that if $\mathcal{V} \subseteq \mathcal{U}$ is a cover of $\mathrm{Ext}_X(D^{\mathbb{N}})$ with no finite subcover, then by recursively applying the contrapositive of Lemma 3.2, we obtain a sequence $\mathbf{d} \in D^{\mathbb{N}}$ such that for no $n \in \mathbb{N}$ is there a finite subset of $\mathcal{V}$ covering $\mathcal{N}_{\mathbf{d} \restriction n,X}$, contradicting the fact that $\mathbf{d} \in \bigcup \mathcal{V}$. □

PROPOSITION 3.3. *Suppose that* X *is a (complete) ultrametric space and* $D \subseteq X$ *is discrete and open. Then* $\mathrm{Ext}_X(D^{\mathbb{N}})$ *has a compatible (complete) ultrametric.*

PROOF. By replacing ρ_X with $\rho_X/(1+\rho_X)$, we can assume that $\rho_X < 1$. Fix real numbers $\epsilon_d > 0$ such that $\sup_{d \in D} \epsilon_d < 1$ and $\mathcal{B}(d, \epsilon_d) = \{d\}$ for all $d \in D$, and define $\rho \colon \mathrm{Ext}_X(D^{\mathbb{N}}) \times \mathrm{Ext}_X(D^{\mathbb{N}}) \to \mathbb{R}$ by

$$\rho(\mathbf{c}, \mathbf{d}) = \begin{cases} 0 & \text{if } \mathbf{c} = \mathbf{d} \text{ and} \\ \rho_X(\mathbf{c}(n), \mathbf{d}(n)) \prod_{m<n} \epsilon_{\mathbf{c}(m)} & \text{otherwise,} \end{cases}$$

where $n = n(\mathbf{c}, \mathbf{d})$ is the least natural number for which $\mathbf{c}(n) \neq \mathbf{d}(n)$.

To establish that ρ is an ultrametric, it is sufficient to show that if $\mathbf{b}, \mathbf{c}, \mathbf{d} \in \mathrm{Ext}_X(D^{\mathbb{N}})$ are distinct, then $\rho(\mathbf{b}, \mathbf{d}) \leq \max\{\rho(\mathbf{b}, \mathbf{c}), \rho(\mathbf{c}, \mathbf{d})\}$. Setting $n = \max\{n(\mathbf{b}, \mathbf{c}), n(\mathbf{c}, \mathbf{d})\}$, there are three cases to check:

(1) If $n(\mathbf{b}, \mathbf{d}) < n$, then $\rho(\mathbf{b}, \mathbf{d}) \in \{\rho(\mathbf{b}, \mathbf{c}), \rho(\mathbf{c}, \mathbf{d})\}$.
(2) If $n(\mathbf{b}, \mathbf{d}) = n$, then $n(\mathbf{b}, \mathbf{c}) = n(\mathbf{b}, \mathbf{d}) = n(\mathbf{c}, \mathbf{d})$, so

$$\begin{aligned} \rho(\mathbf{b}, \mathbf{d}) &= \rho_X(\mathbf{b}(n), \mathbf{d}(n)) \textstyle\prod_{m<n} \epsilon_{\mathbf{b}(m)} \\ &\leq \max\{\rho_X(\mathbf{b}(n), \mathbf{c}(n)), \rho_X(\mathbf{c}(n), \mathbf{d}(n))\} \textstyle\prod_{m<n} \epsilon_{\mathbf{b}(m)} \\ &= \max\{\rho(\mathbf{b}, \mathbf{d}), \rho(\mathbf{c}, \mathbf{d})\}. \end{aligned}$$

(3) If $n(\mathbf{b}, \mathbf{d}) > n$, then $\rho(\mathbf{b}, \mathbf{d}) < \prod_{m \leq n} \epsilon_{\mathbf{b}(m)} \leq \rho(\mathbf{b}, \mathbf{c})$.

To see that the topology generated by ρ is contained in that of $\mathrm{Ext}_X(D^{\mathbb{N}})$, suppose that $\mathbf{d} \in \mathrm{Ext}_X(D^{\mathbb{N}})$ and $\epsilon > 0$. If $\mathbf{d} \in D^{\mathbb{N}}$, then there exists $n \in \mathbb{N}$ sufficiently large that $\prod_{m<n} \epsilon_{\mathbf{d}(m)} \leq \epsilon$, in which case $\mathcal{N}_{\mathbf{d} \restriction n,X}$ is an open neighborhood of $\mathbf{d}$ contained in $\mathcal{B}(\mathbf{d}, \epsilon)$. Otherwise, there exist $t \in D^{<\mathbb{N}}$ and $x \in \sim D$ for which $\mathbf{d} = t \frown (x)$, in which case $\mathcal{N}_{t,\mathcal{B}_X(x,\epsilon)}$ is an open neighborhood of $\mathbf{d}$ contained in $\mathcal{B}(\mathbf{d}, \epsilon)$.

To see that the topology of $\mathrm{Ext}_X(D^{\mathbb{N}})$ is contained in that generated by ρ, suppose that $t \in D^{<\mathbb{N}}$, $U \subseteq X$ is open, and $\mathbf{d} \in \mathcal{N}_{t,U}$, and fix $0 < \epsilon < 1$ such that $\mathcal{B}_X(\mathbf{d}(|t|), \epsilon) \subseteq U$. Then $\mathcal{B}(\mathbf{d}, \epsilon \prod_{n<|t|} \epsilon_{t(n)}) \subseteq \mathcal{N}_{t,U}$.

To see that the completeness of ρ_X yields that of ρ, suppose that $(\mathbf{d}_k)_{k \in \mathbb{N}}$ is an injective Cauchy sequence of elements of $\mathrm{Ext}_X(D^{\mathbb{N}})$. If there exists $\mathbf{d} \in D^{\mathbb{N}}$ with the property that $\mathbf{d}_k(n) \to \mathbf{d}(n)$ for all $n \in \mathbb{N}$, then $\mathbf{d}_k \to \mathbf{d}$. Otherwise, there is a sequence $t \in D^{<\mathbb{N}}$ of maximal length such that $\mathbf{d}_k(n) \to t(n)$ for all $n < |t|$. By passing to a subsequence of $(\mathbf{d}_k)_{k \in \mathbb{N}}$, we can assume that $n(\mathbf{d}_j, \mathbf{d}_k) = |t|$, thus $\rho(\mathbf{d}_j, \mathbf{d}_k) = \rho_X(\mathbf{d}_j(|t|), \mathbf{d}_k(|t|)) \prod_{n<|t|} \epsilon_{t(n)}$, for all distinct $j, k \in \mathbb{N}$. It follows that

$(\mathbf{d}_k(|t|))_{k \in \mathbb{N}}$ is Cauchy, and therefore convergent, thus $(\mathbf{d}_k)_{k \in \mathbb{N}}$ is convergent as well. $\square$

Let $\mathrm{Cnvg}(X)$ denote the set of convergent sequences $(x_n)_{n \in \mathbb{N}}$ of elements of X, and let $\mathrm{Cnvg}_X(D^{\mathbb{N}})$ denote the $\mathbb{N}$-ary relation on $D^{\mathbb{N}}$ given by $\mathrm{Cnvg}_X(D^{\mathbb{N}}) = \bigcup_{(\mathbf{d},t) \in (\mathrm{Cnvg}(X) \cap D^{\mathbb{N}}) \times D^{<\mathbb{N}}} \prod_{n \in \mathbb{N}} \mathcal{N}_{t \frown (\mathbf{d}(n))}$. In our applications of the box-open D-dimensional dihypergraph dichotomy, the following fact will yield extensions of homomorphisms between dihypergraphs.

PROPOSITION 3.4. *Suppose that X and Y are metric spaces, $D \subseteq X$ is dense, discrete, and open, and $\phi \colon D^{\mathbb{N}} \to Y$ is a continuous homomorphism from* $\mathrm{Cnvg}_X(D^{\mathbb{N}})$ *to* $\mathrm{Cnvg}(Y)$. *Then there is a continuous extension of ϕ to* $\mathrm{Ext}_X(D^{\mathbb{N}})$.

PROOF. Given a point $x \in X$, we say that a sequence $(X_n)_{n \in \mathbb{N}}$ of subsets of X *converges* to x, or $X_n \to x$, if for every open neighborhood $U \subseteq X$ of x, there exists $n \in \mathbb{N}$ for which $\bigcup_{m \geq n} X_m \subseteq U$.

LEMMA 3.5. *Suppose that $t \in D^{<\mathbb{N}}$ and $x \in {\sim}D$. Then there exists $y_{t,x} \in Y$ such that $\mathbf{d}(n) \to x \implies \phi(\mathcal{N}_{t \frown (\mathbf{d}(n))}) \to y_{t,x}$ for all $\mathbf{d} \in D^{\mathbb{N}}$.*

PROOF. Note that if $\mathbf{d} \in D^{\mathbb{N}}$, $\mathbf{d}(n) \to x$, and $y_n \in \phi(\mathcal{N}_{t \frown (\mathbf{d}(n))})$ for all $n \in \mathbb{N}$, then there exists $y_{t,x} \in Y$ for which $y_n \to y_{t,x}$. If there exists $\mathbf{c} \in D^{\mathbb{N}}$ such that $\mathbf{c}(n) \to x$ and $\phi(\mathcal{N}_{t \frown (\mathbf{c}(n))}) \not\to y_{t,x}$, then there exist an infinite set $N \subseteq \mathbb{N}$, an open neighborhood $V \subseteq Y$ of $y_{t,x}$, and points $y_n \in \phi(\mathcal{N}_{t \frown (\mathbf{c}(n))}) \setminus V$ for all $n \in N$. By thinning down N, we can assume that it is co-infinite. Set $\mathbf{b} = (\mathbf{c} \upharpoonright N) \cup (\mathbf{d} \upharpoonright {\sim}N)$, and observe that $\mathbf{b}(n) \to x$ but $(\phi(\mathbf{b}(n)))_{n \in \mathbb{N}}$ does not converge, a contradiction. $\square$

To see that the extension given by $\overline{\phi}(t \frown (x)) = y_{t,x}$ is continuous, suppose that $\mathbf{d} \in \mathrm{Ext}_X(D^{\mathbb{N}})$ and $V \subseteq Y$ is an open neighborhood of $\overline{\phi}(\mathbf{d})$, and fix an open neighborhood $W \subseteq Y$ of $\overline{\phi}(\mathbf{d})$ whose closure is contained in V. If $\mathbf{d} \in D^{\mathbb{N}}$, then the continuity of ϕ yields $n \in \mathbb{N}$ for which $\phi(\mathcal{N}_{\mathbf{d} \upharpoonright n}) \subseteq W$, in which case $\overline{\phi}(\mathcal{N}_{\mathbf{d} \upharpoonright n, X}) \subseteq \overline{\phi(\mathcal{N}_{\mathbf{d} \upharpoonright n})} \subseteq \overline{W} \subseteq V$. Otherwise, there exists $t \in D^{<\mathbb{N}}$ and $x \in {\sim}D$ for which $\mathbf{d} = t \frown (x)$, so Lemma 3.5 yields an open neighborhood $U \subseteq X$ of x for which $\phi(\mathcal{N}_{t,U} \cap D^{\mathbb{N}}) \subseteq W$, in which case $\overline{\phi}(\mathcal{N}_{t,U}) \subseteq \overline{\phi(\mathcal{N}_{t,U})} \subseteq \overline{W} \subseteq V$. $\square$

4. Countable unions of hyperrectangles

Given sequences $(X_d)_{d \in D}$ and $(Y_d)_{d \in D}$ of topological spaces, a *hyperrectangular homomorphism* from a pair (R_X, S_X) of subsets of $\prod_{d \in D} X_d$ to a pair (R_Y, S_Y) of subsets of $\prod_{d \in D} Y_d$ is a function ϕ of the form $\prod_{d \in D} \phi_d$, where $\phi_d \colon \mathrm{proj}_d(R_X \cup S_X) \to Y_d$ for all $d \in D$, such that $\phi(R_X) \subseteq R_Y$ and $\phi(S_X) \subseteq S_Y$.

Let $\mathbb{N}_* = \mathbb{N} \cup \{\infty\}$ denote the one-point compactification of $\mathbb{N}$, and $\mathbb{H}_{(D \times \mathbb{N})^{\mathbb{N}},\infty}$ the D-dimensional dihypergraph on $\mathrm{Ext}_{D \times \mathbb{N}_*}((D \times \mathbb{N})^{\mathbb{N}})$ consisting of all sequences $(t \frown ((d, \infty)))_{d \in D}$, where $t \in (D \times \mathbb{N})^{<\mathbb{N}}$.

Given a sequence $(X_d)_{d \in D}$ and disjoint sets $R, S \subseteq \prod_{d \in D} X_d$, let $H_{R,S}$ denote the $(D \times \mathbb{N})$-dimensional dihypergraph on R consisting of all sequences $((x_{c,d,n})_{c \in D})_{(d,n) \in D \times \mathbb{N}}$ of elements of R with the property that $\overline{x}_d = \lim_{n \to \infty} x_{d,d,n}$ exists for all $d \in D$ and $(\overline{x}_d)_{d \in D} \in S$.

PROPOSITION 4.1. *Suppose that D is a discrete set, $(X_d)_{d \in D}$ is a sequence of metric spaces, and $R, S \subseteq \prod_{d \in D} X_d$ are disjoint.*

(1) *The $(D \times \mathbb{N})$-dimensional dihypergraph $H_{R,S}$ is box open.*
(2) *There is an $\aleph_0$-coloring of $H_{R,S}$ if and only if there is a countable union of closed hyperrectangles separating R from S.*
(3) *There is a continuous homomorphism from $\mathbb{H}_{(D\times\mathbb{N})^{\mathbb{N}}}$ to $H_{R,S}$ if and only if there is a continuous hyperrectangular homomorphism from $(\triangle^D((D \times \mathbb{N})^{\mathbb{N}}), \mathbb{H}_{(D\times\mathbb{N})^{\mathbb{N}},\infty})$ to (R, S).*

PROOF. To see (1), note that if $((x_{c,d,n})_{c\in D})_{(d,n)\in D\times\mathbb{N}} \in H_{R,S}$, $\epsilon_n \to 0$, and $U_{d,n} = \{(x_c)_{c\in D} \in R \mid \rho_{X_d}(x_d, x_{d,d,n}) < \epsilon_n\}$ for all $(d,n) \in D \times \mathbb{N}$, then $\prod_{(d,n)\in D\times\mathbb{N}} U_{d,n} \subseteq H_{R,S}$.

To see (2), note first that if $Q \subseteq R$ and $(\overline{x}_d)_{d\in D} \in \prod_{d\in D} \overline{\mathrm{proj}_d(Q)}$, then there are sequences $(x_{d,n})_{n\in\mathbb{N}}$ of elements of $\mathrm{proj}_d(Q)$ such that $x_{d,n} \to \overline{x}_d$ for all $d \in D$, so there are sequences $(x_{c,d,n})_{c\in D} \in Q$ such that $x_{d,d,n} = x_{d,n}$ for all $(d,n) \in D \times \mathbb{N}$, thus $x_{d,d,n} \to \overline{x}_d$ for all $d \in D$. It follows that if Q is $H_{R,S}$-independent, then $\prod_{d\in D} \overline{\mathrm{proj}_d(Q)}$ and S are disjoint, so if $c\colon R \to \mathbb{N}$ is an $\aleph_0$-coloring of $H_{R,S}$, then the union of the closed hyperrectangles $\prod_{d\in D} \overline{\mathrm{proj}_d(c^{-1}(\{n\}))}$ separates R from S. Conversely, suppose that F_d is a closed subset of X_d for all $d \in D$ and $S \cap \prod_{d\in D} F_d = \emptyset$. If $((x_{c,d,n})_{c\in D})_{(d,n)\in D\times\mathbb{N}}$ is a sequence of elements of $\prod_{c\in D} F_c$ such that $\overline{x}_d = \lim_{n\to\infty} x_{d,d,n}$ exists for all $d \in D$, then $(\overline{x}_d)_{d\in D} \in \prod_{d\in D} F_d$, so $(\overline{x}_d)_{d\in D} \notin S$, thus $R \cap \prod_{d\in D} F_d$ is $H_{R,S}$-independent. Hence if there is a countable union of closed hyperrectangles separating R from S, then there is an $\aleph_0$-coloring of $H_{R,S}$.

To see (3), suppose first that $\prod_{d\in D} \phi_d$ is a continuous hyperrectangular homomorphism from $(\triangle^D((D \times \mathbb{N})^{\mathbb{N}}), \mathbb{H}_{(D\times\mathbb{N})^{\mathbb{N}},\infty})$ to (R, S), and define $\phi\colon (D \times \mathbb{N})^{\mathbb{N}} \to \prod_{d\in D} X_d$ by $\phi(\mathbf{d}) = (\phi_d(\mathbf{d}))_{d\in D}$. Clearly ϕ is continuous and $\phi((D \times \mathbb{N})^{\mathbb{N}}) = (\prod_{d\in D} \phi_d)(\triangle^D((D \times \mathbb{N})^{\mathbb{N}})) \subseteq R$. To see that ϕ is a homomorphism from $\mathbb{H}_{(D\times\mathbb{N})^{\mathbb{N}}}$ to $H_{R,S}$, we need only note that $(\lim_{n\to\infty} \phi_d(\mathcal{N}_{t\frown((d,n))}))_{d\in D} = (\phi_d(t \frown ((d,\infty))))_{d\in D}$ for all $t \in (D \times \mathbb{N})^{<\mathbb{N}}$, since $(\prod_{d\in D} \phi_d)(\mathbb{H}_{(D\times\mathbb{N})^{\mathbb{N}},\infty}) \subseteq S$. Conversely, suppose that $\phi\colon (D \times \mathbb{N})^{\mathbb{N}} \to R$ is a continuous homomorphism from $\mathbb{H}_{(D\times\mathbb{N})^{\mathbb{N}}}$ to $H_{R,S}$. Then each of the functions $\mathrm{proj}_d \circ \phi$ is a continuous homomorphism from $\mathrm{Cnvg}_{(D\times\mathbb{N})\cup\{(d,\infty)\}}((D \times \mathbb{N})^{\mathbb{N}})$ to $\mathrm{Cnvg}(X_d)$, so Proposition 3.4 yields continuous extensions $\phi_d\colon \mathrm{Ext}_{(D\times\mathbb{N})\cup\{(d,\infty)\}}((D \times \mathbb{N})^{\mathbb{N}}) \to X_d$ for all $d \in D$, and clearly $\prod_{d\in D} \phi_d$ is a hyperrectangular homomorphism from $(\triangle^D((D \times \mathbb{N})^{\mathbb{N}}), \mathbb{H}_{(D\times\mathbb{N})^{\mathbb{N}},\infty})$ to (R, S). □

In particular, we obtain the following.

THEOREM 4.2 ($\mathtt{OGD}^{\mathbb{N}}(\Gamma)$). *Suppose that D is a countable discrete set, $(X_d)_{d\in D}$ is a sequence of metric spaces, $R, S \subseteq \prod_{d\in D} X_d$ are disjoint, and $R \in \Gamma$. Then exactly one of the following holds:*

(1) *There is a countable union of closed hyperrectangles separating R from S.*
(2) *There is a continuous hyperrectangular homomorphism from $(\triangle^D((D \times \mathbb{N})^{\mathbb{N}}), \mathbb{H}_{(D\times\mathbb{N})^{\mathbb{N}},\infty})$ to (R, S).*

PROOF. This follows from Proposition 4.1. □

Theorems 1.1 and 1.3 together with the special cases of Theorem 4.2 where $D = 1$ and Γ is either the pointclass of analytic subsets of metric spaces or the pointclass of all subsets of analytic metric spaces easily yield the Kechris–Louveau–Woodin generalizations of Hurewicz's characterization of the circumstances under which two sets can be separated by an F_σ set.

Theorem 1.1 and the special case of Theorem 4.2 where $D = 2$ and Γ is the pointclass of analytic subsets of metric spaces easily yield the Lecomte–Zeleny characterization of the circumstances under which an analytic subset of the plane can be separated from another subset of the plane by a countable union of closed rectangles. Theorem 1.3 and the special case of Theorem 4.2 where $D = 2$ and Γ is the pointclass of all subsets of analytic metric spaces easily yield the generalization in which analyticity is weakened to separability under the axiom of determinacy.

PROPOSITION 4.3. *Suppose that D is a discrete set of cardinality at least two, X is a metric space, and H is a D-dimensional dihypergraph on X.*

1. *There is a countable union of closed hyperrectangles separating $\triangle^D(X)$ from H if and only if there is a $\mathbf{\Delta}^0_2$-measurable $\aleph_0$-coloring of H.*
2. *There is a continuous hyperrectangular homomorphism from $(\triangle^D((D \times \mathbb{N})^{\mathbb{N}}), \mathbb{H}_{(D\times\mathbb{N})^{\mathbb{N}},\infty})$ to $(\triangle^D(X), H)$ if and only if there is a continuous homomorphism from $\mathbb{H}_{(D\times\mathbb{N})^{\mathbb{N}},\infty}$ to H.*

PROOF. To see (1), note first that if $(F_{d,n})_{(d,n)\in D\times\mathbb{N}}$ is a sequence of closed subsets of X for which $\bigcup_{n\in\mathbb{N}}\prod_{d\in D} F_{d,n}$ separates $\triangle^D(X)$ from H, then the H-independent closed sets $\bigcap_{d\in D} F_{d,n}$ cover X, so there is a $\mathbf{\Delta}^0_2$-measurable $\aleph_0$-coloring of H. Conversely, if there is a $\mathbf{\Delta}^0_2$-measurable $\aleph_0$-coloring of H, then there is a cover $(F_n)_{n\in\mathbb{N}}$ of X by H-independent closed sets, in which case $\bigcup_{n\in\mathbb{N}} F_n^D$ is a countable union of closed hyperrectangles separating $\triangle^D(X)$ from H.

To see (2), note first that if $\phi\colon \mathrm{Ext}_{D\times\mathbb{N}_*}((D\times\mathbb{N})^{\mathbb{N}}) \to X$ is a homomorphism from $\mathbb{H}_{(D\times\mathbb{N})^{\mathbb{N}},\infty}$ to H, and ϕ_d is the restriction of ϕ to $\mathrm{Ext}_{(D\times\mathbb{N})\cup\{(d,\infty)\}}((D \times \mathbb{N})^{\mathbb{N}})$ for all $d \in D$, then $\prod_{d\in D}\phi_d$ is a hyperrectangular homomorphism from $(\triangle^D((D\times\mathbb{N})^{\mathbb{N}}), \mathbb{H}_{(D\times\mathbb{N})^{\mathbb{N}},\infty})$ to $(\triangle^D(X), H)$. Conversely, observe that if $\prod_{d\in D}\phi_d$ is a continuous hyperrectangular homomorphism from $(\triangle^D((D\times\mathbb{N})^{\mathbb{N}}), \mathbb{H}_{(D\times\mathbb{N})^{\mathbb{N}},\infty})$ to $(\triangle^D(X), H)$, then the function $\phi = \bigcup_{d\in D}\phi_d$ is a continuous homomorphism from $\mathbb{H}_{(D\times\mathbb{N})^{\mathbb{N}},\infty}$ to H. □

In particular, we obtain the following.

THEOREM 4.4 ($\mathtt{OGD}^{\mathbb{N}}(\Gamma)$). *Suppose that D is a countable discrete set of cardinality at least two, X is a metric space in Γ, and H is a D-dimensional dihypergraph on X. Then exactly one of the following holds:*

1. *There is a $\mathbf{\Delta}^0_2$-measurable $\aleph_0$-coloring of H.*
2. *There is a continuous homomorphism from $\mathbb{H}_{(D\times\mathbb{N})^{\mathbb{N}},\infty}$ to H.*

PROOF. This follows from Propositions 4.1 and 4.3. □

Theorem 1.1 and the special case of Theorem 4.4 where $D = 2$ and Γ is the pointclass of analytic subsets of metric spaces easily yield the Lecomte–Zeleny characterization of the circumstances under which a graph on an analytic metric space has a $\mathbf{\Delta}^0_2$-measurable $\aleph_0$-coloring. Theorem 1.3 and the special case of Theorem 4.4 where $D = 2$ and Γ is the pointclass of all subsets of analytic metric spaces easily yield the generalization in which analyticity is weakened to separability under the axiom of determinacy.

5. Sigma-continuous functions with closed witnesses

Given a function $\pi\colon X \to Y$, let H_π denote the $\mathbb{N}$-dimensional dihypergraph on graph(π) consisting of all sequences $((x_n, y_n))_{n\in\mathbb{N}}$ of elements of graph(π) with the property that $\overline{x} = \lim_{n\to\infty} x_n$ exists but $\pi(\overline{x}) \notin \overline{\{y_n \mid n \in \mathbb{N}\}}$.

PROPOSITION 5.1. *Suppose that X and Y are metric spaces and $\pi\colon X \to Y$.*

(1) *The dihypergraph H_π is box open.*
(2) *There is an $\aleph_0$-coloring of H_π if and only if π is σ-continuous with closed witnesses.*
(3) *There is a continuous homomorphism from $\mathbb{H}_{\mathbb{N}^\mathbb{N}}$ to H_π if and only if there is a continuous function $\psi\colon \mathrm{Ext}_{\mathbb{N}_*}(\mathbb{N}^\mathbb{N}) \to X$ with the property that $\pi \circ \psi$ is a reduction of $\mathbb{N}^\mathbb{N}$ to a closed set and $(\pi \circ \psi) \upharpoonright \mathbb{N}^\mathbb{N}$ is continuous.*

PROOF. To see (1), note that if $((x_n, y_n))_{n\in\mathbb{N}} \in H_\pi$ and $\overline{x} = \lim_{n\to\infty} x_n$, then there exists $\epsilon > 0$ such that $\rho_Y(\pi(\overline{x}), y_n) \geq 2\epsilon$ for all $n \in \mathbb{N}$. Fix positive real numbers $\epsilon_n \to 0$, and observe that the intersection of graph(π) with $\prod_{n\in\mathbb{N}} \mathcal{B}_X(x_n, \epsilon_n) \times \mathcal{B}_Y(y_n, \epsilon)$ is contained in H_π.

To see (2), note first that if $F \subseteq X$ is a closed set on which π is continuous, then graph$(\pi) \cap (F \times Y)$ is H_π-independent, so if π is σ-continuous with closed witnesses, then there is a Borel $\aleph_0$-coloring of H_π. Conversely, if $R \subseteq$ graph(π) and $(\overline{x}_n)_{n\in\mathbb{N}}$ is a convergent sequence of elements of the closure of $\mathrm{proj}_X(R)$, then there are sequences $(x_{m,n})_{m\in\mathbb{N}}$ of elements of $\mathrm{proj}_X(R)$ such that $x_{m,n} \to \overline{x}_n$ for all $n \in \mathbb{N}$. If R is H_π-independent, then $\pi(x_{m,n}) \to \pi(\overline{x}_n)$ for all $n \in \mathbb{N}$, so there is a function $f\colon \mathbb{N} \to \mathbb{N}$ such that $\rho_X(x_{f(n),n}, \overline{x}_n) \to 0$ and $\rho_Y(\pi(x_{f(n),n}), \pi(\overline{x}_n)) \to 0$. It follows that $(x_{f(n),n})_{n\in\mathbb{N}}$ converges, so the fact that R is H_π-independent ensures that $\pi(\lim_{n\to\infty} x_{f(n),n}) = \lim_{n\to\infty} \pi(x_{f(n),n})$, in which case $\pi(\lim_{n\to\infty} \overline{x}_n) = \lim_{n\to\infty} \pi(\overline{x}_n)$, thus $\pi \upharpoonright \overline{\mathrm{proj}_X(R)}$ is continuous. In particular, it follows that if there is an $\aleph_0$-coloring of H_π, then π is σ-continuous with closed witnesses.

To see (3), note first that if $\psi\colon \mathrm{Ext}_{\mathbb{N}_*}(\mathbb{N}^\mathbb{N}) \to X$ is continuous, $\pi \circ \psi$ is a reduction of $\mathbb{N}^\mathbb{N}$ to a closed set, $(\pi \circ \psi) \upharpoonright \mathbb{N}^\mathbb{N}$ is continuous, and $(\mathbf{d}_n)_{n\in\mathbb{N}} \in \mathbb{H}_{\mathbb{N}^\mathbb{N}}$, then $(\pi \circ \psi)(\lim_{n\to\infty} \mathbf{d}_n) \notin \overline{\{(\pi \circ \psi)(\mathbf{d}_n) \mid n \in \mathbb{N}\}}$, so the continuity of ψ ensures that $((\psi(\mathbf{d}_n), (\pi \circ \psi)(\mathbf{d}_n)))_{n\in\mathbb{N}} \in H_\pi$, thus $(\psi \upharpoonright \mathbb{N}^\mathbb{N}) \times ((\pi \circ \psi) \upharpoonright \mathbb{N}^\mathbb{N})$ is a homomorphism from $\mathbb{H}_{\mathbb{N}^\mathbb{N}}$ to H_π.

Conversely, suppose that $\phi\colon \mathbb{N}^\mathbb{N} \to$ graph(π) is a continuous homomorphism from $\mathbb{H}_{\mathbb{N}^\mathbb{N}}$ to H_π, and set $\phi_X = \mathrm{proj}_X \circ \phi$ and $\phi_Y = \mathrm{proj}_Y \circ \phi$. As the definition of H_π ensures that ϕ_X is a homomorphism from $\mathrm{Cnvg}_{\mathbb{N}_*}(\mathbb{N}^\mathbb{N})$ to $\mathrm{Cnvg}(X)$, Proposition 3.4 yields a continuous extension $\overline{\phi}_X\colon \mathrm{Ext}_{\mathbb{N}_*}(\mathbb{N}^\mathbb{N}) \to X$.

If there exists $t \in \mathbb{N}^{<\mathbb{N}}$ for which $\phi_Y \upharpoonright \mathcal{N}_t$ is constant, then the function $\psi\colon \mathrm{Ext}_{\mathbb{N}_*}(\mathbb{N}^\mathbb{N}) \to X$ given by $\psi(\mathbf{d}) = \overline{\phi}_X(t \frown \mathbf{d})$ is continuous and $\pi \circ \psi$ is a reduction of $\mathbb{N}^\mathbb{N}$ to a singleton, so we can assume that

$$(\dagger) \qquad \forall t \in \mathbb{N}^{<\mathbb{N}}\ \phi_Y(\mathcal{N}_t) \text{ is infinite.}$$

If there exists $t \in \mathbb{N}^{<\mathbb{N}}$ for which $(\pi \circ \overline{\phi}_X)(\mathcal{N}_{t,\mathbb{N}_*} \setminus \mathbb{N}^\mathbb{N})$ is finite, then by appealing to ($\dagger$) and extending t, we can ensure that $\overline{\phi_Y(\mathcal{N}_t)}$ and $(\pi \circ \overline{\phi}_X)(\mathcal{N}_{t,\mathbb{N}_*} \setminus \mathbb{N}^\mathbb{N})$ are disjoint, so the function $\psi\colon \mathrm{Ext}_{\mathbb{N}_*}(\mathbb{N}^\mathbb{N}) \to X$ given by $\psi(\mathbf{d}) = \overline{\phi}_X(t \frown \mathbf{d})$ is continuous, $\pi \circ \psi$ is a reduction of $\mathbb{N}^\mathbb{N}$ to $\overline{\phi_Y(\mathcal{N}_t)}$, and $(\pi \circ \psi)(\mathbf{d}) = \phi_Y(t \frown \mathbf{d})$ for all $\mathbf{d} \in \mathbb{N}^\mathbb{N}$. We can therefore additionally assume that

$$(\ddagger) \qquad \forall t \in \mathbb{N}^{<\mathbb{N}}\ (\pi \circ \overline{\phi}_X)(\mathcal{N}_{t,\mathbb{N}_*} \setminus \mathbb{N}^\mathbb{N}) \text{ is infinite.}$$

Fix positive real numbers $\epsilon_n \to 0$, as well as an enumeration $(t_k)_{k \in \mathbb{N}}$ of $\mathbb{N}^{<\mathbb{N}}$ such that $t_j \sqsubseteq t_k \implies j \leq k$ for all $j, k \in \mathbb{N}$. We say that a sequence $(X_n)_{n \in \mathbb{N}}$ of subsets of X is *closed and discrete* if there are no convergent sequences in $\prod_{n \in \mathbb{N}} X_n$. We will recursively construct functions $f \colon \mathbb{N}^{<\mathbb{N}} \to \mathbb{N}^{<\mathbb{N}}$ and $g \colon \mathbb{N}^{<\mathbb{N}} \setminus \{\emptyset\} \to \mathbb{N}^{<\mathbb{N}}$ such that, for all $k \in \mathbb{N}$, the following conditions hold:

(a) $\forall n \in \mathbb{N}\ f(t_k) \sqsubset g(t_k \frown (n))$.
(b) $\forall m \neq n\ g(t_k \frown (m))(|f(t_k)|) \neq g(t_k \frown (n))(|f(t_k)|)$.
(c) $(\phi_Y(\mathcal{N}_{g(t_k \frown (n))}))_{n \in \mathbb{N}}$ is either closed-and-discrete or convergent.
(d) $g(t_{k+1}) \sqsubseteq f(t_{k+1})$.
(e) $\forall j \leq k\ (\pi \circ \overline{\phi}_X)(f(t_j) \frown (\infty)) \notin \overline{\phi_Y(\mathcal{N}_{f(t_{k+1})})}$.
(f) $\forall j \leq k\ (\pi \circ \overline{\phi}_X)(f(t_{k+1}) \frown (\infty)) \neq \lim_{n \to \infty} \phi_Y(\mathcal{N}_{g(t_j \frown (n))})$.

We begin by setting $f(\emptyset) = \emptyset$.

Suppose now that $k \in \mathbb{N}$ and we have already found $f(t_k)$, as well as $g(t_j \frown (n))$ for all $j < k$ and $n \in \mathbb{N}$, and fix extensions $\mathbf{d}_{k,n} \in \mathbb{N}^\mathbb{N}$ of $f(t_k) \frown (n)$ for all $n \in \mathbb{N}$. Then there is an injective sequence $(i_{k,n})_{n \in \mathbb{N}}$ of natural numbers such that $(\{\phi_Y(\mathbf{d}_{k,i_{k,n}})\})_{n \in \mathbb{N}}$ is either closed-and-discrete or convergent, and continuity yields extensions $g(t_k \frown (n)) \sqsubset \mathbf{d}_{k,i_{k,n}}$ of $f(t_k) \frown (i_{k,n})$ with the property that $\phi_Y(\mathcal{N}_{g(t_k \frown (n))}) \subseteq \mathcal{B}_Y(\phi_Y(\mathbf{d}_{k,i_{k,n}}), \epsilon_n)$ for all $n \in \mathbb{N}$. By (†), there is an extension $\mathbf{d}_k \in \mathbb{N}^\mathbb{N}$ of $g(t_{k+1})$ such that $\phi_Y(\mathbf{d}_k) \neq (\pi \circ \overline{\phi}_X)(f(t_j) \frown (\infty))$ for all $j \leq k$, and therefore an extension $u_k \sqsubset \mathbf{d}_k$ of $g(t_{k+1})$ such that $(\pi \circ \overline{\phi}_X)(f(t_j) \frown (\infty)) \notin \overline{\phi_Y(\mathcal{N}_{u_k})}$ for all $j \leq k$. By (‡), there is an extension $f(t_{k+1})$ of u_k satisfying condition (f), thereby completing the recursive construction.

Conditions (a), (b), and (d) ensure that the function $\phi_X \circ \phi'$ is a homomorphism from $\mathrm{Cnvg}_{\mathbb{N}_*}(\mathbb{N}^\mathbb{N})$ to $\mathrm{Cnvg}(X)$, where $\phi' \colon \mathbb{N}^\mathbb{N} \to \mathbb{N}^\mathbb{N}$ is given by $\phi'(\mathbf{d}) = \bigcup_{n \in \mathbb{N}} f(\mathbf{d} \upharpoonright n)$. Proposition 3.4 therefore yields a continuous extension $\psi \colon \mathrm{Ext}_{\mathbb{N}_*}(\mathbb{N}^\mathbb{N}) \to X$. As $(\pi \circ \psi) \upharpoonright \mathbb{N}^\mathbb{N} = \phi_Y \circ \phi'$, it only remains to show that $\pi \circ \psi$ is a reduction of $\mathbb{N}^\mathbb{N}$ to a closed set.

Suppose, towards a contradiction, that there exists $k \in \mathbb{N}$ with the property that $(\pi \circ \psi)(t_k \frown (\infty)) \in \overline{(\pi \circ \psi)(\mathbb{N}^\mathbb{N})}$. As conditions (a), (b), and (d) ensure that

$$\begin{aligned}(\pi \circ \psi)(t_k \frown (\infty)) &= \pi(\lim_{n \to \infty} (\phi_X \circ \phi')(\mathcal{N}_{t_k \frown (n)})) \\ &= \pi(\lim_{n \to \infty} \phi_X(\mathcal{N}_{f(t_k \frown (n))})) \\ &= (\pi \circ \overline{\phi}_X)(f(t_k) \frown (\infty)),\end{aligned}$$

it follows that $(\pi \circ \overline{\phi}_X)(f(t_k) \frown (\infty)) \in \overline{\phi_Y(\mathbb{N}^\mathbb{N})}$. As condition (e) ensures that $(\pi \circ \overline{\phi}_X)(f(t_k) \frown (\infty)) \notin \overline{\phi_Y(\mathcal{N}_{f(t_j)})}$ for all $j > k$, and condition (f) implies that if $i < k$ and $(\pi \circ \overline{\phi}_X)(f(t_k) \frown (\infty)) \in \overline{\phi_Y(\mathcal{N}_{f(t_i)})}$ then there exists $j > i$ for which $(\pi \circ \overline{\phi}_X)(f(t_k) \frown (\infty)) \in \overline{\phi_Y(\mathcal{N}_{f(t_j)})}$, it follows that $(\pi \circ \overline{\phi}_X)(f(t_k) \frown (\infty)) \in \overline{\phi_Y(\mathcal{N}_{f(t_k)})} \setminus \bigcup_{n \in \mathbb{N}} \overline{\phi_Y(\mathcal{N}_{f(t_k \frown (n))})}$. Conditions (a), (b), and (d) therefore yield extensions $\mathbf{c}_n \in \mathbb{N}^\mathbb{N}$ of $f(t_k) \frown (n)$ such that $\pi(\lim_{n \to \infty} \phi_X(\mathbf{c}_n)) \in \overline{\{(\pi \circ \phi_X)(\mathbf{c}_n) \mid n \in \mathbb{N}\}}$, contradicting the fact that ϕ is a homomorphism from $\mathbb{H}_{\mathbb{N}^\mathbb{N}}$ to H_π. □

In particular, we obtain the following.

THEOREM 5.2 ($\mathtt{OGD}^\mathbb{N}(\Gamma)$). *Suppose that X and Y are separable metric spaces, $X \in \Gamma$, and $\pi \colon X \to Y$ is Borel. Then exactly one of the following holds:*

(1) *The function π is σ-continuous with closed witnesses.*

(2) *There is a continuous function* $\psi \colon \mathrm{Ext}_{\mathbb{N}_*}(\mathbb{N}^\mathbb{N}) \to X$ *with the property that* $\pi \circ \psi$ *is a reduction of* $\mathbb{N}^\mathbb{N}$ *to a closed set and* $(\pi \circ \psi) \restriction \mathbb{N}^\mathbb{N}$ *is continuous.*

PROOF. This follows from Proposition 5.1. □

Theorem 1.1 and the special case of Theorem 5.2 where Γ is the pointclass of analytic subsets of metric spaces easily yield the Jayne–Rogers characterization of the circumstances under which a function from an analytic metric space to a separable metric space is σ-continuous with closed witnesses. Theorem 1.3 and the special case of Theorem 5.2 where Γ is the pointclass of all subsets of analytic metric spaces easily yield the generalization in which analyticity is weakened to separability under the axiom of determinacy.

6. Cardinal invariants and chromatic numbers

In this section, we explore connections between cardinal invariants and the chromatic numbers of box-open dihypergraphs.

PROPOSITION 6.1. *The chromatic number of* $\mathbb{H}'_{\mathbb{N}^\mathbb{N}}$ *is* $\mathfrak{d}$.

PROOF. This follows from the fact that a subset of $\mathbb{N}^\mathbb{N}$ is $\mathbb{H}'_{\mathbb{N}^\mathbb{N}}$-independent if and only if its closure is compact. □

We say that an $\mathbb{N}$-dimensional dihypergraph is *hereditary* if it is closed under subsequences.

PROPOSITION 6.2 ($\mathtt{OGD}^\mathbb{N}(\Gamma)$). *Suppose that* X *is a Hausdorff space in* Γ *and* H *is a box-open hereditary* $\mathbb{N}$*-dimensional dihypergraph on* X*. Then either* $\chi(H) \leq \aleph_0$ *or* $\chi(H) \geq \mathfrak{d}$.

PROOF. This follows from Proposition 6.1 and the fact that every homomorphism from $\mathbb{H}_{\mathbb{N}^\mathbb{N}}$ to H is a homomorphism from $\mathbb{H}'_{\mathbb{N}^\mathbb{N}}$ to H. □

As every analytic space is trivially the union of $\mathfrak{d}$-many compact sets, Theorem 1.1 and the special case of Proposition 6.2 where Γ is the pointclass of analytic subsets of metric spaces and H is of the form H_X ensure that if an analytic subset of a metric space is not contained in a K_σ set, then $\mathfrak{d}$ is the least cardinal κ for which it is contained in a union of κ-many compact sets.

As every analytic space is trivially the union of $\mathfrak{d}$-many closed sets, Theorem 1.1 and the special case of Proposition 6.2 where Γ is the pointclass of analytic subsets of metric spaces, H is of the form $H_{R,S}$, and $D = 1$ ensure that if an analytic subset of a metric space cannot be separated from another set by an F_σ set, then $\mathfrak{d}$ is the least cardinal κ for which it can be separated from the other set by a union of κ-many closed sets.

As every analytic graph of a function is trivially the union of $\mathfrak{d}$-many compact sets that are graphs of functions, Theorem 1.1 and the special case of Proposition 6.2 where Γ is the pointclass of analytic subsets of metric spaces and H is of the form H_π ensure that if a Borel function π from an analytic metric space to a separable metric space is not σ-continuous with closed witnesses, then $\mathfrak{d}$ is the least cardinal κ for which X is the union of κ-many closed sets on which π is continuous.

The special case of the last fact for Baire-class-one functions is due to Solecki (see [**Sol98**]). He established this by noting that the two elements of his basis for Baire-class-one functions that are not σ-continuous have the desired property. The above argument shows that this more sophisticated basis theorem is unnecessary to obtain the desired result; one need only note that σ-continuity with closed witnesses can be characterized using a box-open hereditary $\mathbb{N}$-dimensional dihypergraph.

We next turn our attention to the computation of the chromatic number of $\mathbb{H}_{\mathbb{N}^\mathbb{N}}$ itself. For each partial function $f \colon D^{<\mathbb{N}} \rightharpoonup D$, define $\mathbf{D}_f = \{\mathbf{d} \in D^\mathbb{N} \mid \forall n \in \mathbb{N}\ (\mathbf{d} \upharpoonright n \in \operatorname{dom}(f) \implies \mathbf{d}(n) \neq f(\mathbf{d} \upharpoonright n))\}$.

Proposition 6.3. *Suppose that D is a set of cardinality at least two. Then $\chi(\mathbb{H}_{D^\mathbb{N}}) = \min\{|\mathcal{F}| \mid \mathcal{F} \subseteq D^{D^{<\mathbb{N}}}$ and $D^\mathbb{N} = \bigcup_{f \in \mathcal{F}} \mathbf{D}_f\}$.*

Proof. This follows from the fact that a set $X \subseteq D^\mathbb{N}$ is $\mathbb{H}_{D^\mathbb{N}}$-independent if and only if there is a function $f \colon D^{<\mathbb{N}} \to D$ for which $X \subseteq \mathbf{D}_f$. □

We now establish an analog of Proposition 6.2 without the assumption that H is hereditary.

Proposition 6.4 ($\mathtt{OGD}^D(\Gamma)$). *Suppose that D is a countable set of cardinality at least two, X is a Hausdorff space in Γ, and H is a box-open D-dimensional dihypergraph on X. Then either $\chi(H) \leq \aleph_0$ or $\chi(H) \geq \operatorname{cov}(\mathcal{M})$.*

Proof. It is sufficient to show that $\chi(\mathbb{H}_{D^\mathbb{N}}) \geq \operatorname{cov}(\mathcal{M})$, which follows from Proposition 6.3 and the observation that if $f \colon D^{<\mathbb{N}} \to D$, then $\mathbf{D}_f$ is meager with respect to the usual topology on $D^\mathbb{N}$. □

Alternatively, one can obtain the above result by noting that if H is a box-open D-dimensional dihypergraph on a Hausdorff space, $\mathcal{I}_H$ is the σ-ideal generated by the family of closed H-independent sets, and $\chi(H) > \aleph_0$, then $\chi(H) = \operatorname{cov}(\mathcal{I}_H)$.

Conversely, given a σ-ideal $\mathcal{I}$ on a topological space X, let $H_\mathcal{I}$ be the $\mathbb{N}$-dimensional dihypergraph on X consisting of all sequences $(x_n)_{n \in \mathbb{N}}$ of elements of X for which $\overline{\{x_n \mid n \in \mathbb{N}\}} \notin \mathcal{I}$.

Proposition 6.5. *Suppose that X is a separable metric space and $\mathcal{I}$ is a σ-ideal on X generated by a family of closed subsets of X.*

(1) *If $\mathcal{I}$ covers X, then $H_\mathcal{I}$ is box open.*
(2) *If $\operatorname{cov}(\mathcal{I}) > \aleph_0$, then $\chi(H_\mathcal{I}) = \operatorname{cov}(\mathcal{I})$.*

Proof. To see (1), suppose that $(x_n)_{n \in \mathbb{N}} \in H_\mathcal{I}$, fix positive real numbers $\epsilon_n \to 0$, and observe that if $(y_n)_{n \in \mathbb{N}} \in \prod_{n \in \mathbb{N}} \mathcal{B}_X(x_n, \epsilon_n)$, then $\overline{\{x_n \mid n \in \mathbb{N}\}} \subseteq \{x_n \mid n \in \mathbb{N}\} \cup \overline{\{y_n \mid n \in \mathbb{N}\}}$, thus $(y_n)_{n \in \mathbb{N}} \in H_\mathcal{I}$.

To see (2), note first that every $H_\mathcal{I}$-independent set is in $\mathcal{I}$, so $\operatorname{cov}(\mathcal{I}) \leq \chi(H_\mathcal{I})$. Conversely, as every set in $\mathcal{I}$ is contained in the union of countably-many closed sets in $\mathcal{I}$, the fact that every closed set in $\mathcal{I}$ is $H_\mathcal{I}$-independent ensures that $\chi(H_\mathcal{I}) \leq \operatorname{cov}(\mathcal{I}) \cdot \aleph_0 = \operatorname{cov}(\mathcal{I})$. □

As a corollary, we also obtain an analog of Proposition 6.1 without the assumption that H is hereditary.

Proposition 6.6. *The chromatic number of $\mathbb{H}_{\mathbb{N}^\mathbb{N}}$ is $\operatorname{cov}(\mathcal{M})$.*

Proof. Proposition 6.4 ensures that $\chi(\mathbb{H}_{\mathbb{N}^\mathbb{N}}) \geq \operatorname{cov}(\mathcal{M})$. As the open dihypergraph dichotomy yields a homomorphism from $\mathbb{H}_{\mathbb{N}^\mathbb{N}}$ to $H_\mathcal{M}$, Proposition 6.5 implies that $\chi(\mathbb{H}_{\mathbb{N}^\mathbb{N}}) \leq \operatorname{cov}(\mathcal{M})$. □

Stronger bounds can be obtained when D is finite. In the special case that $D = 2$, this is trivial.

PROPOSITION 6.7 ($\mathtt{OGD}(\Gamma)$). *Suppose that X is a Hausdorff space in Γ and G is an open graph on X. Then either $\chi(G) \leq \aleph_0$ or $\chi(G) = \mathfrak{c}$.*

PROOF. It is sufficient to show that $\chi(\mathbb{H}_{2^\mathbb{N}}) = \mathfrak{c}$, which follows from the fact that $\mathbb{H}_{2^\mathbb{N}} = \mathbb{K}_{2^\mathbb{N}}$. □

The case that $2 < |D| < \aleph_0$ is substantially subtler, and related to the adaptive global prediction numbers considered in [**Bla10**, §10]. Our first observation is a straightforward analog of Proposition 6.4.

PROPOSITION 6.8 ($\mathtt{OGD}^D(\Gamma)$). *Suppose that D is a finite set of cardinality at least two, X is a Hausdorff space in Γ, and H is an open D-dimensional dihypergraph on X. Then either $\chi(H) \leq \aleph_0$ or $\chi(H) \geq \operatorname{cov}(\mathcal{N})$.*

PROOF. It is sufficient to show that $\chi(\mathbb{H}_{D^\mathbb{N}}) \geq \operatorname{cov}(\mathcal{N})$, which follows from Proposition 6.3 and the observation that if $f\colon D^{<\mathbb{N}} \to D$ and μ is a strictly positive probability measure on D, then $\mathbf{D}_f$ is $\mu^\mathbb{N}$-null. □

We next establish an analog of the main result of [**Bre03**]. We say that a partial function $u\colon \mathbb{N} \rightharpoonup D$ is a *D-Silver condition* if its domain is co-infinite, and we say that a set $X \subseteq D^\mathbb{N}$ is *D-Silver null* if every D-Silver condition $u\colon \mathbb{N} \rightharpoonup D$ extends to a D-Silver condition $v\colon \mathbb{N} \rightharpoonup D$ for which $\mathcal{N}_v$ is disjoint from X.

THEOREM 6.9. *Suppose that D is a finite set of cardinality at least two, $X \subseteq D^\mathbb{N}$, and $\chi(\mathbb{H}_{D^\mathbb{N}} \upharpoonright X) < \mathfrak{b}$. Then X is D-Silver null.*

PROOF. By Proposition 6.3, it is sufficient to show that if $\mathcal{F} \subseteq D^{D^{<\mathbb{N}}}$ has cardinality strictly less than $\mathfrak{b}$, then $\bigcup_{f\in\mathcal{F}} \mathbf{D}_f$ is D-Silver null. By a straightforward recursive construction of length $|D|$, we need only show that for all $d \in D$, every D-Silver condition $u\colon \mathbb{N} \rightharpoonup D$ extends to a D-Silver condition $v\colon \mathbb{N} \rightharpoonup D$ with the property that

$$\forall f \in \mathcal{F} \forall \mathbf{d} \in \mathbf{D}_f \cap \mathcal{N}_v \forall^\infty n \in {\sim}\operatorname{dom}(v)\ d \neq f(\mathbf{d} \upharpoonright n).$$

It is clearly sufficient to handle the special case that $u = \emptyset$.

For all $f \in \mathcal{F}$, fix a function $g_f\colon \mathbb{N} \to \mathbb{N}$ such that for all $n \in \mathbb{N}$ and $t \in D^n$, if there exists $i \in \mathbb{N}$ for which $d = f(t \frown (d)^i)$, then there is such an $i < g_f(n)$. Fix a function $g\colon \mathbb{N} \to \mathbb{N}$ eventually dominating g_f for all $f \in \mathcal{F}$, set $h'(0) = 0$, and recursively define $h(n) = g(h'(n)+1)$ and $h'(n+1) = h'(n)+1+h(n)$ for all $n \in \mathbb{N}$. Let v be the function on ${\sim}\{h'(n+1) \mid n \in \mathbb{N}\}$ with constant value d.

To see that v is as desired, note first that if $\mathbf{d} \in \mathcal{N}_v$, then a straightforward inductive argument reveals that $\mathbf{d} \upharpoonright h'(n+1)$ is the concatenation of $\mathbf{d} \upharpoonright (h'(n)+1)$ and $(d)^{h(n)}$, for all $n \in \mathbb{N}$. It follows that if $f \in \mathcal{F}$, $\mathbf{d} \in \mathbf{D}_f$, and $n \in \mathbb{N}$ is sufficiently large that $g_f(h'(n)+1) \leq g(h'(n)+1)$, then $d \neq f(\mathbf{d} \upharpoonright h'(n+1))$. □

THEOREM 6.10 ($\mathtt{OGD}^D(\Gamma)$). *Suppose that D is a finite set of cardinality at least two, X is a Hausdorff space in Γ, and H is an open D-dimensional dihypergraph on X. Then either $\chi(H) \leq \aleph_0$ or $\chi(H) \geq \mathfrak{b}$.*

PROOF. It is sufficient to show that $\chi(\mathbb{H}_{D^\mathbb{N}}) \geq \mathfrak{b}$, which follows from Theorem 6.9. □

We now establish an analog of the main result of [**Kam00**].

THEOREM 6.11. *It is consistent that* $\chi(\mathbb{H}_{D^{\mathbb{N}}}) > \operatorname{cof}(\mathcal{N})$ *for every finite set* D *of cardinality at least two.*

PROOF. Given a natural number $D \geq 2$, we say that a subtree T of $D^{<\mathbb{N}}$ is *D-perfect* if every element of T has an extension $t \in T$ with the property that $\forall d \in D\ t \frown (d) \in T$. Let S_D denote the set of such trees. Let $\leq$ denote the partial order on S_D with respect to which $S \leq T$ if and only if $S \subseteq T$, and define $\mathbb{S}_D = (S_D, \leq)$. When $D = 2$, this is just Sacks forcing, and the usual proof that the latter is proper and has the Sacks property works just as well for every $\mathbb{S}_D$.

LEMMA 6.12. *Suppose that* $D \geq 2$ *is a natural number,* G *is* $\mathbb{S}_D$*-generic over* V, $\mathbf{d}$ *is the unique branch through* $\bigcap G$, *and* $f \colon D^{<\mathbb{N}} \to D$ *is in* V. *Then there exists* $n \in \mathbb{N}$ *for which* $\mathbf{d}(n) = f(\mathbf{d} \upharpoonright n)$.

PROOF. This follows from the fact that there are $\leq$-densely many trees $T \in S_D$ for which there exists $t \in D^{<\mathbb{N}}$ such that $t \frown (f(t))$ is $\sqsubseteq$-comparable with every element of T. $\square$

Suppose now that CH holds in V, and fix a sequence $(D_\beta)_{\beta<\omega_2}$ of natural numbers that are at least two for which every such natural number appears cofinally often, as well as a countable support iteration $(\mathbb{P}_\alpha, \dot{\mathbb{Q}}_\beta)_{\alpha\leq\omega_2,\beta<\omega_2}$ such that $\Vdash_{\mathbb{P}_\beta} \dot{\mathbb{Q}}_\beta = (\mathbb{S}_{D_\beta})^{V^{\mathbb{P}_\beta}}$ for all $\beta < \omega_2$.

As $\mathbb{P}_{\omega_2}$ has the Sacks property, it follows that $\operatorname{cof}(\mathcal{N}) = \aleph_1$ in $V^{\mathbb{P}_{\omega_2}}$ (see, for example, [**Bla10**, §11.5]). To see that $\chi(\mathbb{H}_{D^{\mathbb{N}}}) > \aleph_1$ in $V^{\mathbb{P}_{\omega_2}}$ for all natural numbers $D \geq 2$, note that if $\mathcal{F} \subseteq D^{D^{<\mathbb{N}}}$ is in $V^{\mathbb{P}_{\omega_2}}$ and has cardinality at most $\aleph_1$ in $V^{\mathbb{P}_{\omega_2}}$, then there exists $\alpha < \omega_2$ for which $\mathcal{F} \in V^{\mathbb{P}_\alpha}$. Fix $\beta \geq \alpha$ for which $D_\beta = D$, note that $D^{\mathbb{N}} \nsubseteq \bigcup_{f\in\mathcal{F}} \mathbf{D}_f$ in $V^{\mathbb{P}_{\beta+1}}$ by Lemma 6.12, and appeal to Proposition 6.3. $\square$

THEOREM 6.13 ($\mathtt{OGD}^D(\Gamma)$). *It is consistent that whenever* D *is a finite set of cardinality at least two,* X *is a Hausdorff space in* Γ, *and* H *is an open* D*-dimensional dihypergraph on* X, *either* $\chi(H) \leq \aleph_0$ *or* $\chi(H) > \operatorname{cof}(\mathcal{N})$.

PROOF. It is sufficient to establish the consistency of $\chi(\mathbb{H}_{D^{\mathbb{N}}}) > \operatorname{cof}(\mathcal{N})$ for all finite sets D of cardinality at least two, which follows from Theorem 6.11. $\square$

We finally establish a consistent upper bound as well.

THEOREM 6.14. *It is consistent that* $\chi(\mathbb{H}_{D^{\mathbb{N}}}) < \mathfrak{d}$ *for every finite set* D *of cardinality at least three.*

PROOF. Given a natural number $D \geq 3$, let P_D denote the set of pairs $(\mathbf{D}, f)$ with the property that $f \colon D^{<n} \to D$ for some $n \in \mathbb{N}$, $\mathbf{D} \subseteq \mathbf{D}_f$, and $\mathbf{c} \upharpoonright n \neq \mathbf{d} \upharpoonright n$ for all distinct $\mathbf{c}, \mathbf{d} \in \mathbf{D}$. Let $\leq$ denote the partial order on P_D with respect to which $(\mathbf{C}, f) \leq (\mathbf{D}, g)$ if and only if $\mathbf{C} \subseteq \mathbf{D}$ and $f \sqsubseteq g$, and define $\mathbb{P}_D = (P_D, \leq)$.

LEMMA 6.15. *Suppose that* $n \in \mathbb{N}$ *and* $f \colon D^{<n} \to D$. *Then any two elements of* P_D *of the form* $(\mathbf{C}, f)$ *and* $(\mathbf{D}, f)$ *are* $\leq$*-compatible.*

PROOF. Fix $m \geq n$ sufficiently large that $c \upharpoonright m \neq d \upharpoonright m$ for all distinct $c \in \mathbf{C}$ and $d \in \mathbf{D}$. As $D \geq 3$, there exists $g \colon D^{<m} \to D$ such that $\mathbf{C} \cup \mathbf{D} \subseteq \mathbf{D}_g$ and $f \sqsubseteq g$, in which case $(\mathbf{C} \cup \mathbf{D}, g)$ is a common $\leq$-extension of $(\mathbf{C}, f)$ and $(\mathbf{D}, f)$. $\square$

Fix a sequence $(D_n)_{n\in\mathbb{N}}$ of natural numbers that are at least three for which every such natural number appears cofinally often, and let $\mathbb{P} = (P, \leq)$ denote the finite support product of $(\mathbb{P}_{D_n})_{n\in\mathbb{N}}$. Lemma 6.15 ensures that each of the partial orders $\mathbb{P}_D$ is σ-linked, thus so too is $\mathbb{P}$.

LEMMA 6.16. *Suppose that $D \geq 3$ is a natural number, G is $\mathbb{P}$-generic over V, and $\mathbf{d} \in (D^{\mathbb{N}})^V$. Then there exists $k \in \mathbb{N}$ such that $D_k = D$ and $\mathbf{d}(n) \neq f(\mathbf{d} \upharpoonright n)$ for all $n \in \mathbb{N}$, where $f = \bigcup\{f_k \mid (\mathbf{D}_n, f_n)_{n\in\mathbb{N}} \in G\}$.*

PROOF. This follows from the fact that there are $\leq$-densely many sequences $(\mathbf{D}_n, f_n)_{n\in\mathbb{N}} \in P$ for which there exists $k \in \mathbb{N}$ such that $D = D_k$ and $\mathbf{d} \in \mathbf{D}_k$. □

LEMMA 6.17. *Suppose that $\dot{\mathbf{d}}$ is a $\mathbb{P}$-name for an element of $\mathbb{N}^{\mathbb{N}}$, $(k_n)_{n\in\mathbb{N}} \in \mathbb{N}^{\mathbb{N}}$, and $(f_n)_{n\in\mathbb{N}}$ is in the finite support product of $({D_n}^{{D_n}^{<k_n}})_{n\in\mathbb{N}}$. Then $\forall i \in \mathbb{N} \exists j \in \mathbb{N} \forall (\mathbf{D}_n, f_n)_{n\in\mathbb{N}} \in P$ $(\mathbf{D}_n, f_n)_{n\in\mathbb{N}} \not\Vdash_{\mathbb{P}} \dot{\mathbf{d}}(i) \geq j$.*

PROOF. Suppose, towards a contradiction, that there exist $i \in \mathbb{N}$ and $(\mathbf{D}_{j,n}, f_n)_{n\in\mathbb{N}} \in P$ such that $(\mathbf{D}_{j,n}, f_n)_{n\in\mathbb{N}} \Vdash_{\mathbb{P}} \dot{\mathbf{d}}(i) \geq j$ for all $j \in \mathbb{N}$. By passing to a subsequence, we can assume that there are sequences $(I_n)_{n\in\mathbb{N}}$ of finite sets and $(\mathbf{d}_{i,j,n})_{(i,j,n)\in I_n\times\mathbb{N}\times\mathbb{N}}$ of elements of $\mathbb{N}^{\mathbb{N}}$ such that $\mathbf{D}_{j,n} = \{\mathbf{d}_{i,j,n} \mid i \in I_n\}$ for all $j, n \in \mathbb{N}$ and $\mathbf{d}_{i,n} = \lim_{j\to\infty} \mathbf{d}_{i,j,n}$ exists for all $i, n \in \mathbb{N}$. Set $\mathbf{D}_n = \{\mathbf{d}_{i,n} \mid i \in I_n\}$. Then $(\mathbf{D}_n, f_n)_{n\in\mathbb{N}} \in P$, so there exist $k \in \mathbb{N}$ and an extension of $(\mathbf{D}_n, f_n)_{n\in\mathbb{N}} \in P$ for which $(\mathbf{D}'_n, f'_n)_{n\in\mathbb{N}} \Vdash_{\mathbb{P}} \dot{\mathbf{d}}(i) = k$. As $(\mathbf{D}_{j,n}, f'_n)_{n\in\mathbb{N}} \in P$ for all sufficiently large $j \in \mathbb{N}$, Lemma 6.15 ensures that $(\mathbf{D}_{j,n}, f_n)_{n\in\mathbb{N}}$ and $(\mathbf{D}'_n, f'_n)_{n\in\mathbb{N}}$ are compatible for all sufficiently large $j > k$, the desired contradiction. □

Suppose now that $\mathfrak{d} > \aleph_1$ in V, and fix a finite support iteration $(\mathbb{P}_\alpha, \dot{\mathbb{Q}}_\beta)_{\alpha\leq\omega_1,\beta<\omega_1}$ such that $\Vdash_{\mathbb{P}_\beta} \dot{\mathbb{Q}}_\beta = \mathbb{P}^{V^{\mathbb{P}_\beta}}$ for all $\beta < \omega_1$.

Proposition 6.3 and Lemma 6.16 easily imply that $\chi(\mathbb{H}_{D^{\mathbb{N}}}) = \aleph_1$ in $V^{\mathbb{P}_{\omega_1}}$. To see that $\mathfrak{d} > \aleph_1$ in $V^{\mathbb{P}_{\omega_1}}$, note that if $(\dot{\mathbf{d}}_\alpha)_{\alpha<\omega_1} \in V$ is a sequence of $\mathbb{P}_{\omega_1}$-names for elements of $\mathbb{N}^{\mathbb{N}}$, then Lemma 6.17 yields a sequence $(\mathbf{d}_\alpha)_{\alpha<\omega_1} \in V$ of elements of $\mathbb{N}^{\mathbb{N}}$ such that $\Vdash_{\mathbb{P}_{\omega_1}} \dot{\mathbf{d}}_\alpha \leq^* \mathbf{d}_\alpha$ for all $\alpha < \omega_1$. The fact that $\mathfrak{d} > \aleph_1$ in V then yields $\mathbf{d} \in (\mathbb{N}^{\mathbb{N}})^V$ such that $\mathbf{d} \not\leq^* \mathbf{d}_\alpha$ for all $\alpha < \omega_1$, so $\Vdash_{\mathbb{P}_{\omega_1}} \mathbf{d} \not\leq^* \dot{\mathbf{d}}_\alpha$ for all $\alpha < \omega_1$. □

References

[Bla10] Andreas Blass, *Combinatorial cardinal characteristics of the continuum*, Handbook of set theory. Vols. 1, 2, 3, Springer, Dordrecht, 2010, pp. 395–489, DOI 10.1007/978-1-4020-5764-9_7. MR2768685

[Bre03] Jörg Brendle, *Evasion and prediction. III. Constant prediction and dominating reals*, J. Math. Soc. Japan **55** (2003), no. 1, 101–115, DOI 10.2969/jmsj/1196890844. MR1939187

[Dav64] Morton Davis, *Infinite games of perfect information*, Advances in game theory, Princeton Univ. Press, Princeton, N.J., 1964, pp. 85–101. MR0170727

[Fen93] Qi Feng, *Homogeneity for open partitions of pairs of reals*, Trans. Amer. Math. Soc. **339** (1993), no. 2, 659–684, DOI 10.2307/2154292. MR1113695

[Hur28] Witold Hurewicz, *Relativ perfekte teile von punktmengen und mengen (a)*, Fund. Math. **12** (1928), 78–109.

[JR82] J. E. Jayne and C. A. Rogers, *First level Borel functions and isomorphisms*, J. Math. Pures Appl. (9) **61** (1982), no. 2, 177–205. MR673304

[Kam00] Shizuo Kamo, *Cardinal invariants associated with predictors*, Logic Colloquium '98 (Prague), Lect. Notes Log., vol. 13, Assoc. Symbol. Logic, Urbana, IL, 2000, pp. 280–295. MR1743265

[Kec77] Alexander S. Kechris, *On a notion of smallness for subsets of the Baire space*, Trans. Amer. Math. Soc. **229** (1977), 191–207, DOI 10.2307/1998505. MR450070

[KLW87] A. S. Kechris, A. Louveau, and W. H. Woodin, *The structure of σ-ideals of compact sets*, Trans. Amer. Math. Soc. **301** (1987), no. 1, 263–288, DOI 10.2307/2000338. MR879573

[LZ14] Dominique Lecomte and Miroslav Zeleny, *Baire-class ξ colorings: the first three levels*, Trans. Amer. Math. Soc. **366** (2014), no. 5, 2345–2373, DOI 10.1090/S0002-9947-2014-05876-5. MR3165641

[Sol98] Sławomir Solecki, *Decomposing Borel sets and functions and the structure of Baire class 1 functions*, J. Amer. Math. Soc. **11** (1998), no. 3, 521–550, DOI 10.1090/S0894-0347-98-00269-0. MR1606843

[Sou17] M. Ya. Souslin, *Sur une définition des ensembles mesurables B sans nombres transfinis*, C. R. Math. Acad. Sci. Paris **164** (1917), 88–91.

[SR75] Jean Saint-Raymond, *Approximation des sous-ensembles analytiques par l'intérieur* (French, with English summary), C. R. Acad. Sci. Paris Sér. A-B **281** (1975), no. 2-3, Aii, A85–A87. MR377840

Kurt Gödel Research Center for Mathematical Logic, Universität Wien, Währinger Strasse 25, 1090 Wien, Austria

Email address: raphael.carroy@univie.ac.at

URL: http://www.logique.jussieu.fr/~carroy/indexeng.html

Kurt Gödel Research Center for Mathematical Logic, Universität Wien, Währinger Strasse 25, 1090 Wien, Austria

Email address: benjamin.miller@univie.ac.at

URL: http://www.logic.univie.ac.at/benjamin.miller

Kurt Gödel Research Center for Mathematical Logic, Universität Wien, Währinger Strasse 25, 1090 Wien, Austria

Email address: daniel.soukup@univie.ac.at

URL: http://www.logic.univie.ac.at/~soukupd73

Contemporary Mathematics
Volume **752**, 2020
https://doi.org/10.1090/conm/752/15128

An introduction to combinatorics of determinacy

William Chan

1. Introduction

This article is an introduction to combinatorics under the axiom of determinacy. The main topics are partition properties and ∞-Borel codes. To illustrate the important ideas, the article will focus on the simplest settings. This will mean that one will work at simple pointclass such as $\boldsymbol{\Sigma}_1^1$, small cardinals such as ω_1 or ω_2, or very natural models of determinacy such as $L(\mathbb{R})$.

Despite the simplicity, there are still many interesting combinatorial questions in these settings. One purpose of this article is to serve as a reference for some of the notations and backgrounds that will be helpful for some results of Jackson, Trang, and the author on combinatorics around ω_1, ω_2, and $\mathbb{R}$ under AD and AD^+. (Section 2 to 6 were written for [**4**] and [**7**]. Section 7 and 8 will be helpful for reading [**3**] and [**5**].)

The article should be accessible with very basic knowledge of descriptive set theory and general set theory. Some familiarity with determinacy and games, the pointclass $\boldsymbol{\Sigma}_1^1$, and the bounding principle for $\boldsymbol{\Sigma}_1^1$ should suffice for most of the article. Knowledge of club sets, ultrafilters, measures, and basic constructibility will also be helpful. There will be occasional mentions of the theory of prewellorderings and scales such as the Moschovakis coding lemma or the Kunen-Martin theorem. In a few places, one will refer to the result of Kechris that AD implies $L(\mathbb{R}) \models \mathsf{DC}_{\mathbb{R}}$. The last two sections will require more familiarity with topics in general set theory such as HOD and iterated forcing.

The article has essentially two parts. The first part deals with various partition properties which are very powerful tools of AD. The second part introduces the ∞-Borel code and the Vopénka forcing which are useful AD^+ tool. Except possibly for a few minor observations, no results in this article are due to the author.

Section 2 introduces the partition property in its various forms. It also defines the associated measures induced by the partition properties.

Section 3 develops the notation and theory around good codings of functions. This section proves Martin's criterion for establishing partition properties.

Section 4 gives some examples of good coding of functions on ω_1. This is used to give a simple proof of the weak partition property on ω_1 and two proofs of

2010 *Mathematics Subject Classification.* Primary 03E60 (Determinacy Principles), 03E15 (Descriptive Set Theory), and 03E02 (Partition Relations).

The author was supported by NSF grant DMS-1703708.

the strong partition property on ω_1. One will give Martin's original proof of the strong partition property using sharps. Then one gives a proof due to Kechris of the strong partition property on ω_1 which uses category arguments and the simple generic coding function for ω_1.

Section 5 develops the theory around the Kunen function which is named after the eponymous Kunen tree. Here it will be shown that every function from ω_1 to ω_1 has a Kunen function by using the original Kunen tree. The Kunen functions are especially useful for choosing representatives of certain ultrapowers. This will be used to establish the identity of ω_2 as the ultrapower ${}^{\omega_1}\omega_1/\mu$, where μ is the club ultrafilter on ω_1. Then one will prove a useful sliding lemma and use this to establish the weak partition property on ω_2.

Section 6 is dedicated to proving a result of Martin and Paris which shows that ω_2 is not a strong partition cardinal. However, this section presents a proof of Jackson which produces an explicit partition without a homogeneous subset.

Section 7 proves some results of Woodin about the structure of $L(\mathbb{R})$ and the nature of ∞-Borel codes in $L(\mathbb{R})$. In particular, it will be shown that in $L(\mathbb{R})$, there is an "ultimate" ∞-Borel code that can be used to generate all other ∞-Borel codes for sets of reals. This is quite useful for cardinality results in $L(\mathbb{R})$.

Section 8 will present some further descriptive set theoretic applications due to Woodin of the Vopénka forcing. The countable section uniformization and the Woodin's perfect set dichotomy theorem will be proved here.

2. Partition Properties

The following is the usual partition property:

DEFINITION 2.1. Let κ be an ordinal, $\lambda \leq \kappa$, and $\gamma < \kappa$, then $\kappa \to (\kappa)^\lambda_\gamma$ indicates that for every $P : [\kappa]^\lambda \to \gamma$, there is some $\beta < \gamma$ and some $A \subseteq \kappa$ with $|A| = \kappa$ so that for all $f \in [A]^\lambda$, $P(f) = \beta$.

The most frequent situation is when $\gamma = 2$.

κ is a strong partition cardinal if and only if $\kappa \to (\kappa)^\kappa_2$.

κ is a weak partition cardinal if and only if $\kappa \to (\kappa)^\alpha_2$ for all $\alpha < \kappa$.

DEFINITION 2.2. Let κ be an ordinal. Let $S : \kappa \to \kappa$. Let $Y^S = \{(\alpha, \beta) : \alpha < \kappa \wedge \beta < S(\alpha)\}$.

Let $f : \kappa \to \kappa$. f is said to have uniform cofinality S if and only if there is a function $g : Y^S \to \kappa$ with the following properties

(1) For all $\alpha < \kappa$ and all $\beta < \gamma < S(\alpha)$, $g(\alpha, \beta) < g(\alpha, \gamma)$

(1) For all $\alpha < \kappa$, $f(\alpha) = \sup\{g(\alpha, \beta) : \beta < S(\alpha)\}$.

Let μ be a measure on κ. $f : \kappa \to \kappa$ is said to have uniform cofinality S μ-almost everywhere if and only if there is a $g : Y^S \to \kappa$ as above so that for μ-almost all $\alpha < \kappa$, $f(\alpha) = \sup\{g(\alpha, \beta) : \beta < S(\alpha)\}$.

A function $f : \kappa \to \kappa$ has uniform cofinality ω if and only if f has uniform cofinality S where $S(\alpha) = \omega$ for all $\alpha < \kappa$.

DEFINITION 2.3. Let $f : \kappa \to \kappa$ be a function. f is discontinuous everywhere if and only if for all $\alpha < \kappa$, $f(\alpha) > \sup\{f(\beta) : \beta < \alpha\}$.

(Jackson) A function $f : \kappa \to \kappa$ has the correct type if and only if f has uniform cofinality ω and is discontinuous everywhere.

Let $[\kappa]^\kappa_*$ denote the subset of $[\kappa]^\kappa$ consisting of functions of the correct type.

There are partition properties formulated for functions of the correct type in which the homogeneous set can be chosen to be club. Club homogeneous sets are conceptually useful in various construction. Functions of the correct type seem to naturally appear in many proofs establishing the partition properties under AD.

It should be noted that throughout the survey an asterisk, $*$, will be used to denote the corresponding concept that involves functions of the correct type when there is an ordinary version of this concept.

DEFINITION 2.4. Let κ be an ordinal. Let $\lambda \leq \kappa$ and $\gamma < \kappa$. Let $\kappa \to_* (\kappa)^\lambda_\gamma$ assert that for all $P : [\kappa]^\lambda_* \to \gamma$, there exists a club $C \subseteq \kappa$ and a $\beta < \lambda$ so that for all $f \in [C]^\lambda_*$, $P(f) = \beta$. Such a club C is said to be homogeneous for P taking value β (for functions of the correct type).

DEFINITION 2.5. Let κ be an ordinal. If $A \subseteq \kappa$ is such that $|A| = \kappa$, then let $\mathsf{enum}_A : \kappa \to A$ denote the increasing enumeration of A. (In context, when one writes enum_A, it should be clear what κ is.)

FACT 2.6. *Let κ be an ordinal and $\lambda \leq \kappa$. Then $\kappa \to_* (\kappa)^\lambda_2$ implies $\kappa \to (\kappa)^\lambda_2$. $\kappa \to (\kappa)^{\omega\cdot\lambda}_2$ implies $\kappa \to_* (\kappa)^\lambda_2$.*

PROOF. Assume $\kappa \to_* (\kappa)^\lambda_2$.

Let $P : [\kappa]^\lambda \to 2$. Then there is some $C \subseteq \omega_1$ club so that C is homogeneous for P for function of the correct type. Let $D = \{\alpha < \kappa : (\exists \beta \in \mathrm{Lim})(\alpha = \mathsf{enum}_C(\beta + \omega))\}$, where Lim refers to the class of limit ordinals. Note $|D| = \kappa$.

The next claim is that every $f \in [D]^\lambda$ is a function of the correct type. Let $g : \kappa \times \omega \to \kappa$ be defined as follows: Suppose $\gamma < \kappa$. Let β_γ be the unique limit ordinal β so that $f(\gamma) = \mathsf{enum}_C(\beta_\gamma + \omega)$. Then for each $n \in \omega$, define $g(\gamma, n) = \mathsf{enum}_C(\beta_\gamma + n)$. Then it is clear that $f(\gamma) = \sup\{g(\gamma, n) : n \in \omega\}$. This shows that f has uniform cofinality ω. Suppose $\gamma < \kappa$. There is a unique limit ordinal β_γ so that $f(\gamma) = C(\beta_\gamma + \omega)$. Then for all $\epsilon < \gamma$, $f(\epsilon) \leq \mathsf{enum}_C(\beta_\gamma) < f(\gamma)$. That is, $\sup\{f(\epsilon) : \epsilon < \gamma\} \leq \mathsf{enum}_C(\beta_\gamma) < f(\gamma)$. f is discontinuous everywhere.

Thus every $f \in [D]^\lambda$ belongs to $[C]^\lambda_*$. Since C is homogeneous for P for functions of the correct type, D is homogeneous for P in the ordinary sense. This establishes $\kappa \to (\kappa)^\lambda_2$.

Now suppose $\kappa \to (\kappa)^{\omega\cdot\lambda}_2$.

Suppose $P : [\kappa]^\lambda \to 2$. Let $\mathsf{block} : [\kappa]^{\omega\cdot\lambda} \to [\kappa]^\lambda$ be defined by $\mathsf{block}(f)(\gamma) = \sup\{\omega \cdot \gamma + n : n \in \omega\}$. Define $P' : [\kappa]^{\omega\cdot\lambda} \to 2$ by $P'(f) = P(\mathsf{block}(f))$. Let $A \subseteq \omega_1$ be such that $|A| = \kappa$ and A is homogeneous for P' in the ordinary sense. Without loss of generality, suppose $P'(h) = 0$ for all $h \in [A]^{\omega\cdot\lambda}$. Let C be the collection of limit points of A. That is, $C = \{\alpha \in \kappa : \alpha = \sup(C \cap \alpha)\}$. C is a club subset of ω_1.

Suppose $f \in [C]^\lambda_*$. Since f is of the correct type, let $g : \lambda \times \omega \to \omega_1$ witness that it has uniform cofinality ω. Let $\gamma < \kappa$. Since f is discontinuous everywhere, $\sup\{f(\alpha) : \alpha < \gamma\} < f(\gamma)$. Therefore there is some $N_\gamma \in \omega$ so that for all $n \geq N_\gamma$, $g(\gamma, n) > \sup\{f(\alpha) : \alpha < \gamma\}$. Since $f(\gamma)$ is a limit point of A, define by recursion, $h(\omega \cdot \gamma + n)$ to be the least element of A greater than

$$\max\{g(\gamma, N_\gamma + n), h(\omega \cdot \gamma + k) : k < n\}.$$

Then $h \in [A]^{\omega\cdot\gamma}$ and $\mathsf{block}(h) = f$. Thus $P(f) = P'(h) = 0$.

It has been shown that for all $f \in [C]^\lambda_*$, $P(f) = 0$. $\kappa \to_* (\kappa)^\lambda_2$ has been established. $\square$

Partition properties on a cardinal κ yield other very interesting properties on κ. Much of the material of the remainder of this section can be found in [**20**].

FACT 2.7. *$\kappa \to (\kappa)^2_2$ implies that κ is regular.*

PROOF. Suppose $\eta < \kappa$ and $h : \eta \to \kappa$ is a cofinal function. Define $P : [\kappa]^2 \to 2$ by

$$P(\alpha, \beta) = \begin{cases} 0 & (\exists \gamma < \eta)(\alpha \leq h(\gamma) \leq \beta) \\ 1 & \text{otherwise} \end{cases}.$$

Let $A \subseteq \omega_1$ with $|A| = \kappa$ be homogeneous for P. If A is homogeneous for P taking value 0, then one can show that $\text{ot}(\text{rang}(h)) = \kappa$, which is a contradiction. If A is homogeneous for 1, then one can show that $\text{rang}(h) \subseteq \min A$. This violates $\kappa \to (\kappa)^2_2$. □

DEFINITION 2.8. Let κ be a regular cardinal and $\eta < \kappa$ be a limit ordinal. A set $C \subseteq \kappa$ is η-closed if and only if for all $f : \eta \to C$ increasing, $\sup(f) \in C$. $C \subseteq \kappa$ is a η-club if and only if C is η-closed and unbounded.

Let W^κ_η denote the filter of sets containing an η-club as a subset.

FACT 2.9. *Let κ be a regular cardinal and $\eta < \kappa$ be a limit ordinal. Let $\delta < \kappa$ and $\langle C_\alpha : \alpha < \delta\rangle$ be a sequence of η-clubs. Then $\bigcap_{\alpha \in \delta} C_\alpha$ is an η-club.*

PROOF. Clearly $\bigcap_{\gamma < \delta} C_\gamma$ is η-closed. One needs to show that $\bigcap_{\gamma<\delta} C_\gamma$ is unbounded.

Fix $\epsilon < \kappa$. Let β^0_0 be the least element of C_0 greater than ϵ.

Suppose for $\alpha < \eta$ and $\gamma < \delta$, β^ξ_α has been defined for all $\xi < \gamma$. Note $\sup\{\beta^\xi_\alpha : \xi < \gamma\} < \kappa$ by the regularity of κ. Let β^γ_α be the least element of C_γ which is larger than $\sup\{\beta^\xi_\alpha : \xi < \gamma\}$.

Suppose for $\alpha < \eta$, β^γ_ν has been defined for all $\nu < \alpha$, and $\gamma < \delta$. By the regularity of κ, $\sup\{\beta^\gamma_\nu : \nu < \alpha \wedge \gamma < \delta\} < \kappa$. Let β^0_α be the least element of C_0 greater than $\sup\{\beta^\gamma_\nu : \nu < \alpha \wedge \gamma < \delta\}$.

Note that for all $\gamma_0, \gamma_1 \in \delta$, $\sup\{\beta^{\gamma_0}_\alpha : \alpha < \eta\} = \sup\{\beta^{\gamma_1}_\alpha : \alpha < \eta\}$. Let λ denote this common value. $\lambda \in C_\gamma$ for all $\gamma < \delta$ since C_γ is an η-club. Thus $\lambda \in \bigcap_{\gamma<\delta} C_\gamma$ and $\lambda > \epsilon$. This shows that $\bigcap_{\gamma<\delta} C_\gamma$ is unbounded. □

In ZF, Fact 2.9 does not imply that W^κ_η is κ-complete. Suppose $\langle A_\alpha : \alpha < \delta\rangle$ where $\delta < \kappa$ is a sequence in W^κ_η. For each A_α, there is an η-club $C \subseteq A_\alpha$. To apply Fact 2.9, one would need to produce a sequence of η-clubs, $\langle C_\alpha : \alpha < \delta\rangle$, so that $C_\alpha \subseteq A_\alpha$ for each $\alpha < \delta$. This appears to require some choice principle.

It will be shown next that appropriate partition properties imply that W^κ_γ is an ultrafilter, is κ-complete, and is normal.

FACT 2.10. *Let κ be a regular cardinal and $\eta < \kappa$ be a limit ordinal. Let $A \subseteq \kappa$ be an unbounded set. Let* $\text{Lim}^\eta(A) = \{\sup(f) : f \in [A]^\eta\}$. *That is,* $\text{Lim}^\eta(A)$ *is the collection of all $\alpha \in \kappa$ which are the supremum of an η-increasing sequence through A.*

Then $\text{Lim}^\eta(A)$ *is an η-club.*

PROOF. $\text{Lim}^\eta(A)$ is clearly unbounded since A is unbounded.

Suppose $f : \eta \to \text{Lim}^\eta(A)$ is an increasing function. For each $\alpha < \eta$, let $g(\alpha)$ be the least element $\gamma \in A$ so that $f(\alpha) < \gamma < f(\alpha + 1)$ which exists since

$f(\alpha + 1) = \sup(h)$ for some $h \in [A]^\eta$. Thus $\sup(f) = \sup(g)$ and $g \in [A]^\eta$. Thus $\sup(f) \in \mathrm{Lim}^\eta(A)$. $\mathrm{Lim}^\eta(A)$ is an η-club. □

FACT 2.11. *Let κ be a regular cardinal and $\eta < \kappa$ be a limit ordinal. Suppose $\kappa \to (\kappa)^\eta_2$. Then W^κ_η is an ultrafilter.*

PROOF. Let $A \subseteq \kappa$. Define the partition $P : [\kappa]^\eta \to 2$ by $P(f) = 1 \Leftrightarrow \sup(f) \in A$.

By $\kappa \to (\kappa)^\eta_2$, let $B \subseteq \kappa$ with $|B| = \kappa$ be homogeneous for this partition. Without loss of generality, suppose B is homogeneous taking value 1. $\mathrm{Lim}^\eta(B)$ is an η-club by Fact 2.10. Let $\alpha \in \mathrm{Lim}^\eta(B)$. Let $f \in [B]^\eta$ be such that $\sup(f) = \alpha$. Then $P(f) = 1$ implies that $\alpha = \sup(f) \in A$. This shows that $\mathrm{Lim}^\eta(B) \subseteq A$ which implies $A \in W^\kappa_\eta$.

If B was homogeneous for P taking value 0, then the same argument would have shown $\kappa \setminus A \in W^\kappa_\eta$. □

FACT 2.12. *Assume κ is a regular cardinal and $\eta < \kappa$ is a limit ordinal. Let $\lambda < \kappa$ be infinite. If $\kappa \to (\kappa)^\eta_\lambda$ holds, then W^κ_η is λ^+-complete.*

PROOF. Let $\langle A_\alpha : \alpha < \lambda\rangle$ be a sequence in W^κ_η. Suppose $\bigcap_{\alpha<\lambda} A_\alpha \notin W^\kappa_\eta$. Since W^κ_η is an ultrafilter by Fact 2.11, one may assume that $\bigcap_{\alpha<\lambda} A_\alpha = \emptyset$ by adding one further set to the sequence.

Let $P : [\kappa]^\eta \to \lambda$ be defined by $P(f)$ is the least $\alpha < \lambda$ so that $\sup(f) \notin A_\alpha$. By $\kappa \to (\kappa)^\eta_\lambda$, there is some α^* and a $B \subseteq \kappa$ so that $|B| = \kappa$ and $P(f) = \alpha^*$ for all $f \in [B]^\eta$.

By Fact 2.10, $\mathrm{Lim}^\eta(B) \in W^\kappa_\eta$. However, $\mathrm{Lim}^\eta(B) \cap A_{\alpha^*} = \emptyset$. This contradicts $A_{\alpha^*} \in W^\kappa_\eta$. □

FACT 2.13. *Assume κ is a regular cardinal and $\eta < \kappa$ is a limit ordinal. $\kappa \to (\kappa)^{\eta+\eta}_2$ implies $\kappa \to (\kappa)^\eta_\lambda$ for all $\lambda < \kappa$. Therefore, $\kappa \to (\kappa)^{\eta+\eta}_2$ implies W^κ_η is a κ-complete measure on κ.*

If κ has the weak partition property, then W^κ_η is a κ-complete measure on κ for all limit ordinals $\eta < \kappa$.

PROOF. Let $P : [\kappa]^\eta \to \lambda$. Define $Q : [\kappa]^{\eta+\eta} \to 2$ by $Q(f_0\hat{\ }f_1) = 0$ if and only if $P(f_0) = P(f_1)$. By $\kappa \to (\kappa)^{\eta+\eta}_2$, let $A \subseteq \kappa$ with $|A| = \kappa$ be homogeneous for Q.

Suppose A is homogeneous for Q taking value 1. Let $Q' : [A]^{\eta+\eta} \to 2$ be defined by

$$Q'(f_0\hat{\ }f_1) = \begin{cases} 0 & P(f_0) < P(f_1) \\ 1 & P(f_0) > P(f_1) \end{cases}$$

Again by $\kappa \to (\kappa)^{\eta+\eta}_2$, there is a $B \subseteq A$ with $|B| = \kappa$ which is homogeneous for Q'. One can check that if B is homogeneous for Q' taking value 1, then one would have an infinite descending sequences of ordinals. If B is homogeneous for Q' taking value 0, then one can produce an injection of κ into $\lambda < \kappa$. B can not be homogeneous. Contradiction. Therefore, A must have been homogeneous for Q taking value 0.

Now the claim is that A is homogeneous for P: Let $f_0, f_1 \in [A]^\eta$. Find $f_2 \in [A]^\eta$ such that $\min(f_2)$ is larger than both $\sup(f_0)$ and $\sup(f_1)$. Then $f_0\hat{\ }f_2 \in [A]^{\eta+\eta}$ and $f_1\hat{\ }f_2 \in [A]^{\eta+\eta}$. Since A is homogeneous for Q taking value 0, $Q(f_0\hat{\ }f_2) = 0$ and $Q(f_1\hat{\ }f_2) = 0$. This implies that $P(f_0) = P(f_2) = P(f_1)$. Thus A is homogeneous for P. □

FACT 2.14. *Let κ be a regular cardinal and $\eta < \kappa$ be a limit ordinal. Suppose $\kappa \to (\kappa)^{\eta+\eta}_2$. Then κ is a normal κ-complete ultrafilter.*

If κ has the weak partition property, then W^κ_η is a normal κ-complete measure on κ for each limit ordinal $\eta < \kappa$.

PROOF. By Fact 2.13, W^κ_η is a κ-complete ultrafilter.

Let $F : \kappa \to \kappa$ be a regressive function. That is, $A' = \{\alpha \in \kappa : F(\alpha) < \alpha\} \in W^\kappa_\eta$. Let $A \subseteq A'$ be a η-club set. Define $P : [\kappa]^\eta \to 2$ by

$$P(f) = \begin{cases} 0 & F(\sup(f)) < \min f \\ 1 & \text{otherwise} \end{cases}$$

Since $\kappa \to (\kappa)^{\eta+\eta}_2$ implies $\kappa \to (\kappa)^\eta_2$, let $B \subseteq A$ with $|B| = \kappa$ be homogeneous for P.

Let $\tilde{B} = \{\mathsf{enum}_B(\eta \cdot \alpha + \eta) : \alpha < \kappa\}$, where $\mathsf{enum}_B : \kappa \to \kappa$ is the increasing enumeration of B. Suppose $f \in [\tilde{B}]^\eta$. Since $\tilde{B} \subseteq B \subseteq A$ and A is η-closed, $\sup(f) \in A \subseteq A'$. Therefore, $F(\sup(f)) < \sup(f)$. Let $\gamma < \eta$ be least so that $F(\sup(f)) < f(\gamma)$. Since $f(\gamma + 1) \in \tilde{B}$, let $f_0 : \gamma + 1 \to B$ be such that for all $\nu < \gamma+1$, $f(\gamma) < f_0(\nu) < f(\gamma+1)$. Let $f_1 : (\eta - (\gamma+1)) \to \tilde{B}$ be the tail of f above $f(\gamma)$. Note that $f_2 = f_0\hat{\ } f_1$ is an η-sequence in B with the property that $\min(f_2) > f(\gamma) > F(\sup(f))$ and $\sup(f_2) = \sup(f)$. Thus $F(\sup(f_2)) = F(\sup(f)) < \min(f_2)$. Since $f_2 \in [B]^\eta$, one has that B must be homogeneous for P taking value 0.

Let $f \in [B]^\eta$. Let $f' : \eta \to B$ be the increasing enumeration of $\{\min B\} \cup \mathrm{rang}(f)$. Thus $P(f') = 0$ implies that $F(\sup(f)) = F(\sup(f')) < \min(f') = \min(B)$. It has been shown that for all $f \in [B]^\eta$, $F(\sup(f)) < \min(B)$. Thus for all $\alpha \in \mathrm{Lim}^\eta(B)$, $F(\alpha) < \min(B)$. Since $\mathrm{Lim}^\eta(B) \in W^\kappa_\eta$ and W^κ_η is κ-complete, there is some $\xi < \min(B)$ and $C \in W^\kappa_\eta$ so that $F[C] = \xi$. It has been shown that F is constant W^κ_η-almost everywhere. Normality has been established. □

Under suitable circumstances on κ, one can determine all the normal κ-complete measures on κ.

FACT 2.15. *Let κ be a regular cardinal and $\eta < \kappa$ be a limit ordinal. Assume $\kappa \to (\kappa)^\eta_2$. Then $W^\kappa_\eta = W^\kappa_{\mathrm{cof}(\eta)}$.*

Suppose that $\eta_0 < \eta_1$ are two infinite regular cardinals less than κ. Then $W^\kappa_{\eta_0} \neq W^\kappa_{\eta_1}$.

Suppose the collection of infinite regular cardinals below κ has cardinality less than κ. Let μ be any κ-complete normal measure on κ. There is some infinite regular cardinal $\eta < \kappa$ so that μ is equivalent to W^κ_η.

PROOF. The partition property $\kappa \to (\kappa)^\eta_2$ implies that W^κ_η and $W^\kappa_{\mathrm{cof}(\eta)}$ are both ultrafilters.

Suppose $A \subseteq \kappa$ is a $\mathrm{cof}(\eta)$-club. Let $f \in [A]^\eta$. Let $\rho : \mathrm{cof}(\eta) \to \eta$ be a cofinal increasing sequence. Note that $f \circ \rho \in [A]^{\mathrm{cof}(\eta)}$. Since A is a $\mathrm{cof}(\eta)$-club, $\sup(f \circ \rho) = \sup(f) \in A$. Thus A is also an η-club. This shows that $W^\kappa_{\mathrm{cof}(\eta)} \subseteq W^\kappa_\eta$.

Suppose that $\neg(W^\kappa_\eta \subseteq W^\kappa_{\mathrm{cof}(\eta)})$. Let $A \in W^\kappa_\eta$ be such that $A \notin W^\kappa_{\mathrm{cof}(\eta)}$. Since $W^\kappa_{\mathrm{cof}(\eta)}$ is an ultrafilter, $\kappa \setminus A \in W^\kappa_{\mathrm{cof}(\eta)}$. It has already been shown that $W^\kappa_{\mathrm{cof}(\eta)} \subseteq W^\kappa_\eta$. Therefore, $\kappa \setminus A \in W^\kappa_\eta$. However, $A, \kappa \setminus A \in W^\kappa_\eta$ contradicts the fact that W^κ_η is an ultrafilter.

It has been established that $W^\kappa_\eta = W^\kappa_{\mathrm{cof}(\eta)}$.

Now suppose $\eta_0 < \eta_1$ are two regular cardinals less than κ. Let $A_i = \{\alpha < \kappa : \mathrm{cof}(\alpha) = \eta_i\}$. For $i \in 2$, A_i is an η_i-club. Therefore, $A_i \in W^\kappa_{\eta_i}$. If $W^\kappa_{\eta_0} = W^\kappa_{\eta_1}$, then $A_0 \cap A_1 = \emptyset \in W^\kappa_{\eta_1}$. This contradicts $W^\kappa_{\eta_1}$ is a filter.

Now suppose the collection K of infinite regular cardinals below κ has cardinality less than κ. Let μ be a κ-complete normal measure on κ.

Let T be the set of limit ordinals below κ. Since κ and T are in bijection, μ is equivalent to a measure concentrating on T. Therefore, assume for simplicity that $T \in \mu$. For each $\eta \in K$, let $A_\eta = \{\alpha \in \kappa : \mathrm{cof}(\alpha) = \eta\}$. $T = \bigcup_{\eta \in K} A_\eta$. Since μ is κ-complete and $|K| < \kappa$, there is some η_0 so that $A_{\eta_0} \in \mu$.

The claim is that $\mu = W^\kappa_{\eta_0}$: Let $B \in W^\kappa_{\eta_0}$. Let $C \subseteq B$ be an η_0-club set. Suppose $B \notin \mu$. Then $\kappa \setminus B \in \mu$. Let $Q = A_{\eta_0} \cap (\kappa \setminus B)$. Note $Q \in \mu$.

Let $\alpha \in Q$. Note that $\mathrm{cof}(\alpha) = \eta_0$ and $\alpha \notin B$ so $\alpha \notin C$. $\sup(C \cap \alpha) < \alpha$ since otherwise $\alpha \in C$ since $\mathrm{cof}(\alpha) = \eta_0$ and C is an η_0-club. Let $G : Q \to \kappa$ be defined $G(\alpha) = \sup(C \cap \alpha)$. G is a regressive function on Q. Since μ is normal, there is some $\gamma < \kappa$ so that $J = \{\alpha : G(\alpha) = \gamma\} \in \mu$. For any $\alpha \in J$, $C \cap \alpha \subseteq \gamma$. Since $J \in \mu$ implies that J is unbounded, this implies that $C \subseteq \gamma$. This is impossible since C is unbounded. □

3. Good Coding of Functions

DEFINITION 3.1. Let κ be a regular cardinal and $\lambda \leq \kappa$ be an ordinal. A good coding system for ${}^\lambda\kappa$ consists of Γ, decode, and $\mathsf{GC}_{\beta,\gamma}$ for each $\beta < \lambda$ and $\gamma < \kappa$ with the following properties:

(1) Γ is a pointclass closed under continuous substitution and $\exists^\mathbb{R}$. Let $\check{\Gamma}$ denote the dual pointclass. Let $\Delta = \Gamma \cap \check{\Gamma}$.

(2) $\mathsf{decode} : \mathbb{R} \to \mathscr{P}(\lambda \times \kappa)$. For all $f \in {}^\lambda\kappa$, there is some $x \in \mathbb{R}$ so that $\mathsf{decode}(x) = f$.

(3) For all $\beta < \lambda$ and $\gamma < \kappa$, $\mathsf{GC}_{\beta,\gamma} \subseteq \mathbb{R}$, $\mathsf{GC}_{\beta,\gamma} \in \Delta$, and $\mathsf{GC}_{\beta,\gamma}$ is defined by $x \in \mathsf{GC}_{\beta,\gamma}$ if and only if

$$\mathsf{decode}(x)(\beta,\gamma) \wedge (\forall \gamma' < \kappa)(\mathsf{decode}(x)(\beta,\gamma') \Rightarrow \gamma = \gamma').$$

For each $\beta < \lambda$, let $\mathsf{GC}_\beta = \bigcup_{\gamma<\kappa} \mathsf{GC}_{\beta,\gamma}$.

(4) (Boundedness property) Suppose $A \in \exists^\mathbb{R}\Delta$ and $A \subseteq \mathsf{GC}_\beta$, then there exists some $\delta < \kappa$ so that $A \subseteq \bigcup_{\gamma<\delta} \mathsf{GC}_{\beta,\gamma}$.

(5) Δ is closed under less than κ length wellordered unions.

Suppose $x \in \mathbb{R}$, let $\mathsf{fail}(x)$ be the least $\beta < \lambda$ so that $x \notin \mathsf{GC}_\beta$ if it exists. Otherwise, let $\mathsf{fail}(x) = \infty$.

Let $\mathsf{GC} = \bigcap_{\beta<\lambda} \mathsf{GC}_\beta$. Note that if $x \in \mathsf{GC}$, then $\mathsf{decode}(x)$ is the graph of a function in ${}^\lambda\kappa$. If $x \in \mathsf{GC}$, then one will use function notations such as $\mathsf{decode}(x)(\beta) = \gamma$ to indicate $(\beta,\gamma) \in \mathsf{decode}(x)$.

Assuming AD, (5) follows from the other four conditions. This comes from a pointclass argument. See the end of the proof of Theorem 2.34 of [**12**]. Later, one will apply this to ω_1 with the associated pointclass $\mathbf{\Sigma}^1_1$. It is clear that (5) holds in this setting since $\mathbf{\Delta}^1_1$ is closed under countable unions.

REMARK 3.2. The meaning of $x \in \mathsf{GC}_{\beta,\gamma}$ is that x is good at β in the sense that $\mathsf{decode}(x)$ has successfully mapped β to γ. One interprets $x \in \mathsf{GC}_\beta$ to be mean that x is good at β in the sense that $\mathsf{decode}(x)$ has successfully mapped β to some value. So a reals x belongs to GC means that x is a good code in the sense that $\mathsf{decode}(x)$ is truly a function from λ to κ.

DEFINITION 3.3. Let κ be a regular cardinal and let $\lambda \leq \kappa$ be an ordinal. Let $(\Gamma, \mathsf{decode}, \mathsf{GC}_{\beta,\gamma} : \beta < \lambda, \gamma < \kappa)$ be a coding system for ${}^\lambda\kappa$.

Let S^1 consists of reals r coding a Lipschitz continuous function $\Xi_r : \mathbb{R} \to \mathbb{R}$ which has the following property:

$$(\forall y)[\mathsf{fail}(y) \leq \mathsf{fail}(\Xi_r(y)) \wedge (\mathsf{fail}(y) < \infty \Rightarrow \mathsf{fail}(y) < \mathsf{fail}(\Xi_r(y)))].$$

Let S^2 consists of Player 2 strategies s so that its associated Lipschitz continuous function $\Xi_s : \mathbb{R} \to \mathbb{R}$ has the property that $(\forall x)(\mathsf{fail}(x) \leq \mathsf{fail}(\Xi_s(x))$.

FACT 3.4. *Let κ be a regular cardinal and $\lambda \leq \kappa$ be an ordinal. Let $(\Gamma, \mathsf{decode}, \mathsf{GC}_{\beta,\gamma} : \beta < \lambda, \gamma < \kappa)$ be a coding system for ${}^\lambda\kappa$.*

If $r \in S^1$, then there is a club $C \subseteq \kappa$ (obtained uniformly from r) so that for all $\delta \in C$, for all $\beta < \min\{\lambda, \delta\}$ and $\gamma < \delta$,

$$\Xi_r\left[\bigcap_{\beta'<\beta}\bigcup_{\gamma'<\gamma} \mathsf{GC}_{\beta',\gamma'}\right] \subseteq \bigcap_{\beta'\leq\beta}\bigcup_{\gamma'<\delta} \mathsf{GC}_{\beta',\gamma'}$$

Suppose $s \in S^2$, then there is a club $C \subseteq \kappa$ so that for all $\delta \in C$, for all $\beta < \min\{\lambda, \delta\}$, for all $\gamma < \delta$,

$$\Xi_s\left[\bigcap_{\beta'\leq\beta}\bigcup_{\gamma'<\gamma} \mathsf{GC}_{\beta',\gamma'}\right] \subseteq \bigcap_{\beta'\leq\beta}\bigcup_{\gamma'<\delta} \mathsf{GC}_{\beta',\gamma'}.$$

PROOF. Suppose $r \in S^1$. For each $\beta < \lambda$ and $\gamma < \kappa$, let

$$R_{\beta,\gamma} = \bigcap_{\beta'<\beta}\bigcup_{\gamma'<\gamma} \mathsf{GC}_{\beta',\gamma'}.$$

So if $x \in R_{\beta,\gamma}$, then $\mathsf{decode}(x)$ is good up to β by successfully mapping each ordinal less than β to some value less than γ. By property (5) of the good coding system, Δ is closed under less than κ wellordered union and also intersections. Property (3) states that $\mathsf{GC}_{\beta',\gamma'} \in \Delta$ for all $\beta' < \lambda$ and $\gamma' < \kappa$. Thus $R_{\beta,\gamma} \in \Delta$. Then $\Xi_r[R_{\beta,\gamma}]$ is $\exists^{\mathbb{R}}\Delta$.

Claim 1: For all $\beta' \leq \beta$, $\Xi_r[R_{\beta,\gamma}] \subseteq \mathsf{GC}_{\beta'}$.

To see Claim 1: Note that $x \in R_{\beta,\gamma}$ implies that $\mathsf{fail}(x) \geq \beta$. Since $r \in S^1$, $\mathsf{fail}(\Xi_r(x)) > \beta$. Thus for all $\beta' \leq \beta$, $\Xi_r(x) \in \mathsf{GC}_{\beta'}$. This shows Claim 1.

Therefore $\Xi_r[R_{\beta,\gamma}]$ is a $\exists^{\mathbb{R}}\Delta$ subset of $\mathsf{GC}_{\beta'}$ for each $\beta' \leq \beta$. By the boundedness property (4) of a good coding system, there is some $\epsilon_{\beta'} < \kappa$ so that $\Xi_r[R_{\beta,\gamma}] \subseteq \bigcup_{\gamma'<\epsilon_{\beta'}} \mathsf{GC}_{\beta',\gamma'}$.

Since κ is regular, let $\Upsilon(\beta, \gamma)$ be the least $\epsilon < \kappa$ so that $\beta < \epsilon$, $\gamma < \epsilon$, and for all $\beta' \leq \beta$, $\Xi_r[R_{\beta,\gamma}] \subseteq \bigcup_{\gamma'<\epsilon} \mathsf{GC}_{\beta',\gamma'}$. Thus $\Upsilon : \lambda \times \kappa \to \kappa$ is a well defined function.

Let $C = \{\delta : (\forall\beta < \delta)(\forall\gamma < \delta)(\Upsilon(\min\{\lambda, \beta\}, \gamma) < \delta)\}$.

Claim 2: C is a club.

To see Claim 2: Let $\alpha < \kappa$. Let $\alpha_0 = \alpha$. If α_n has been defined, let $\alpha_{n+1} = \Upsilon(\min\{\lambda, \alpha_n\}, \alpha_n)$. Let $\alpha_\infty = \sup\{\alpha_n : n \in \omega\}$. By definition of Υ, $\alpha_n < \alpha_{n+1}$ for all n. Let $\beta < \min\{\lambda, \alpha_\infty\}$ and $\gamma < \alpha_\infty$. Then there is some n so that $\beta < \alpha_n$ and $\gamma < \alpha_n$. Then $\Upsilon(\min\{\lambda, \beta\}, \gamma) \leq \Upsilon(\min\{\lambda, \alpha_n\}, \alpha_n) = \alpha_{n+1} < \alpha_\infty$. This shows that $\alpha_\infty \in C$. As $\alpha < \alpha_\infty$, C is unbounded. It is straightforward to show that C is closed. Claim 2 has been shown.

Fix a $\delta \in C$. Pick a $\beta < \min\{\lambda, \delta\}$ and a $\gamma < \delta$. Suppose $x \in R_{\beta,\gamma}$. Let $\beta' \leq \beta$. Then

$$\Xi_r(x) \in \bigcup_{\gamma' < \Upsilon(\beta,\gamma)} \mathsf{GC}_{\beta',\gamma'} \subseteq \bigcup_{\gamma'<\delta} \mathsf{GC}_{\beta',\gamma'}.$$

Since $\beta' \leq \beta$ was arbitrary,

$$\Xi_r(x) \in \bigcap_{\beta' \leq \beta} \bigcup_{\gamma'<\delta} \mathsf{GC}_{\beta',\gamma'}.$$

This completes the proof of the result for S^1.

The argument for S^2 is similar. □

REMARK 3.5. Let $r \in S^1$. Let C be the club produced by Fact 3.4. This club C is called the club on which the opposite player (Player 2) has taken control of the output.

This means the following: In most applications, r codes a Lipschitz function coming from a Player 1 winning strategy. (Hence the notation S^1.) Let $\delta \in C$. Fact 3.4 states that for any code y so that $\mathsf{decode}(y)$ successfully defines a function up to some $\beta < \min\{\lambda, \delta\}$ and takes value some γ strictly below δ, then the Player 1 strategy coded by r when played against y produces some code x which successfully defines a function at and below β and still takes value below δ.

This observation is used in a game to show that although Player 1 may have a winning strategy which determines that a final sequence will land in a certain payoff set, there is a club C so that if the opposite player (Player 2) plays suitable codes for functions through this club, Player 2 actually determines the value of the final sequence. A similar statement holds if Player 2 has the winning strategy in this game.

This is the main idea of the following game of Martin to establish the partition properties and of an earlier game of Solovay to show the club filter is an ultrafilter.

DEFINITION 3.6. Suppose κ is a regular cardinal and λ is such that $\omega \cdot \lambda < \kappa$. Suppose $f \in {}^{\omega\cdot\lambda}\kappa$. Let $\mathsf{block} : {}^{\omega\cdot\lambda}\kappa \to {}^{\lambda}\kappa$ be defined by $\mathsf{block}(f)(\alpha) = \sup\{f(\omega \cdot \alpha + k) : k \in \omega\}$.

Suppose $f, g \in {}^{\omega\cdot\lambda}\kappa$. Let $\mathsf{joint} : {}^{\omega\cdot\lambda}\kappa \times {}^{\omega\cdot\lambda}\kappa \to {}^{\lambda}\kappa$ be defined by

$$\mathsf{joint}(f, g)(\alpha) = \sup\{f(\omega \cdot \alpha + k), g(\omega \cdot \alpha + k) : k \in \omega\}.$$

THEOREM 3.7. *(Martin) Assume* $\mathsf{ZF} + \mathsf{AD}$. *Suppose* λ, κ *are ordinals such that* $\omega \cdot \lambda \leq \kappa$. *Suppose there is a good coding system* $(\Gamma, \mathsf{decode}, \mathsf{GC}_{\beta,\gamma} : \beta \in \omega \cdot \lambda, \gamma < \kappa)$ *for* ${}^{\omega\cdot\lambda}\kappa$. *Then* $\kappa \to_* (\kappa)^\lambda_2$ *holds.*

PROOF. Let $P : [\kappa]^\lambda_* \to 2$. Consider the following game

$$G \quad \begin{array}{lcccccccccc} \mathrm{I} & x_0 & & x_1 & & x_2 & & x_3 & & & x \\ \mathrm{II} & & y_0 & & y_1 & & y_2 & & y_3 & & y \end{array}$$

where Player 1 and 2 take turns playing integers. Player 1 produces the real x, and Player 2 produces the real y. Player 1 wins the game if and only the conjunction of the following two conditions hold:

(1) $\mathsf{fail}(x) > \mathsf{fail}(y) \vee \mathsf{fail}(x) = \mathsf{fail}(y) = \infty$.

(2) $(\mathsf{fail}(x) = \mathsf{fail}(y) = \infty) \Rightarrow P(\mathsf{joint}(\mathsf{decode}(x), \mathsf{decode}(y))) = 0$.

(Case I) Suppose Player 1 has a winning strategy σ in this game.

The strategy σ induces a Lipschitz function Ξ_σ. Explicitly, $\Xi_\sigma(y)$ is just the response of Player 1 using σ when Player 2 plays using y. Note that $\sigma \in S^1$.

Let C be the club from Fact 3.4. On C, Player 2 takes control of the output in the sense below. Let D be the limit points of C.

Let $f \in [D]^\lambda_*$. Since f is of the correct type, let $F : \lambda \times \omega \to \kappa$ be a witness to f having uniform cofinality ω. For each $\alpha < \lambda$, let $\nu_\alpha = \sup\{f(\xi) : \xi < \alpha\}$. Let $g(\omega \cdot \alpha)$ be the least element of C greater than $\max\{\nu_\alpha, F(\alpha, 0)\}$. Suppose $g(\omega \cdot \alpha + k)$ has been defined for some $k \in \omega$. Let $g(\omega \cdot \alpha + k + 1)$ be the least element of C greater than $\max\{g(\omega \cdot \alpha + k), F(\alpha, k + 1)\}$. Now $g \in [C]^{\omega\cdot\lambda}$, g is discontinuous everywhere, and $\mathsf{block}(g) = f$.

By property (2) of the good coding system, there is some $y \in \mathsf{GC}$ so that $\mathsf{decode}(y) = g$. Let Player 2 play this y against Player 1 using σ. Since $y \in \mathsf{GC}$, $\mathsf{fail}(y) = \infty$. Thus $\mathsf{fail}(\Xi_\sigma(r)) = \infty$.

For $\beta < \omega \cdot \lambda$, let $\epsilon_\beta = \sup\{g(\alpha) : \alpha < \beta\}$. Since g is discontinuous everywhere, $\epsilon_\beta < g(\beta)$. Then

$$\Xi_\sigma(y) \in \Xi_\sigma\left[\bigcap_{\beta' < \beta} \bigcup_{\gamma' < \epsilon_\beta} \mathsf{GC}_{\beta',\gamma'}\right] \subseteq \bigcap_{\beta' \leq \beta} \bigcup_{\gamma' < g(\beta)} \mathsf{GC}_{\beta',\gamma'}$$

by the definition of C. Hence $\mathsf{decode}(\Xi_\sigma(y))(\beta) < g(\beta)$. Thus for all $\alpha < \lambda$,

$$\sup\{\mathsf{decode}(\Xi_\sigma(y))(\omega\cdot\alpha+k), \mathsf{decode}(y)(\omega\cdot\alpha+k) : k \in \omega\} = \sup\{g(\omega\cdot\alpha+k) : k \in \omega\} = f(\alpha).$$

This shows that $\mathsf{joint}(\mathsf{decode}(\Xi_\sigma(y)), \mathsf{decode}(y)) = f$. Since σ is a Player 1 winning strategy, one has that $P(f) = 0$. Since $f \in [D]^\lambda_*$ was arbitrary, D is homogeneous for P taking value 0.

(Case II) Suppose Player 2 has a winning strategy τ.

Note that Player 2 wins if and only if the disjunction of the following holds:

(1) $\mathsf{fail}(x) < \mathsf{fail}(y) \vee (\mathsf{fail}(x) = \mathsf{fail}(y) < \infty)$.

(2) $\mathsf{fail}(x) = \mathsf{fail}(y) = \infty \wedge P(\mathsf{joint}(\mathsf{decode}(x), \mathsf{decode}(y))) = 1$.

Therefore, $\tau \in S^2$. Let C be the club coming from Fact 3.4. C is the club for which Player 1 takes control of the output. One may assume that C consists of only limit ordinals. Let D be the limit points of C.

Let $f \in [D]^\lambda_*$. As before, there is a $g \in [C]^{\omega\cdot\lambda}$ such that $\mathsf{block}(g) = f$. Let $x \in \mathsf{GC}$ be such that $\mathsf{decode}(x) = g$. Since $\mathsf{fail}(x) = \infty$, $\mathsf{fail}(\Xi_\tau(x)) = \infty$ and $\Xi_\tau(x) \in \mathsf{GC}$ as well.

Let $\beta < \omega \cdot \lambda$. Note that $g(\beta) + 1 < g(\beta + 1)$

$$\Xi_\tau(x) \in \Xi_\tau\left[\bigcap_{\beta' \leq \beta} \bigcup_{\gamma' < g(\beta)+1} \mathsf{GC}_{\beta',\gamma'}\right] \subseteq \bigcap_{\beta' \leq \beta} \bigcup_{\gamma' < g(\beta+1)} \mathsf{GC}_{\beta',\gamma'}$$

by definition of C. Hence $\mathsf{decode}(\Xi_\tau(x))(\beta) < g(\beta + 1)$. Thus for all $\alpha < \lambda$,

$$\sup\{\mathsf{decode}(x)(\omega\cdot\alpha+k), \mathsf{decode}(\Xi_\tau(x))(\omega\cdot\alpha+k) : k \in \omega\} = \sup\{g(\omega\cdot\alpha+k) : k \in \omega\} = f(\alpha).$$

This shows that $\mathsf{joint}(\mathsf{decode}(x), \mathsf{decode}(\Xi_\tau(x))) = f$. Since τ is a Player 2 winning strategy, one has that $P(f) = 1$. D is homogeneous for P taking value 1.

The proof is complete. $\square$

THEOREM 3.8. *(Almost everywhere uniformization on codes) Assume* $\mathsf{ZF}+\mathsf{AD}$. *Let* κ *be a regular cardinal and* $\lambda < \kappa$. *Let* $(\Gamma, \mathsf{decode}, \mathsf{GC}_{\beta,\gamma} : \beta < \omega \cdot \lambda, \gamma < \kappa)$ *be a good coding system for* ${}^{\omega\cdot\lambda}\kappa$. *Let* $R \subseteq [\kappa]^\lambda_* \times \mathbb{R}$ *be a relation.*

There is a club $C \subseteq \kappa$ *and a Lipschitz continuous function* $F : \mathbb{R} \to \mathbb{R}$ *so that for all* $x \in \mathsf{GC}$ *with* $\mathsf{decode}(x) \in [C]^{\omega\cdot\lambda}$ *and* $\mathsf{block}(\mathsf{decode}(x)) \in [C]^\lambda_* \cap \mathrm{dom}(R)$, $R(\mathsf{block}(\mathsf{decode}(x)), F(x))$.

PROOF. Define a game G as follows:

	I	x_0		x_1		x_2		x_3		x
G										
	II		y_0, z_0		y_1, z_1		y_2, z_2		y_3, z_3	y, z

Player 2 wins if and only if the disjunction of the following hold:
(1) $\mathsf{fail}(x) < \mathsf{fail}(y) \vee \mathsf{fail}(x) = \mathsf{fail}(y) < \infty$.
(2) $(\mathsf{fail}(x) = \mathsf{fail}(y) = \infty) \wedge (\mathsf{joint}(\mathsf{decode}(x), \mathsf{decode}(y)) \in \mathrm{dom}(R) \Rightarrow R(\mathsf{joint}(\mathsf{decode}(x), \mathsf{decode}(y)), z))$.

Claim 1: Player 2 has a winning strategy.

To prove Claim 1: Suppose not. By AD, Player 1 has a winning strategy σ. Note that Player 1 winning condition is the conjunction of the follow:
(1) $\mathsf{fail}(y) < \mathsf{fail}(x) \vee \mathsf{fail}(x) = \mathsf{fail}(y) = \infty$.
(2)$\mathsf{fail}(x) = \mathsf{fail}(y) = \infty \Rightarrow [\mathsf{joint}(\mathsf{decode}(x), \mathsf{decode}(y)) \in \mathrm{dom}(R) \wedge \neg R(\mathsf{joint}(\mathsf{decode}(x), \mathsf{decode}(y)), z)]$.

Let Ξ_σ be the associated continuous function. Note that

$$(\forall y)(\forall z)[\mathsf{fail}(y) \leq \mathsf{fail}(\Xi_\sigma(y,z)) \wedge (\mathsf{fail}(y) < \infty \Rightarrow \mathsf{fail}(y) < \mathsf{fail}(\Xi_\sigma(y,z)))].$$

That is, if one fixes any z, the associated function $F_z(y) = \Xi_\sigma(y, z)$ belongs to S^1. With a small modification to the argument of Fact 3.4, there is a club C with the property that for all $\delta \in C$, for all $z \in \mathbb{R}$, for all $\beta < \min\{\lambda, \delta\}$, and $\gamma < \delta$

$$F_z\left[\bigcap_{\beta'<\beta}\bigcup_{\gamma'<\gamma} \mathsf{GC}_{\beta',\gamma'}\right] \subseteq \bigcap_{\beta'\leq\beta}\bigcup_{\gamma'<\gamma} \mathsf{GC}_{\beta',\gamma'}.$$

That is, C is a club for which Player 2 controls the output against F_z for every $z \in \mathbb{R}$.

Let D be the limit points of C. Let $f \in [D]^\lambda_*$. As in the proof of Theorem 3.7, let $g \in [D]^{\omega\cdot\lambda}$ be such that $\mathsf{block}(g) = f$. Let $y \in \mathsf{GC}$ be such that $\mathsf{decode}(y) = g$. By the same argument as in Theorem 3.7, one has that for all z, $\mathsf{joint}(\mathsf{decode}(\Xi_\sigma(y,z)), \mathsf{decode}(y)) = f$. Since σ is a Player 1 winning strategy, one must have that $f \in \mathrm{dom}(R)$. Since $f \in \mathrm{dom}(R)$, there is some z^* so that $R(f, z^*)$. If Player 2 plays (y, z^*), then Player 1 loses using σ. This contradicts σ being a Player 1 winning strategy. Claim 1 has been shown.

Let τ be a Player 2 winning strategy in this game. Let $G(x) = \pi_1[\Xi_\tau(x)]$, where π_1 refers to the projection onto the first coordinate. Then G is a Lipschitz continuous function satisfying the condition from the definition of S^2. By Fact 3.4, let C be the club on which Player 2 takes controls of the output. Again one may assume C consists of only limit ordinals. Let D be the limit points of C.

Let $f \in [D]^\lambda_* \cap \mathrm{dom}(R)$. Let $g \in [C]^{\omega\cdot\lambda}$ be such that $\mathsf{block}(g) = f$. Let $x \in \mathsf{GC}$ be such that $\mathsf{decode}(x) = g$. By the argument in Theorem 3.7,

$$\mathsf{joint}(\mathsf{decode}(x), \mathsf{decode}(G(x))) = f.$$

Since τ is a Player 2 winning strategy, one must have that $R(f, \pi_2(\Xi_r(x)))$.

Define $F(x) = \pi_2(\Xi_r(x))$. Then F is a Lipschitz function with the desired uniformization property. □

THEOREM 3.9. *Let κ be a regular cardinal and $\lambda \leq \kappa$. Suppose $(\Gamma, \mathsf{decode}, \mathsf{GC}_{\beta,\gamma} : \beta < \lambda, \gamma < \kappa)$ is a good coding system for ${}^\lambda\kappa$. Let $M \models \mathsf{AD}$ be an inner model containing all the reals and within M, $(\Gamma, \mathsf{decode}, \mathsf{GC}_{\beta,\gamma} : \beta < \lambda, \gamma < \kappa)$ is a good coding system.*

Then for any $\Phi : [\kappa]^\lambda \to \kappa$, there is a club D, necessarily in M by the Moschovakis coding lemma, so that $\Phi \restriction [D]^\lambda_* \in M$.

PROOF. The hypothesis implies that $\kappa < \Theta^M$. Let $\pi : \mathbb{R} \to \kappa$ be a surjection in M. Define a relation $R \subseteq [\kappa]^\lambda_* \times \mathbb{R}$ by

$$R(f, x) \Leftrightarrow \Phi(f) = \pi(x)$$

Let the Lipschitz function $F : \mathbb{R} \to \mathbb{R}$ and club $C \subseteq \kappa$ be the objects given by Theorem 3.8. Let D be the set of limit points of C.

Let $f \in [D]^\lambda_*$. Let $x \in \mathsf{GC}$ be any function such that $\mathsf{decode}(x) \in [C]^{\omega\cdot\lambda}_*$ and $\mathsf{block}(\mathsf{decode}(x)) = f$. Then $R(\mathsf{block}(\mathsf{decode}(x)), F(x))$. This means that $\Phi(f) = \pi(F(x))$.

Note that for any x and x' so that
$\mathsf{decode}(x), \mathsf{decode}(x') \in [C]^{\omega\cdot\lambda}_*$ and $\mathsf{block}(\mathsf{decode}(x)) = \mathsf{block}(\mathsf{decode}(x'))$,
one has that $\pi(F(x)) = \pi(F(x')) = \Phi(\mathsf{block}(\mathsf{decode}(x)))$.

By the Moschovakis coding lemma and the fact that $\pi \in M$, C and hence D belongs to M. So within M, one can define $\Phi \restriction [D]^\lambda_*$ as follows, $\Phi(f) = \gamma$ if and only if there exists some $x \in \mathsf{GC}$ so that $\mathsf{decode}(x) \in [C]^{\omega\cdot\lambda}_*$, $\mathsf{block}(\mathsf{decode}(x)) = f$, and $\pi(F(x)) = \gamma$. By the above, this is well defined and works. □

One will say that a cardinal κ is "reasonable" if one is in the situation where there exists a good coding system:

DEFINITION 3.10. (Jackson) Let κ be a regular cardinal and $\lambda \leq \kappa$. κ is λ-reasonable if and only if there is a good coding system for ${}^\lambda\kappa$.

4. Reasonableness at ω_1

DEFINITION 4.1. Fix some recursive pairing function $\pi : \omega \times \omega \to \omega$. A real $x \in {}^\omega 2$ codes a relation R_x defined as follows: $R_x(m, n) \Leftrightarrow x(\pi(m, n)) = 1$.

Let the domain of x be $\mathrm{dom}(x) = \{n : (\exists m)(R_x(m, n) \vee R_x(m, n))\}$. Let LO be the set of reals x so that R_x is a linear ordering on its domain. Let WO be the set of reals x so that R_x is a well ordering. LO is an arithmetic set of reals. WO is Π^1_1.

If $x \in \mathrm{WO}$, then $\mathrm{ot}(x)$ is the order type of x. If $\beta < \mathrm{ot}(x)$, let n^x_β denote the unique element of $\mathrm{dom}(x)$ whose rank according to R_x is β.

If $\alpha < \omega_1$, then let $\mathrm{WO}_\alpha = \{x \in \mathrm{WO} : \mathrm{ot}(x) = \alpha\}$. $\mathrm{WO}_{<\alpha} = \{x \in \mathrm{WO} : \mathrm{ot}(x) < \alpha\}$. $\mathrm{WO}_{\leq\alpha}$, $\mathrm{WO}_{>\alpha}$, and $\mathrm{WO}_{\geq\alpha}$ are defined similarly. WO_α, $\mathrm{WO}_{<\alpha}$, and $\mathrm{WO}_{\leq\alpha}$ are $\mathbf{\Delta}^1_1$ (in any element of WO_α).

DEFINITION 4.2. The ordertype function $\mathrm{ot} : \mathrm{WO} \to \omega_1$ is a Π^1_1 norm. Let $\preceq$ denote the induced prewellordering: $x \preceq y$ if and only if $\mathrm{ot}(x) \leq \mathrm{ot}(y)$.

Being a Π_1^1-norm implies that there is a Σ_1^1 relation $\leq_{\Sigma_1^1}$ and a Π_1^1 relation $\leq_{\Pi_1^1}$ so that

$$y \in \mathrm{WO} \Rightarrow (\forall x)[(x \in \mathrm{WO} \wedge x \preceq y) \Leftrightarrow (x \leq_{\Sigma_1^1} y) \Leftrightarrow (x \leq_{\Pi_1^1} y)].$$

It is useful to have a concrete coding of a nice collection of club subsets of ω_1.

FACT 4.3. *Let τ be a Player 2 strategy with the property that for all $w \in \mathrm{WO}$, $\tau(w) \in \mathrm{WO} \wedge \mathrm{ot}(w) < \mathrm{ot}(\tau(w))$. Let $C_\tau = \{\eta : (\forall w \in \mathrm{WO}_{<\eta})(\mathrm{ot}(\tau(w)) < \eta)\}$. C_τ is a club.*

PROOF. Let $R_\alpha = \{\tau(w) : w \in \mathrm{WO}_{<\alpha}\}$. $R_\alpha \subseteq \mathrm{WO}$ is $\boldsymbol{\Sigma}_1^1$. By boundedness, $\sup\{\mathrm{ot}(v) : v \in R_\alpha\} < \omega_1$.

Let $\Phi(\alpha) = \sup\{\mathrm{ot}(\tau(w)) : w \in \mathrm{WO}_{<\alpha}\} + 1$. By the above, $\Phi(\alpha)$ is defined.

It is clear that C_τ is closed.

Let $\alpha < \omega_1$. Let $\alpha_0 = \alpha$. Let $\alpha_{n+1} = \Phi(\alpha_n)$. Let $\eta = \sup\{\alpha_n : n \in \omega\}$. Let $\beta < \eta$. There is some n so that $\beta < \alpha_n$. Let $w \in \mathrm{WO}_\beta \subseteq \mathrm{WO}_{<\alpha_n}$. Thus $\mathrm{ot}(\tau(w)) < \Phi(\alpha_n) = \alpha_{n+1} < \eta$. So $\eta \in C_\tau$. This shows that C_τ is unbounded. □

DEFINITION 4.4. Let $\mathsf{clubcode}_{\omega_1}$ be the set of Player 2 strategies so that for all $w \in \mathrm{WO}$, $\tau(w) \in \mathrm{WO} \wedge \mathrm{ot}(w) < \mathrm{ot}(\tau(w))$.

Note that $\mathsf{clubcode}_{\omega_1}$ is a Π_2^1 set of reals.

FACT 4.5. *Assume that $\tau \in \mathsf{clubcode}_{\omega_1}$. The relation $S(w)$ defined by $w \in \mathrm{WO} \wedge \mathrm{ot}(w) \in C_\tau$ is $\boldsymbol{\Pi}_1^1$.*

Assume $\mathsf{AC}_\omega^\mathbb{R}$. If $\alpha < \omega_1$, then the relation $T_\alpha(w)$ defined by $w \in \mathrm{WO}_{<\alpha} \wedge \mathrm{ot}(w) \in C_\tau$ is $\boldsymbol{\Delta}_1^1$.

PROOF. Note that $S(w)$ holds if and only if $w \in \mathrm{WO} \wedge (\forall v)(v <_{\Sigma_1^1} w \Rightarrow \tau(v) <_{\Pi_1^1} w)$, where $<_{\Sigma_1^1}$ and $<_{\Pi_1^1}$ are Σ_1^1 and Π_1^1 relations, respectively, coming from the ordertype function, ot, being a Π_1^1-norm.

Note that $\alpha \cap C_\tau$ is a countable set. Using $\mathsf{AC}_\omega^\mathbb{R}$, let $r \in \omega$ be such that $\{\mathrm{ot}(r_n) : n \in \omega\} = \alpha \cap C_\tau$. (Here $r_n(m) = r(\langle n, m\rangle)$, where $\langle\cdot,\cdot\rangle : \omega \times \omega \to \omega$ is a recursive bijective pairing function.) One can see that T_α is $\boldsymbol{\Delta}_1^1$ using as parameters r and some element of WO_α. □

FACT 4.6. *Assume $\mathsf{ZF} + \mathsf{AD}$. Let $C \subseteq \omega_1$ be a club. There is a $\tau \in \mathsf{clubcode}_{\omega_1}$ so that $C_\tau \subseteq C$.*

PROOF. Consider the game G_C:

	I	w_0		w_1		w_2		w_3		w
G_C										
	II		v_0		v_1		v_2		v_3	v

Player 2 wins this game if and only if

$$w \in \mathrm{WO} \Rightarrow (v \in \mathrm{WO} \wedge \mathrm{ot}(v) > \mathrm{ot}(w) \wedge \mathrm{ot}(v) \in C).$$

Claim 1: Player 2 has the winning strategy.

Suppose σ is a Player 1 winning strategy. Note that $\sigma[\mathbb{R}] \subseteq \mathrm{WO}$ and is $\boldsymbol{\Sigma}_1^1$. By boundedness, $\gamma = \sup\{\mathrm{ot}(w) : w \in \sigma[\mathbb{R}]\} < \omega_1$. Since C is unbounded, let $\delta \in C$ be such that $\delta > \gamma$. Let $v \in \mathrm{WO}_\delta$. If Player 2 plays v against σ, $v \in \mathrm{WO}$, $\mathrm{ot}(v) = \delta \in C$, and $\mathrm{ot}(\sigma(v)) \leq \gamma < \delta = \mathrm{ot}(v)$. Player 2 has won. This contradicts σ being a Player 1 winning strategy.

Let τ be a Player 2 winning strategy. It is clear that $\tau \in \mathsf{clubcode}_{\omega_1}$.

Claim 2: $C_\tau \subseteq C$.

Suppose $\eta \in C_\tau$. Let $\beta < \eta$. Let $w \in \mathrm{WO}_\beta$. Then $\beta < \mathrm{ot}(\tau(w)) < \eta$ and $\mathrm{ot}(\tau(w)) \in C$. Since $\beta < \eta$ was arbitrary and C is closed, $\eta \in C$. □

FACT 4.7. *Suppose $A \subseteq \mathsf{clubcode}_{\omega_1}$ is $\mathbf{\Sigma}^1_1$. Then uniformly from A, there is a club C so that for all $\tau \in A$, $C \subseteq C_\tau$.*

PROOF. For each $\beta < \omega_1$, let $R_\beta = \{\tau(w) : \tau \in A \wedge w \in \mathrm{WO}_{<\beta}\}$. $R_\beta \subseteq \mathrm{WO}$ and is $\mathbf{\Sigma}^1_1$. By the boundedness principle, $\sup\{\mathrm{ot}(v) : v \in R_\beta\} < \omega_1$.

Let $\Phi(\beta) = \sup\{\mathrm{ot}(v) : v \in R_\beta\} + 1$.

Let $C = \{\eta : (\forall \beta < \eta)(\Phi(\beta) < \eta)\}$. As before, one can check that C is a club subset of ω_1.

Fix $\tau \in A$. Let $\eta \in C$. Suppose $\beta < \eta$. Let $w \in \mathrm{WO}_\beta$. Then $\beta < \mathrm{ot}(\tau(w)) < \Phi(\beta) < \eta$. Thus $\eta \in C_\tau$. □

FACT 4.8. *Assume $\mathsf{ZF} + \mathsf{AD}$. Let $\langle \mathcal{A}_\alpha : \alpha < \omega_1\rangle$ be such that each $\mathcal{A}_\alpha$ is a nonempty $\subseteq$-downward closed collection of club subsets of ω_1. Then there is a sequence $\langle C_\alpha : \alpha < \omega_1\rangle$ with each $C_\alpha \subseteq \omega_1$ a club subset of ω_1 and $C_\alpha \in \mathcal{A}_\alpha$.*

In particular: Let μ be the club measure on ω_1. Let $\langle A_\alpha : \alpha < \omega_1\rangle$ be a sequence of sets in μ, that is each A_α contains a club subset of ω_1. Then there is a sequence $\langle C_\alpha : \alpha < \omega_1\rangle$ with each $C_\alpha \subseteq \omega_1$ a club subset of ω_1 so that $C_\alpha \subseteq A_\alpha$.

PROOF. Consider the game

$$G \quad \begin{array}{lcccccccccc} \mathrm{I} & w(0) & & w(1) & & w(2) & & w(3) & & & w \\ \mathrm{II} & & z(0) & & z(1) & & z(2) & & z(3) & & z \end{array}$$

where Player 2 wins if and only if $w \in \mathrm{WO} \Rightarrow (z \in \mathrm{clubcode}_{\omega_1} \wedge C_z \in \mathcal{A}_{\mathrm{ot}(w)})$.

Claim 1: Player 2 has a winning strategy in this game.

To prove this, suppose otherwise that Player 1 has a winning strategy σ. Note that $\sigma[\mathbb{R}] \subseteq \mathrm{WO}$. Since $\sigma[\mathbb{R}]$ is a $\mathbf{\Sigma}^1_1$ subset of WO, by the boundedness principle, there is a $\zeta \in \omega_1$ so that for all $w \in \sigma[\mathbb{R}]$, $\mathrm{ot}(w) < \zeta$. By $\mathsf{AC}^{\mathbb{R}}_\omega$, pick $\langle C_\alpha : \alpha < \zeta\rangle$ with the property that for all $\alpha < \zeta$, $C_\alpha \in \mathcal{A}_\alpha$. Let $C = \bigcap_{\alpha<\zeta} C_\alpha$ which is also a club. By Fact 4.6, there is some $z \in \mathrm{clubcode}_{\omega_1}$ so that $C_z \subseteq C$. Note that since for each $\alpha < \zeta$, $C_z \subseteq C_\alpha$, $C_\alpha \in \mathcal{A}_\alpha$, and $\mathcal{A}_\alpha$ is $\subseteq$-downward closed, one has that $C_z \in \mathcal{A}_\alpha$ for each $\alpha < \zeta$. Thus Player 2 wins against σ by playing z. Contradiction.

Thus let τ be a Player 2 winning strategy in this game. For each $\alpha < \omega_1$, let $P_\alpha = \tau[\mathrm{WO}_\alpha]$. Note that $P_\alpha \subseteq \mathrm{clubcode}_{\omega_1}$ with the property that for all $z \in P_\alpha$, $C_z \in \mathcal{A}_\alpha$. Note that P_α is $\mathbf{\Sigma}^1_1$. By Fact 4.7, there is a uniform procedure to obtain from the set P_α, a club C_α with the property that $C_\alpha \subseteq C_z$ for all $z \in P_\alpha$. In particular since $\mathcal{A}_\alpha$ is $\subseteq$-downward closed, $C_\alpha \in \mathcal{A}_\alpha$. This completes the proof. □

FACT 4.9. *(Martin) Assume $\mathsf{ZF} + \mathsf{AD}$. Let $\lambda < \omega_1$. Then ω_1 is λ-reasonable.*

PROOF. Note that ω_1 is regular by $\mathsf{AC}^{\mathbb{R}}_\omega$. Let $\Gamma = \mathbf{\Sigma}^1_1$. Note that $\mathbf{\Delta}^1_1$ is closed under countable unions. Thus (1) and (5) hold.

For each $x \in \mathbb{R}$, let $x_n \in \mathbb{R}$ be defined by $x_n(k) = x(\langle n, k\rangle)$, where $\langle \cdot, \cdot \rangle : \omega \times \omega \to \omega$ denotes a fixed recursive bijective pairing function.

Fix some $w \in \mathrm{WO}_\lambda$ with $\mathrm{dom}(w) = \omega$.

Define $\mathsf{decode}(x)(\alpha, \beta)$ if and only if $x_{n^w_\alpha} \in \mathrm{WO}_\beta$. Suppose $f \in {}^\lambda\omega_1$. By $\mathsf{AC}^{\mathbb{R}}_\omega$, for each $\beta < \lambda$, let $u_\beta \in \mathrm{WO}_{f(\beta)}$. Let $x \in \mathbb{R}$ be such that $x_{n^w_\beta} = u_\beta$. Thus $\mathsf{decode}(x) = f$. This shows (2).

Let $\mathsf{GC}_{\beta,\gamma} = \{x : x_{n^w_\beta} \in \mathrm{WO}_\gamma\}$. Note that $\mathsf{GC}_{\beta,\gamma}$ is $\mathbf{\Delta}^1_1$. It satisfies the property in (3).

Suppose $A \in \mathbf{\Sigma}^1_1$ and $A \subseteq \mathsf{GC}_\beta$. Let $R = \{x_{n^w_\beta} : x \in A\}$. Then R is a $\mathbf{\Sigma}^1_1$ subset of WO. By the usual boundedness principle, there is some $\delta < \omega_1$ so that for all $v \in R$, $\mathrm{ot}(v) < \delta$. Then $A \subseteq \bigcup_{\gamma<\delta} \mathsf{GC}_{\beta,\gamma}$.

It has been shown that $(\mathbf{\Sigma}^1_1, \mathsf{decode}, \mathsf{GC}_{\beta,\gamma} : \beta < \lambda, \gamma < \omega_1)$ is a good coding system for ${}^\lambda\omega_1$. □

COROLLARY 4.10. *(Martin) Assume* $\mathsf{ZF} + \mathsf{AD}$. ω_1 *is a weak partition cardinal, i.e. for all* $\lambda < \omega_1$, $\omega_1 \to_* (\omega_1)^\lambda_2$.

COROLLARY 4.11. *Assume* $\mathsf{ZF} + \mathsf{AD}$. *The club filter* $\mu = W^{\omega_1}_\omega$ *is the unique countably complete normal measure on* ω_1.

PROOF. By Fact 2.15. □

FACT 4.12. *Let* $\alpha < \omega_1$. *Fix a good coding system for* ${}^{\omega\cdot\alpha}\omega_1$. *Suppose* $z \in \mathsf{clubcode}_{\omega_1}$. *Let* D *be the set of limit points of* C_z. *Let* $f \in [D]^\alpha_*$. *The set of* $\{x \in \mathbb{R} : \mathsf{decode}(x) \in [C_z]^{\omega\cdot\alpha} \wedge \mathsf{block}(\mathsf{decode}(x)) = f\}$ *is a* $\mathbf{\Delta}^1_1$ *set*

PROOF. Fix any x^* so that $\mathsf{block}(\mathsf{decode}(x^*)) = f$. Since ω_1 is regular, $\sup(f) = \gamma < \omega_1$. Using x^* as a parameter, one can now check that the desired set is $\mathbf{\Delta}^1_1$ by unraveling the definition of the coding system in Theorem 4.9 and using Fact 4.5. □

The original method of establishing that ω_1 is ω_1-reasonable involves sharps for reals which will be concisely reviewed below. See [**23**] for more details. The reader who would prefer to avoid sharps can skip ahead to Theorem 4.26 to see the argument of Kechris that uses the Banach-Mazur game and category.

Let $\mathscr{L} = \{\dot\in, \dot E\}$ be a language where $\dot\in$ is a binary relation symbol and $\dot E$ is a unary relation symbol.

For each $\mathscr{L}$-formula $\varphi(w, v_1, ..., v_k)$, whose free variable are exactly those listed, then let $h_\varphi(v_1, ..., v_k)$ be a formal function which will be called the Skolem function associated to φ.

Define the language $\mathscr{L}^I = \{\dot\in, \dot E, \dot c_n : n \in \omega\}$ where for each $n \in \omega$, $\dot c_n$ is a distinct constant symbol.

Let skolem denote the smallest class of function $h(v_1, ..., v_k)$ containing h_φ for all φ and closed under composition.

$\mathscr{L}^S$ consist of $\dot\in$, $\dot E$, constants c_n for all n, and new distinct constant symbols $h(c_{i_1}, ..., c_{i_k})$ for all $i_1 < ... < i_k$ in ω and $h \in \mathsf{skolem}$

One will assume one has fixed a recursive coding of formulas and terms of $\mathscr{L}^S$. One will identify terms or formulas with their associated integer code.

Let $\dot\prec$ denote the $\mathscr{L}$-formula that defines the canonical $L[\dot E]$ wellordering. Every formula ϕ of $\mathscr{L}^S$ can be converted into a $\mathscr{L}^I$ formula $\tilde\phi$ that roughly amounts to recursively replacing each Skolem term $h_\varphi(v_1, ..., v_k)$ with its intended meaning that it should represent the $\dot\prec$-least solution to φ if it exists and $\emptyset$ otherwise.

DEFINITION 4.13. Let $A(T)$ assert the following:

(1) T is the set of integer code for a complete and consistent theory extending $\mathsf{ZF} + V = L[\dot{E}]$. The statement $(\forall x)(x \in \dot{E} \Rightarrow x \in \omega)$ belongs to T.
(2) (Indiscernibility) For each $\mathscr{L}$-formula $\varphi(v_1, ..., v_k)$ and increasing sequences of integers $(a_1, ..., a_k)$ and $(b_1, ..., b_k)$, the following sentence belongs to T:

$$\text{“}\varphi(c_{a_1}, ..., c_{a_k}) \Leftrightarrow \varphi(c_{b_1}, ..., c_{b_k})''.$$

(3) (Unbounded) Let $h(v_1, ..., v_k) \in \mathsf{skolem}$. The following sentence belongs to T:

$$\text{“}h(c_1, ..., c_k) \in \mathrm{ON} \Rightarrow h(c_1, ..., c_k) < c''_{k+1}.$$

(This is actually a $\mathscr{L}^S$-sentence ϕ. Precisely, one means that $\tilde{\phi} \in T$.)
(4) (Remarkability) Suppose $h(v_0, ..., v_k, w_1, ..., v_j) \in \mathsf{skolem}$. Then the following sentence belongs to T:

$$\begin{aligned} \text{“}[h(c_0, ..., c_k, c_{a_1}, ..., c_{a_j}) \in \mathrm{ON} \wedge h(c_0, ..., c_k, c_{a_1}, ..., c_{a_j}) < c_{a_1}] \\ \Rightarrow h(c_1, ..., c_k, c_{a_1}, ..., c_{a_j}) = h(c_1, ..., c_k, c_{b_1}, ..., c_{b_k})'' \end{aligned}$$

for all increasing sequences of integers $(a_1, ..., a_j)$ and $(b_1, ..., b_j)$ such that $k < a_1$ and $k < b_1$.

Note that $A(T)$ is an arithmetical statement asserting syntactical conditions on the $\mathscr{L}^I$-theory T.

Now assume T is a set of integers so that $A(T)$.

Let K be a linear ordering. For each $a \in K$, let c_a be distinct new constant symbols. Let $\mathscr{L}^{S,K}$ consists of $\dot{\in}$, $\dot{E}$, and new constant symbols $h(c_{a_1}, ..., c_{a_k})$ for each $h \in \mathsf{skolem}$ and increasing tuple $(a_1, ..., a_k)$ in K.

An equivalence relation is defined on the new constants of $\mathscr{L}^{S,K}$ by have $h_1(c_{a_1}, ..., c_{a_n}) \sim h_2(c_{a_1}, ..., c_{a_n})$ if and only $\tilde{\phi} \in T$ where ϕ is the $\mathscr{L}^S$ formula $h_1(c_1, ..., c_n) = h_2(c_1, ..., c_n)$.

The membership relation can be defined on Skolem constant by referring to T in a similar manner.

By taking quotients, one obtains a structure denoted $\Gamma(T, K)$.

K embeds into $\Gamma(T, K)$ by the map $j(a) = [c_a]_\sim$.

FACT 4.14. *Let $T \subseteq \omega$ be such that $A(T)$. Then $\Gamma(T, \alpha)$ is wellfounded for all ordinals α if and only if $\Gamma(T, \alpha)$ is wellfounded for all $\alpha < \omega_1$.*

Suppose $w \in \mathrm{WO}$ and $(\omega, <_w)$ is its associated wellordering relation on ω. One can check that in this case one can find a structure on ω which is isomorphic to $\Gamma(T, \mathrm{ot}(w))$. This structure is recursive in T and w and produced uniformly from T and w. In such case that $w \in \mathrm{WO}$, one will denote $\Gamma(T, w)$ to be this structure on ω recursive in T and w.

Suppose M is an $\dot{\in}$ structure on ω. For each $k \in \omega$, there is a recursive function ST so that $\mathsf{ST}(M, k)$ is a structure on ω isomorphic to $(k, \dot{\in}^M)$.

Since $\Gamma(T, w)$ is considered as a structure on ω, for each $k \in \Gamma(T, w)$ such that $\Gamma(T, w) \models k$ is an ordinal, $\mathsf{ST}(\Gamma(T, w), k)$ gives (uniformly) a structure on ω isomorphic to ordinal k in $\Gamma(T, w)$.

Note that for each $\beta < \omega_1$, the set of (T, w) such that β is an initial segment of the ordinal of $\Gamma(T, w)$ is a $\boldsymbol{\Delta}^1_1$ set (using any element of WO_β as a parameter).

Consider formulas $\varphi(v, w)$ stating $v \in \mathrm{ON} \wedge \psi(v, w)$ where ψ is some other $\mathscr{L}$-formula. After fixing an enumeration of such formulas, one has an enumeration $\langle t_n : n \in \omega \rangle$ of Skolem constants whose intention is to name ordinals.

For each $\beta < \omega_1$ and $\gamma < \omega_1$, the set of (T, w, n) such that $\beta + \omega$ is an initial segment of $\Gamma(T, w)$ and $t_n^{\Gamma(T,w)}(\beta) < \gamma$ is a $\mathbf{\Delta}_1^1$ set (using any elements of WO_β and WO_γ). Similarly for $t_n^{\Gamma(T,w)}(\beta) = \gamma$.

Let $\mathrm{B}(T)$ assert that for all $\alpha < \omega_1$, $\Gamma(T, \alpha)$ is wellfounded. This is equivalent to

$$(\forall w)(w \in \mathrm{WO} \Rightarrow \Gamma(T, w) \text{ is wellfounded}).$$

This is Π_2^1.

Definition 4.15. A set T is a sharp of a real if and only if $A(T) \wedge B(T)$. Thus the statement that T is a sharp of a real is Π_2^1.

T is the sharp of a real x if and only if T is a sharp of a real and for all $n \in \omega$, "$n \in \dot{E}$" $\in T$ if and only if $n \in x$.

It can be shown that if there is a T such that T is a sharp of x, then this T is unique. Therefore $x^\sharp$ will denote this unique T for the real x.

Solovay showed that every subset of ω_1 is constructible from a real under AD:

Fact 4.16. *(Solovay) Assume* $\mathsf{ZF} + \mathsf{AD}$. *There is an single formula $\theta(x, \alpha)$ so that for all $A \subseteq \omega_1$, there is an $x \in \mathbb{R}$ so that $\alpha \in A \Leftrightarrow L[x] \models \theta(x, \alpha)$.*

Proof. Let $A \subseteq \omega_1$. Consider the game

S_A											
I	$x(0)$		$x(1)$		$x(2)$		$x(3)$				x
II		$y(1)$		$y(2)$		$y(3)$		$y(4)$			y

Player 2 wins if the disjunction of the following holds
(1) $x \notin \mathrm{WO}$.
(2) $x \in \mathrm{WO} \Rightarrow (\forall n)((y_n \in \mathrm{WO}) \wedge (\exists \gamma \geq \mathrm{ot}(x))(\{\mathrm{ot}(y_n) : n \in \omega\} = A \cap \gamma))$. (Here y_n is defined by $y_n(k) = y(\langle n, k\rangle)$.)

Suppose Player 1 has a winning strategy σ. Then for all $y \in \mathbb{R}$, $\sigma(y) \in \mathrm{WO}$. Then $E = \{v : (\exists y)(v = \sigma(y))\}$ is an $\mathbf{\Sigma}_1^1$ subset of WO. Thus there is some $\gamma < \omega_1$ so that $\mathrm{ot}(v) < \gamma$ for all $v \in E$. Let $y \in \mathbb{R}$ be such that for all n, $y_n \in \mathrm{WO}$ and $\{\mathrm{ot}(y_n) : n \in \omega\} = A \cap \gamma$. Then Player 1 using σ loses if Player 2 plays this y.

By AD, Player 2 must have a winning strategy τ. Note that $\alpha \in A$ if and only if $L[\tau]$ satisfies that $1_{\mathrm{Coll}(\omega, \alpha+1)}$ forces that for all $w \in \mathrm{WO}_{\alpha+1}$, there is some n so that $\mathrm{ot}(\tau(w)_n) = \alpha$. The formula θ is defined to assert this statement. The real associated to A is of course τ. □

Fact 4.17. *Let $f : \omega_1 \to \omega_1$. There is some $x \in \mathbb{R}$ and an $n \in \omega$ so that for all $\alpha < \omega_1$, $f(\alpha) = t_n^{\Gamma(x^\sharp, \beta)}(\alpha)$ whenever $\beta > \alpha$.*

Proof. Consider $A = \{\langle \alpha, \beta\rangle : f(\alpha) = \beta\}$ where $\langle \cdot, \cdot \rangle$ denotes some constructible pairing function. By Fact 4.16, there is some real x so that $\langle \alpha, \beta\rangle \in A$ if and only if $L[x] \models \theta(x, \langle \alpha, \beta\rangle)$.

Consider the formula $\psi(\beta, \alpha)$ if and only if $\theta(\dot{E}, \langle \alpha, \beta\rangle)$. There is some n so that t_n is the Skolem function associated to this formula. Fix two ordinals $\alpha < \beta$. Since $\Gamma(x^\sharp, \beta)$ is of the form $L_{\beta'}[x]$ for some $\beta' \geq \beta$ and $\Gamma(x^\sharp, \beta) \prec_\omega L[x]$, $t_n^{\Gamma(x^\sharp, \beta)}(\alpha) = f(\alpha)$. □

It should be noted that coding functions by simply coding their graph as a subset of ω_1 is generally not good enough. For each α, if one had to search for the corresponding β which is the value of $f(\alpha)$, the coding system will be too complex. In this case, one uses the Skolem term to output the value of f when given α. This needs to be handled in the proof of Kechris as well. See Remark 4.25. There it is handled by a modified version of the Solovay game used above.

THEOREM 4.18. *(Martin) Assume* $\mathsf{ZF} + \mathsf{AD}$. ω_1 *is* ω_1*-reasonable.*

PROOF. If $x \in \mathbb{R} = {}^\omega\omega$, let $\mathsf{cut}(x) \in {}^\omega\omega$ be defined by $\mathsf{cut}(x)(k) = x(k+1)$.
One will now define a good coding system for ${}^{\omega_1}\omega_1$.
The pointclass of the coding system is $\mathbf{\Sigma}^1_1$.
For each $x \in \mathbb{R}$, let $\mathsf{decode}(x)(\alpha, \beta)$ holds if and only if

$$A(\mathsf{cut}(x)) \wedge t_{x(0)}^{\Gamma(\mathsf{cut}(x), \alpha+\omega)}(\alpha) = \beta.$$

Define $\mathsf{GC}_{\beta,\gamma}$ by $x \in \mathsf{GC}_{\beta,\gamma}$ if and only if $\mathsf{decode}(x)(\beta, \gamma)$ holds.

Suppose $f : \omega_1 \to \omega_1$. Let z be the real and n be the natural number obtained by applying Fact 4.17 to f. Let $x \in \mathbb{R}$ be such that $x(0) = n$ and $\mathsf{cut}(x) = z^\sharp$. Then $\mathsf{decode}(x) = f$. Such an x will be called a sharp code for f.

Let $\beta < \omega_1$ and $\gamma < \omega_1$, note that $\mathsf{GC}_{\beta,\gamma}$ consists of those $x \in \mathbb{R}$ such that the following holds
(i) $A(\mathsf{cut}(x))$
(ii) β is in the wellfounded part of $\Gamma(\mathsf{cut}(x), \beta + \omega)$
(iii) $t_{x(0)}^{\Gamma(\mathsf{cut}(x), \beta+\omega)}(\beta) = \gamma$.

By the discussion earlier, $\mathsf{GC}_{\beta,\gamma}$ is $\mathbf{\Delta}^1_1$. Note that since $t_{x(0)}$ is a Skolem function, $\tau_{x(0)}^{\Gamma(\mathsf{cut}(x), \beta+\omega)}$ is a function on its domain.

Now suppose $A \subseteq \mathsf{GC}_\beta$ is $\mathbf{\Sigma}^1_1$. Let $E = \{v \in \mathbb{R} : (\exists x)(x \in A \wedge v = \mathsf{ST}(\Gamma(\mathsf{cut}(x), \beta + \omega), t_{x(0)}^{\Gamma(\mathsf{cut}(x), \beta+\omega)})\}$. Then E is a $\mathbf{\Sigma}^1_1$ subset of WO. By the boundedness lemma, there is some $\delta < \omega_1$ so that for all $v \in E$, $\mathrm{ot}(v) < \delta$. Thus $A \subseteq \bigcup_{\gamma < \delta} \mathsf{GC}_{\beta,\gamma}$.

It has been shown that $(\mathbf{\Sigma}^1_1, \mathsf{decode}, \mathsf{GC}_{\beta,\gamma} : \beta < \omega_1, \gamma < \omega_1)$ is a good coding system for ${}^{\omega_1}\omega_1$. □

Next, one will give the argument of Kechris that ω_1 is a strong partition cardinal. This proof uses category and the Kechris-Woodin generic coding idea. However at ω_1, the generic coding function is essentially trivial and exists without even AD. For many other purposes, the generic coding function is very useful.

Note that in Martin's proof, the indiscernibility models are used to make complexity computations. In the argument of Kechris, category quantifiers will be used to ensure sets have the correct complexity.

First some results on the Banach-Mazur games. See [**16**] Section 21.C and 21.D for the proofs of these results:

FACT 4.19. *(Banach-Mazur Game) Assume* ZF. *Let* $A \subseteq {}^\omega\omega$. *Define* $G^*(A)$ *by*

$$G^*_A \quad \begin{array}{c|ccccccc} I & s_0 & & s_2 & & s_4 & & \\ \hline II & & s_1 & & s_3 & & s_5 & \end{array}$$

Player 1 and 2 alternatingly play nonempty strings in ${}^{<\omega}\omega$. Let $x = s_0\hat{}s_1\hat{}s_2...$ be the concatenation of the moves. Player 2 wins G^*_A if and only if $x \in A$.

(1) A is comeager if and only if Player 2 has a winning strategy in G^*_A.

(2) There is some $s \in {}^{<\omega}\omega$ so that ${}^{\omega}\omega \setminus A$ is comeager in N_s if and only if Player 1 has a winning strategy in G^*_A.

FACT 4.20. *(Unfolded Banach-Mazur Game) Assume* ZF. *Let $A \subseteq {}^{\omega}\omega$ and $B \subseteq {}^{\omega}\omega \times {}^{\omega}\omega$ be such that for all $x \in A$, there exists a $y \in {}^{\omega}\omega$ so that $B(x, y)$. Let $G^*_{A,B}$ be the following game*

$$G^*_{A,B} \quad \begin{array}{lcccccc} I & s_0 & & s_2 & & s_4 & \\ II & & s_1, y(0) & & s_3, y(1) & & s_5, y(2) \end{array}$$

Player 1 plays nonempty strings in ${}^{<\omega}\omega$. Player 2 plays nonempty strings in ${}^{<\omega}\omega$ and an element of ω. Let $x = s_0\hat{}s_1\hat{}s_2...$ be the concatenation of the strings played by both player. Let y be the real produced by the extra elements of ω played by Player 2. Player 2 wins $G^*_{A,B}$ if and only if $x \in A$ and $(x, y) \in B$.

If Player 2 wins $G^*_{A,B}$, then Player 2 wins G^*_A. If Player 1 wins $G^*_{A,B}$, then Player 1 wins G^*_A.

COROLLARY 4.21. *Assume* ZF. *Let $A \subseteq {}^{\omega}\omega$ be a $\mathbf{\Sigma}^1_1$ set. Let $B \subseteq {}^{\omega}\omega \times {}^{\omega}\omega$ be a $\mathbf{\Pi}^0_1$ set so that $A = \pi_1[B]$. A is comeager if and only if Player 2 has a winning strategy in $G^*_{A,B}$.*

PROOF. Note that if $A = \pi_1[B]$, then the winning condition for Player 2 in $G^*_{A,B}$ is equivalent to merely $(x, y) \in B$. Thus in this case $G^*_{A,B}$ is a closed game. Thus the determinacy of such games hold under ZF. Now the result follows from Fact 4.19 and 4.20. □

If $\alpha < \omega_1$, one can define a topology on ${}^{\omega}\alpha$ be declaring the basic open sets to be sets of the form $N^{\alpha}_s = \{f \in {}^{\omega}\alpha : f \supseteq s\}$, where $s \in {}^{<\omega}\alpha$. Using this topology, one can define the notions of comeagerness and meagerness for ${}^{\omega}\alpha$. As α is countable, ${}^{\omega}\alpha$ with this topology is homeomorphic to ${}^{\omega}\omega$. Observe that the set $\text{surj}_{\alpha} = \{f \in {}^{\omega}\alpha : f \text{ is a surjection}\}$ is a comeager subset of ${}^{\omega}\alpha$. If $A \subseteq \mathbb{R} \times {}^{\omega}\alpha$, then one writes $(\forall^*_{\alpha} f)A(x, f)$ as an abbreviation for the statement that $A(x, f)$ holds for comeagerly many f in ${}^{\omega}\alpha$.

COROLLARY 4.22. *Assume* ZF. *Let $A \subseteq {}^{\omega}\omega \times {}^{\omega}\omega$ be a $\mathbf{\Sigma}^1_1$ set. Then $A_0(x) \Leftrightarrow (\forall^*_{\omega} y)A(x, y)$ is $\mathbf{\Sigma}^1_1$ and $A_1(x) \Leftrightarrow (\forall^*_{\omega} y)\neg A(x, y)$ is $\mathbf{\Pi}^1_1$.*

PROOF. Let $B \subseteq {}^{\omega}\omega \times {}^{\omega}\omega \times {}^{\omega}\omega$ be such that $A = \{(x, y) : (\exists z)((x, y, z) \in B)\}$ where $B \in \mathbf{\Pi}^0_1$. For each $x \in {}^{\omega}\omega$, let $B_x = \{(y, z) : (x, y, z) \in B\}$ and $A_x = \{y : (x, y) \in A\}$. Note that $B_x \in \mathbf{\Pi}^0_1$ and $\pi_1[B_x] = A_x$.

If σ is a Player 1 strategy and τ is a Player 2 strategy for a game of the form $G^*_{C,D}$ for some appropriate C and D, then let $x_{\sigma * \tau}$ be the real produced by the concatenation of the finite strings played by each player. Let $y_{\sigma * \tau}$ be the auxiliary sequence produced by the moves of τ.

Note that $A_0(x)$ if and only if Player 2 has a winning strategy in $G^*_{A_x, B_x}$ if and only if $(\exists \tau)(\forall \sigma)((x_{\sigma * \tau}, y_{\sigma * \tau}) \in B_x)$ by Fact 4.20. Since $B_x \in \mathbf{\Pi}^0_1$ and $\mathbf{\Pi}^0_1$ is closed under $\forall^{\mathbb{R}}$, the latter expression is $\mathbf{\Sigma}^1_1$.

For $s \in {}^{<\omega}\omega$, let $\Phi_s : N_s \to {}^\omega\omega$ be the canonical homeomorphism between N_s and ${}^\omega\omega$. Let $A^s \subseteq {}^\omega\omega \times {}^\omega\omega$ be defined by $(x, y) \in A^s \Leftrightarrow (x, \Phi_s^{-1}(y)) \in A$. Let $B^s \subseteq {}^\omega\omega \times {}^\omega\omega \times {}^\omega\omega$ be defined by $(x, y, z) \in B^s$ if and only if $(x, \Phi_s^{-1}(y), z) \in B$.

For the second statement, note that $\neg A_1(x)$ if and only if $\neg((\forall^*_\omega y)\neg A(x, y))$. Since $\mathbf{\Pi}^1_1$ sets have Baire property, the latter holds if and only if there exists a $s \in {}^{<\omega}2$ so that A_x is comeager in N_s. Now there exists an $s \in {}^{<\omega}2$ so that A_x is comeager in N_s if and only if there exists an $s \in {}^{<\omega}2$ so that $\Phi_s[A_x \cap N_s]$ is comeager if and only if there exists an $s \in {}^{<\omega}2$ so that Player 2 has a winning strategy in $G^*_{(A^s)_x,(B^s)_x}$. As before, the last expression can be checked to be $\mathbf{\Sigma}^1_1$. Hence A_1 is $\mathbf{\Pi}^1_1$. □

The following is the (essentially trivial) version of the Kechris-Woodin generic coding function for ω_1. (See [**18**].)

FACT 4.23. *There is a continuous function $G : {}^\omega\omega_1 \to \mathrm{WO}$ so that for all $\ell \in {}^\omega\omega_1$ such that $\ell(0) = \{\ell(n+1) : n \in \omega\}$, $G(\ell) \in \mathrm{WO}$ and $\mathrm{ot}(G(\ell)) = \ell(0)$.*

PROOF. For each ℓ as above, let $A_\ell = \{n \in \omega \setminus \{0\} : (\forall m)(\ell(n) = \ell(m) \Rightarrow n \leq m)\}$. Let $G(\ell)$ be the wellordering with domain A_ℓ so that $m <_{G(\ell)} n$ if and only if $\ell(m) < \ell(n)$. $G(\ell) \in \mathrm{WO}$ and $\mathrm{ot}(G(\ell)) = \ell(0)$.

This function is continuous in the sense that for any n, $G(\ell) \upharpoonright n$ is determined by $\ell \upharpoonright m$ for some m. □

FACT 4.24. *Assume $\mathsf{ZF} + \mathsf{AD}$. Let $f : \omega_1 \to \omega_1$. There is a Player 2 strategy so that for all $v \in \mathrm{WO}$, $\{(\alpha, \beta) : (\exists n)((\tau(v)_n)_0 \in \mathrm{WO}_\alpha \wedge (\tau(v)_n)_1 \in \mathrm{WO}_\beta)\} = f \upharpoonright \gamma$ for some $\gamma > \mathrm{ot}(v)$.*

Here $\tau(v)$ is $(v * \tau)_{\mathrm{odd}}$, i.e. the real produced by Player 2 when played against Player 1 playing the bits of v. Also $f \upharpoonright \gamma = \{(\alpha, \beta) \in f : \alpha < \gamma\}$. Recall that if $x \in \mathbb{R}$, x_n is defined by $x_n(k) = x(\langle n, k\rangle)$. This is the nth section of x. This result states that when given $v \in \mathrm{WO}$, the integer sections of $\tau(v)$ codes f up to some ordinal γ greater than $\mathrm{ot}(v)$.

PROOF. Consider the game S_f defined by

$$S_f \quad \begin{array}{lccccccccccc} \mathrm{I} & v(0) & & v(1) & & v(2) & & v(3) & & & & v \\ \mathrm{II} & & r(0) & & r(1) & & r(2) & & r(3) & & & r \end{array}$$

Player 2 wins if and only if $v \in \mathrm{WO}$ implies that $\{(\alpha, \beta) : (\exists n)((r_n)_0 \in \mathrm{WO}_\alpha \wedge (r_n)_1 \in \mathrm{WO}_\beta)\} = f \upharpoonright \gamma$ for some $\gamma > \mathrm{ot}(v)$.

By essentially the same bounding argument of Fact 4.16, one has the Player 2 must have the winning strategy in this game. □

REMARK 4.25. In the above statement, it is very important that some section of $\tau(v)$ contains a code for the image of $f(\alpha)$. To search for $f(\alpha)$, which could be quite large in comparison to α, would push the complexity of $\mathsf{GC}_{\alpha,\beta}$ beyond $\mathbf{\Delta}^1_1$. Instead, a particular winning strategy τ will take a code v for α and ouput $\tau(v)$ which magically contains codes for $f(\alpha)$ among its integer sections, $\{(\tau(v)_n)_1 : n \in \omega\}$.

THEOREM 4.26. *(Kechris' proof) Assume $\mathsf{ZF} + \mathsf{AD}$. ω_1 is ω_1-reasonable.*

PROOF. One will define a good coding system for ${}^{\omega_1}\omega_1$ witnessing ω_1-reasonableness of ω_1.

The associated pointclass is $\mathbf{\Sigma}_1^1$.

Let $\mathsf{add} : \omega_1 \times {}^{\omega}\omega_1 \to {}^{\omega}\omega_1$ be defined by

$$\mathsf{add}(\alpha, g)(n) = \begin{cases} \alpha & n = 0 \\ g(n-1) & n > 0 \end{cases}$$

$\mathsf{add}(\alpha, g)$ simply inserts α at the beginning of g.

In some fix recursive manner, every real x codes a Player 2 strategy τ_x.

Now fix $\alpha, \beta \in \omega_1$. Define the formula $\varphi_\beta^\alpha(x, z)$ to state that

$$(\exists n)([(\tau_x(z)_n)_0 \in \mathrm{WO}_\alpha \wedge (\tau_x(z)_n)_1 \in \mathrm{WO}_\beta] \wedge (\forall m < n)((\tau_x(z)_m)_0 \notin \mathrm{WO}_\alpha)).$$

This formula defines a $\mathbf{\Delta}_1^1$ relation using elements of WO_α and WO_β as parameters.

Let $\phi_\beta^\alpha(x)$ be the statement $(\forall_\alpha^* g)\varphi_\beta^\alpha(x, G(\mathsf{add}(\alpha, g)))$, where G is the generic coding function of Fact 4.23. Note that since surj_α is comeager in

$${}^{\omega}\alpha, (\forall_\alpha^* g)(G(\mathsf{add}(\alpha, g)) \in \mathrm{WO}_\alpha).$$

Now fix a bijection $B : \omega \to \alpha$. Let $T : {}^{\omega}\omega \to {}^{\omega}\alpha$ be defined by $T(r) = B \circ r$. Define $\tilde{G} : {}^{\omega}\omega \to \mathrm{WO}$ by $\tilde{G}(r) = G(\mathsf{add}(\alpha, T(r)))$. Now $\tilde{G}$ is a continuous function with the property that for comeagerly many $r \in {}^{\omega}\omega$ (in the usual sense of comeager), $\tilde{G}(r) \in \mathrm{WO}_\alpha$.

$\phi_\beta^\alpha(x)$ is equivalent to $(\forall_\omega^* r)\varphi_\beta^\alpha(x, \tilde{G}(r))$. Since φ_β^α defines a $\mathbf{\Delta}_1^1$ relation (in parameters from WO_α and WO_β), the collection of (x, r) satisfying $\varphi_\beta^\alpha(x, \tilde{G}(r))$ is $\mathbf{\Delta}_1^1$ using parameters from WO_α, WO_β, and a parameter coding the continuous function $\tilde{G}$. Now the set defined by $\phi_\beta^\alpha(x)$ is $\mathbf{\Delta}_1^1$ by Corollary 4.22. (Note this is not done uniformly in the α and β.)

If $x \in \mathbb{R}$, define $\mathsf{decode}(x)(\alpha, \beta)$ if and only if $\phi_\beta^\alpha(x)$.

Define $\mathsf{GC}_{\alpha,\beta} = \{x \in \mathbb{R} : \phi_\beta^\alpha(x)\}$. $\mathsf{GC}_{\alpha,\beta}$ is $\mathbf{\Delta}_1^1$.

For any $f : \omega_1 \to \omega_1$, Fact 4.24 states that there is some x so that $\mathsf{decode}(x) = f$.

Fix $\alpha < \omega_1$. For each $x \in {}^{\omega}\omega$, let $\psi_0(x, z)$ assert that $(\exists n)((\tau_x(z)_n)_0 \in \mathrm{WO}_\alpha)$. $\psi_0(x, z)$ is $\mathbf{\Delta}_1^1$ using any code of α as a parameter. Let $\psi_1(x, y, z)$ be the conjunction of the following statements

(1) $\psi_0(x, z) \wedge \psi_0(y, z)$.

(2) There exists $w_0, w_1 \in \mathbb{R}$, $n_0, n_1 \in \omega$ so that

(2a) $(\tau_x(z)_{n_0})_0 \in \mathrm{WO}_\alpha$, $(\tau_x(z)_{n_0})_1 = w_0$, and for all $m < n_0$, $(\tau_x(z)_m)_0 \notin \mathrm{WO}_\alpha$.

(2b) $(\tau_y(z)_{n_0})_0 \in \mathrm{WO}_\alpha$, $(\tau_y(z)_{n_1})_1 = w_1$, and for all $m < n_1$, $(\tau_y(z)_m)_0 \notin \mathrm{WO}_\alpha$.

(2c) $w_0 <_{\Sigma_1^1} w_1$, where $<_{\Sigma_1^1}$ is the Σ_1^1 relation witnessing that the ordertype function is a Π_1^1-norm.

Observe that $\psi_1(x, y, z)$ is $\mathbf{\Sigma}_1^1$.

Let $A \subseteq \mathsf{GC}_\alpha$ and A is $\mathbf{\Sigma}_1^1$. Define a relation $\prec$ on ${}^{\omega}\omega$ by $x \prec y$ if and only if

$$x \in A \wedge y \in A \wedge (\forall_\alpha^* g)\psi_1(x, y, G(\mathsf{add}(\alpha, g))).$$

By Corollary 4.22, $\prec$ is a $\mathbf{\Sigma}_1^1$ relation on ${}^{\omega}\omega$.

By checking the various definitions, one can see that for all $x, y \in A$, $\mathsf{decode}(x)(\alpha) < \mathsf{decode}(y)(\alpha)$ if and only if $x \prec y$.

Since $\prec$ is a $\mathbf{\Sigma}_1^1$ relation, it is an ω-Suslin relation. By the Kunen-Martin theorem ([**17**] Section 7), the rank must be less than ω_1. Since ω_1 is regular, there is some $\gamma < \omega_1$ so that $A \subseteq \bigcup_{\beta<\gamma} \mathsf{GC}_\beta$.

(Some presentations of the Kunen-Martin theorem uses $\mathsf{DC}_\mathbb{R}$, but with some care, one can remove the use of $\mathsf{DC}_\mathbb{R}$. Another approach in the case of ω-Suslin relations is to observe that if K denotes a tree witnessing a relation is ω-Suslin, one can prove the Kunen-Martin theorem in $L[K] \models \mathsf{AC}$ and argue in $L[K]$ that the Kunen-Martin tree is wellfounded there and hence in the real world. Then one shows that this tree still works in the real world V. Alternatively, if one assumes Kechris's result that $L(\mathbb{R}) \models \mathsf{AD}$ implies $L(\mathbb{R}) \models \mathsf{DC}_\mathbb{R}$, then one can absorb this problem into $L(\mathbb{R})$ and apply the Kunen-Martin Theorem in $L(\mathbb{R})$.)

It has been shown that $(\mathbf{\Sigma}_1^1, \mathsf{decode}, \mathsf{GC}_{\beta,\gamma} : \beta < \omega_1, \gamma < \omega_1)$ is a good coding system for ${}^{\omega_1}\omega_1$. □

COROLLARY 4.27. *(Martin) Assume* $\mathsf{ZF} + \mathsf{AD}$. $\omega_1 \rightarrow_* (\omega_1)_2^{\omega_1}$.

DEFINITION 4.28. Let $\lambda \leq \omega_1$. Let $\mu_{\omega_1}^\lambda$ consists of those subsets X of $[\omega_1]^\lambda$ which contain a set of the form $[C]_*^\lambda$ where C is a club subset of ω_1. Each $\mu_{\omega_1}^\lambda$ is a countably complete ultrafilter.

COROLLARY 4.29. *Assume* $\mathsf{ZF}+\mathsf{AD}$. *Let* $\lambda \leq \omega_1$. $\prod_{[\omega_1]^\lambda} \omega_1/\mu_{\omega_1}^\lambda = (\prod_{[\omega_1]^\lambda} \omega_1/\mu_{\omega_1}^\lambda)^{L(\mathbb{R})}$.

Since $L(\mathbb{R}) \models \mathsf{DC}$ *by a result of Kechris* [**15**], $\prod_{[\omega_1]^\lambda} \omega_1/\mu_{\omega_1}^\lambda$ *is an ordinal.*

PROOF. Note that the good coding system for ${}^\lambda\omega_1$ constructed above belong to $L(\mathbb{R})$. Now apply Theorem 3.9. □

REMARK 4.30. This section presented two proofs of the strong partition property for ω_1. The original proof of Martin uses indiscernibility. Many early results were proved using this indiscernibility idea. For example, Kunen showed in $\mathsf{AD}+\mathsf{DC}$ that δ_3^1 is a weak partition cardinal and $\delta_3^1 = \aleph_{\omega+1}$.

The second proof of Kechris is perhaps the simplest proof using classical descriptive set theoretic ideas. The ideas of the Kechris-Woodin generic coding function is very useful in various different settings.

However, no proof in the flavor of these two arguments is known to establish the strong partition property at δ_3^1. There is another proof of the strong partition property for ω_1 due to Jackson. It uses ideas such as the Kunen tree and an analysis of all the measures on ω_1 due to Kunen. An exposition can be found in [**10**], [**12**], and [**11**]. More generally, Jackson developed the theory of description. This theory produces the only known proof that δ_3^1 and in fact all δ_{2n+1}^1 are strong partition cardinal under $\mathsf{AD}+\mathsf{DC}$. Jackson has also computed the identity of these cardinals. For example, $\delta_5^1 = \aleph_{\omega^{\omega^\omega}+1}$ and in fact there is a general formula for computing δ_{2n+1}^1. See [**11**] for more information. It should noted that Jackson's proof that ω_1 is a strong partition cardinal, Kunen's result that δ_3^1 is a weak partition cardinal, and all of Jackson's results mentioned above using description theory are proved using DC. The original Martin proof and the Kechris proof of the strong partition property of ω_1 are within $\mathsf{ZF} + \mathsf{AD}$.

5. Kunen Functions and Partition Properties at ω_2

FACT 5.1. *Let* μ *be a normal measure on a cardinal* κ. *Let* $A \in \mu$, *let* $\mathsf{enum}_A : \omega_1 \to A$ *be the increasing enumeration of* A. *Then* $\{\alpha \in A : \mathsf{enum}_A(\alpha) = \alpha\} \in \mu$.

Let $f : \kappa \to \kappa$ and $X \subseteq f[\kappa]$ so that $\{\alpha < \kappa : f(\alpha) \in X\} \in \mu$, then $\mathsf{enum}_X =_\mu f$.

PROOF. Suppose not. Then $B = \{\alpha \in A : \alpha < \mathsf{enum}_A(\alpha)\} \in \mu$. Since enum_A is an order-preserving function, $B = \{\alpha \in A : \mathsf{enum}_A^{-1}(\alpha) < \alpha\}$. By normality, there is some $C \subseteq B$ and $\gamma < \kappa$ so that for all $\alpha \in C$, $\mathsf{enum}_A^{-1}(\alpha) = \gamma$. This contradicts the injectiveness of enum_A.

For the second statement: Let $A = \{\alpha : f(\alpha) \in X\}$. Note that $f \circ \mathsf{enum}_A = \mathsf{enum}_X$. Let $B \subseteq A$ with $B \in \mu$ be such that $\mathsf{enum}_A(\alpha) = \alpha$ for all $\alpha \in B$. Then for all $\gamma \in B$, $\mathsf{enum}_X(\alpha) = f(\mathsf{enum}_A(\alpha)) = f(\alpha)$. Hence $\mathsf{enum}_X =_\mu f$. □

The next result states that an ultrapower of a strong partition cardinal by a measure is a cardinal as long as that ultrapower is well founded.

FACT 5.2. *(Martin) Let κ be a strong partition cardinal and μ be a measure on κ. If ${}^\kappa\kappa/\mu$ is wellfounded, then ${}^\kappa\kappa/\mu$ is a cardinal.*

Assuming $\mathsf{ZF} + \mathsf{AD} + \mathsf{DC}_\mathbb{R}$. If $\kappa < \Theta$ is a strong partition cardinal and μ is a measure on κ, then ${}^\kappa\kappa/\mu$ is a cardinal.

PROOF. Suppose there is an $h \in {}^\kappa\kappa/\mu$ and an injection $\Phi : {}^\kappa\kappa/\mu \to \prod_\kappa h/\mu$, where $\prod_\kappa h/\mu$ is the collection of $[f]_\mu$ such that $f <_\mu h$. Note that $\prod_\kappa h/\mu$ is the initial segment of ${}^\kappa\kappa/\mu$ determined by $[h]_\mu$.

Let $\mathcal{T} = (\kappa \times 2, \sqsubset)$ denote the lexicographic ordering on $\kappa \times 2$. If $F : \kappa \times 2 \to \kappa$ is a $\mathcal{T}$-increasing function, then let $f_0, f_1 : \kappa \to \kappa$ be defined by $f_0(\alpha) = F(\alpha, 0)$ and $f_1(\alpha) = F(\alpha, 1)$. Note that f_0 and f_1 are increasing functions.

Define $P : [\kappa]^\mathcal{T} \to 2$ by

$$P(F) = \begin{cases} 0 & \Phi([f_0]_\mu) < \Phi([f_1]_\mu) \\ 1 & \text{otherwise} \end{cases}$$

Since $\mathrm{ot}(\mathcal{T}) = \kappa$, $\kappa \to (\kappa)^\kappa_2$ implies that there is an $E \subseteq \kappa$ with $|E| = \kappa$ which is homogeneous for this partition.

Suppose P is homogeneous taking value 1. For each $n \in \omega$, define $g_n : \kappa \to \kappa$ by $g_n(\alpha) = \mathsf{enum}_E(\omega \cdot \alpha + n)$. Note that if $m < n$, then $g_m <_\mu g_n$. For any $m < n$, define $G^{m,n} : \mathcal{T} \to \kappa$ by $G^{m,n}(\alpha, 0) = g_m(\alpha)$ and $G^{m,n}(\alpha, 1) = g_n(\alpha)$. Note that $G^{m,n}$ is $\mathcal{T}$-increasing and $G^{m,n} \in [\kappa]^\mathcal{T}$. Since $P(G^{m,n}) = 1$, one has that $\Phi([g_n]_\mu) = \Phi([G_1^{m,n}]_\mu) \leq \Phi([G_0^{m,n}]_\mu) = \Phi([g_m]_\mu)$. Since Φ is an injection, one must actually have $\Phi([g_n]_\mu) < \Phi([g_m]_\mu)$. Thus $\langle \Phi([g_k]_\mu) : k \in \omega \rangle$ is an infinite decreasing sequence in ${}^\kappa\kappa/\mu$. This contradicts the assumption that ${}^\kappa\kappa/\mu$ is wellfounded.

Thus E must be homogeneous for P taking value 0.

Let $\ell \in {}^\kappa\kappa$ be such that $[\ell]_\mu > [h]_\mu$ and $\ell(\alpha) > 0$ for all $\alpha \in \kappa$. Let $U = \{(\alpha, \beta) \in \kappa \times \kappa : \beta < \ell(\alpha)\}$. Let $\mathcal{U}$ denote $(U, \prec)$ where $\prec$ denotes the lexicographic ordering of U. Again, $\mathrm{ot}(\mathcal{U}) = \kappa$.

Let $F = \{\mathsf{enum}_E(\omega \cdot \alpha) : \alpha < \kappa)\}$. Note that $|F| = \kappa$. Pick some $K : \mathcal{U} \to F$ which is order-preserving. For any $f <_\mu \ell$, define $k_f : \kappa \to \kappa$ by

$$k_f(\alpha) = \begin{cases} K(\alpha, f(\alpha)) & f(\alpha) < \ell(\alpha) \\ K(\alpha, 0) & \text{otherwise} \end{cases}.$$

Note that if $f <_\mu g <_\mu \ell$, then $k_f <_\mu k_g$ and that for all $f <_\mu \ell$, $k_f \in [F]^\kappa$.

Define $\Psi : \prod_\kappa \ell/\mu \to \prod_\kappa h/\mu$ as follows: Let $A < [\ell]_\mu$. Pick $f \in A$, i.e. a representative of A. Let $\Psi(A) = \Phi([k_f]_\mu)$. One can check that Ψ is well defined.

The next claim is that Ψ is order preserving: Suppose $A < B < [\ell]_\mu$. Let $f \in A$ and $g \in B$ be representatives for A and B, respectively. Since $k_f <_\mu k_g$, the set $C = \{\alpha \in \kappa : k_g(\alpha) \leq k_f(\alpha)\} \notin \mu$. Define $k'_g : \kappa \to E$ by $k'_g(\alpha) = k_g(\alpha)$ if $\alpha \notin C$ and $k'_g(\alpha)$ is the next element of E greater than $k_f(\alpha)$ if $\alpha \in C$. The purpose of defining F in the manner above was to ensure that between two successive elements of F, there are at least ω many point of E in between. From this, one can verify that $k'_g \in [E]^\kappa$ and $k_f(\alpha) < k'_g(\alpha) < k_f(\alpha+1)$ for all $\alpha < \kappa$. Note that $[k'_g]_\mu = [k_g]_\mu$ since k_g and k'_g agree off of $C \notin \mu$.

Let $F^{k_f,k'_g} : \mathcal{T} \to \kappa$ be defined by

$$F^{k_f,k'_g}(\alpha, i) = \begin{cases} k_f(\alpha) & i = 0 \\ k'_g(\alpha) & i = 1 \end{cases}$$

Again using the main property of F, one can verify that $F^{k_f,k'_g} \in [\kappa]^{\mathcal{T}}$. Thus $P(F^{k_f,k'_g}) = 0$. Thus $\Psi(A) = \Phi([k_f]_\mu) < \Phi([k'_g]_\mu) = \Phi([k_g]_\mu) = \Psi(B)$.

It has been shown that Ψ is order preserving. However, this is not possible since $[h]_\mu < [\ell]_\mu$. This completes the proof. $\square$

FACT 5.3. *Let κ be a measurable cardinal possessing a normal κ-complete measure μ. Let $\epsilon < \kappa$. Let $\mathcal{T}^\epsilon = (\kappa \times \epsilon, \sqsubset)$ where $\sqsubset$ is the lexicographic ordering. For $F : \mathcal{T}^\epsilon \to \kappa$ which is order preserving and $\alpha < \epsilon$, let $F^\alpha \in [\kappa]^\kappa$ be defined by $F^\alpha(\gamma) = F(\gamma, \alpha)$.*

Let $\langle f_\alpha : \alpha < \epsilon \rangle$ be a sequence in $[\kappa]^\kappa$ such that for all $\alpha < \beta < \epsilon$, $f_\alpha <_\mu f_\beta$. Then there is an $F \in [\kappa]^{\mathcal{T}^\epsilon}$ so that for all $\alpha < \epsilon$, $F^\alpha =_\mu f_\alpha$.

Moreover, if $D \subseteq \kappa$ with $|D| = \kappa$ and each $f_\alpha \in [D]^\kappa$, then one can find $F \in [D]^{\mathcal{T}^\epsilon}$ with the above property.

PROOF. Suppose $F(\alpha', \beta')$ has been defined for all $(\alpha', \beta') \sqsubset (\alpha, \beta)$. Then let $F(\alpha, \beta)$ be the least element in the range of f_β greater than $F(\alpha', \beta')$ for all $(\alpha', \beta') \sqsubset (\alpha, \beta)$.

The claim is that F has the desired properties: This is proved by induction on $\beta < \epsilon$.

Consider $\beta = 0$. Suppose $A = \{\alpha < \kappa : F(\alpha, 0) \neq f_0(\alpha)\} \in \mu$. This means for each $\alpha \in A$, there is some $\alpha' < \alpha$, and some $\beta < \epsilon$ so that $F(\alpha', \beta) \geq f_0(\alpha)$. Let $g : \kappa \to \kappa$ be defined by $g(\alpha)$ is the least such α' with the property above if $\alpha \in A$ and 0 if $\alpha \notin A$. Note that $g(\alpha) < \alpha$ for all $\alpha \in A$. By the normality of μ, there is some $A' \subseteq A$ with $A' \in \mu$ and some δ so that $g(\alpha) = \delta$ for all $\alpha \in A'$. This means for all $\alpha \in A' \in \mu$, $f_0(\alpha) \leq \sup\{F(\delta, \beta) : \beta < \epsilon\}$. By the κ-completeness of μ, one may even find a β^* and a $A'' \in \mu$ so that for all $\alpha \in A''$, $f_0(\alpha) \leq F(\delta, \beta^*)$. This is clearly impossible. This shows that $F^0 =_\mu f_0$.

Suppose $\beta < \epsilon$ is such that for all $\beta' < \beta$, $F^{\beta'} = f_{\beta'}$. Suppose $\neg(F^\beta =_\mu f_\beta)$. Then $A = \{\alpha : F^\beta(\alpha) \neq f_\beta(\alpha)\} \in \mu$. Since $f_0(\alpha) = F^0(\alpha)$ for almost all α, it can not be the case that for almost all $\alpha \in A$, there is some $\alpha' < \alpha$ and some $\beta' < \epsilon$ so that $F^{\beta'}(\alpha') \geq f_\beta(\alpha)$. Thus for almost all $\alpha \in A$, there is some $\beta' < \beta$ so that $F^{\beta'}(\alpha) \geq f_\beta(\alpha)$. By the κ-additivity of μ, there is some $\beta^* < \beta$ so that for almost all α, $F^{\beta^*}(\alpha) \geq f_\beta(\alpha)$. By the induction hypothesis, one has that $F^{\beta^*} =_\mu f_{\beta^*}$. One has shown that $f_\beta \leq_\mu f_{\beta^*}$ despite the fact that $\beta^* < \beta$. Contradiction. The result has been established. $\square$

FACT 5.4. *(Martin) Let κ be a strong partition cardinal and μ is a normal κ-complete measure on a cardinal κ. If ${}^\kappa\kappa/\mu$ is wellfounded, then ${}^\kappa\kappa/\mu$ is a regular cardinal.*

Assuming $\mathsf{ZF} + \mathsf{AD} + \mathsf{DC}_{\mathbb{R}}$, if $\kappa < \Theta$ is a strong partition cardinal and μ is a normal κ-complete measure on κ, then ${}^\kappa\kappa/\mu$ is a regular cardinal.

PROOF. By Fact 5.2, $\delta = \text{ot}({}^\kappa\kappa/\mu)$ is a cardinal. Assume that ${}^\kappa\kappa/\mu$ is not a regular cardinal. There is some $\gamma < \delta$ and an increasing function $\Phi : \gamma \to \delta$ which is cofinal.

Let $\mathcal{T}^2$, F^0, F^1 come from the notation of Fact 5.3. Define a partition $P : [\kappa]^{\mathcal{T}^2} \to 2$ by

$$P(F) = \begin{cases} 0 & (\exists \beta < \gamma)([F^0]_\mu < \Phi(\beta) < [F^1]_\mu) \\ 1 & \text{otherwise} \end{cases}$$

Let $D \subseteq \kappa$ with $|D| = \kappa$ be a homogeneous set for P.

(Case I) Assume D is homogeneous for P taking value 0.

By Fact 5.1, every nonconstant function $f \in {}^\kappa D$ has some $g \in [D]^\kappa$ so that $f =_\mu g$. Thus one can show that $[D]^\kappa/\mu$ has ordertype δ.

For each α, let f_α and $f_{\alpha+1}$ be two elements of $[D]^\kappa$ which represents the elements of $[D]^\kappa/\mu$ of rank α and $\alpha + 1$ respectively. By Fact 5.3, there is some $F : [D]^{\mathcal{T}^2} \to \kappa$ so that $F^0 = f_\alpha$ and $F^1 = F_{\alpha+1}$. Then $P(F) = 0$. Hence let ν_α be the least ordinal less than γ so that $[f_\alpha] = [F^\alpha]_\mu < \Phi(\nu_\alpha) < [F^{\alpha+1}]_\mu = [f_{\alpha+1}]_\mu$. One can check that ν_α depends only on α and not on the choice of f_α and $f_{\alpha+1}$. Also if $\alpha \neq \alpha'$, then $\nu_\alpha < \nu_{\alpha'}$. This gives an order preserving map from δ into $\gamma < \delta$ which is impossible.

(Case II) Assume D is homogeneous for P taking value 1.

As argued above, $[D]^\kappa/\mu$ has ordertype δ so $[D]^\kappa/\mu$ is a cofinal subset of κ^κ/μ. Fix $f^* \in [D]^\kappa/\mu$. Since Φ is a cofinal map, there is some $\gamma^* < \gamma$ so that $[f^*]_\mu < \Phi(\gamma^*)$. Then there is some $f' \in [D]^\kappa$ so that $[f^*]_\mu < \Phi(\gamma^*) < [f']_\mu$. By Fact 5.3, there is some $F \in [D]^{\mathcal{T}^2}$ so that $F^0 =_\mu f^*$ and $F^1 =_\mu f'$. Thus $P(F) = 0$. This is impossible since D is homogeneous for P taking value 1.

The failure of both cases would imply P has no homogeneous subset of size κ violating the assumption that κ is a strong partition cardinal. This completes the proof. □

DEFINITION 5.5. Let κ be a cardinal. Let μ be a κ-complete normal ultrafilter on κ.

A function $f : \kappa \to \kappa$ is a block function if and only if $\{\alpha \in \kappa : f(\alpha) < |\alpha|^+\} \in \mu$.

Let $\Xi : \kappa \times \kappa \to \kappa$. For each $\alpha < \kappa$, let $\delta^\Xi_\alpha = \sup\{\Xi(\alpha, \beta) : \beta < \alpha\}$. Let $\Xi_\alpha : \alpha \to \delta^\Xi_\alpha$ be defined by $\Xi_\alpha(\beta) = \Xi(\alpha, \beta)$.

Ξ is a Kunen function for f with respect to μ if and only if $K^\Xi_f = \{\alpha < \kappa : f(\alpha) \leq \delta^\Xi_\alpha \wedge \Xi_\alpha \text{ is a surjection}\} \in \mu$. (Here, when one says that Ξ_α is a surjection, one is considering Ξ_α as a function $\Xi_\alpha : \alpha \to \delta^\Xi_\alpha$.) K^Ξ_f is the set of α on which Ξ provides a bounding for f.

For $\beta < \kappa$, let $\Xi^\beta : \kappa \to \kappa$ be defined by $\Xi^\beta(\alpha) = \Xi(\alpha, \beta)$ where $\alpha > \beta$ and 0 otherwise. Ξ^β is a block function provided that $\{\alpha < \kappa : \Xi_\alpha \text{ is a surjection}\} \in \mu$.

FACT 5.6. *Assume* ZF. *Let μ be a normal measure on a cardinal κ. Suppose $f : \kappa \to \kappa$ is a block function which possesses a Kunen function Ξ with respect to μ. Suppose $G \in \prod_{\alpha\in\kappa} f(\alpha)/\mu$. Then there is a $\beta < \kappa$ so that $[\Xi^\beta]_\mu = G$*

PROOF. Take any $g \in G$. Let $A = \{\alpha \in K_f^\Xi : g(\alpha) < f(\alpha)\}$. Note that for $\alpha \in A$, $g(\alpha) < f(\alpha) < \delta_\alpha^\Xi$. Define $\Phi : A \to \kappa$ by $\Phi(\alpha)$ is the least $\beta < \alpha$ so that $g(\alpha) = \Xi(\alpha, \beta)$ which exists since Ξ_α is a surjection onto δ_α^Ξ. Thus on A, Φ is a regressive function. By normality, there is some $\beta < \kappa$ so that $\Phi(\alpha) = \beta$ for μ-almost all α. Then $\Xi^\beta =_\mu g$. Hence $[\Xi^\beta]_\mu = G$. One can check this β does not depend on the initial choice of $g \in G$. □

DEFINITION 5.7. Let κ be a cardinal. Let μ be a normal measure on κ. Let h be a block function. Suppose h possesses a Kunen function Ξ with respect to μ. An ordinal $\beta < \kappa$ is a minimal code (relative to Ξ) if and only if for all $\gamma < \beta$, $\neg(\Xi^\gamma =_\mu \Xi^\beta)$. Let J_h^Ξ be the collection of β which are minimal codes and $\Xi^\beta <_\mu h$. Define an ordering $\prec_h^\Xi$ on J_h^Ξ by $\alpha \prec_h^\Xi \beta$ if and only if $\Xi^\alpha <_\mu \Xi^\beta$. By Fact 5.6, for every $G < [h]_\mu$, there is a unique $\beta \in J_h^\Xi$ so that $\Xi^\beta \in G$. In this way, one says that β is a minimal code for G or for any $g \in G$ with respect to Ξ.

FACT 5.8. *Let μ be a normal measure on a cardinal κ. Let $f : \kappa \to \kappa$ be a block function possessing a Kunen function Ξ with respect to μ. Then $\prod_{\alpha\in\kappa} f(\alpha)/\mu$, i.e. the initial segment of ${}^\kappa/\kappa/\mu$ determined by $[f]_\mu$, is a wellordering.*

If every block function has a Kunen function, then $\prod_{\alpha<\kappa} |\alpha|^+/\mu$ is wellfounded.

For each $F \in \prod_{\alpha<\kappa} |\alpha^+|/\mu$, $F < \kappa^+$. Thus $\prod_{\alpha<\kappa} |\alpha|^+/\mu \leq \kappa^+$.

PROOF. Let f be a block function possessing a Kunen function Ξ. Every $G \in \prod_{\alpha\in\kappa} f(\alpha)/\mu$ has a unique $\beta \in J_f^\Xi$ so that $[\Xi^\beta]_\mu = G$. There is a bijection of J_f^Ξ with $\prod_{\alpha\in\kappa} f(\alpha)/\mu$ given by $\beta \mapsto [\Xi^\beta]_\mu$. This shows that $|\prod_{\alpha\in\kappa} f(\alpha)/\mu| \leq \kappa$.

Now suppose that $\prod_{\alpha\in\kappa} f(\alpha)/\mu$ is not wellfounded. Then $\prec_f^\Xi$ is an illfounded linear ordering on $J_f^\Xi \subseteq \kappa$. Let β_0 be the least (in the usual ordinal ordering) element of J_f^Ξ so that the initial segment determined by β_0 in $(J_f^\Xi, \prec_h^\Xi)$ has no $\prec$-minimal element. Suppose β_n has been defined, let β_{n+1} be the least ordinal in J_f^Ξ which is $\prec_f^\Xi$-below β_n. This process defines a sequence of ordinals $\langle \beta_n : n \in \omega\rangle$ in J_f^Ξ.

Then $\langle \Xi^{\beta_n} : n \in \omega\rangle$ is a sequence with the property that $[\Xi^{\beta_n}]_\mu$ is an infinite decreasing sequence in $\prod_{\alpha<\kappa} f(\alpha)/\mu$. Let $D_n = \{\alpha \in \kappa : \Xi^{\beta_{n+1}}(\alpha) < \Xi^{\beta_n}(\alpha)\} \in \mu$. Then $\bigcap_{n\in\omega} D_n \in \mu$. Let $\xi \in \bigcap_{n\in\omega} D_n$. Then $\langle \Xi^{\beta_n}(\xi) : n \in \omega\rangle$ is an infinite decreasing sequence of ordinals. This is impossible.

Since f was arbitrary, this shows that $\prod_{\alpha<\kappa} |\alpha|^+/\mu$ is wellfounded. By the first paragraph, each $F \in \prod_{\alpha<\kappa} |\alpha|^+/\mu$ has cardinality less than or equal to κ. Thus $F < \kappa^+$. Thus $\prod_{\alpha<\kappa} |\alpha|^+/\mu \leq \kappa^+$. □

DEFINITION 5.9. Let μ be a normal measure on κ. Let $h : \kappa \to \kappa$ be a block function. Let Ξ be a Kunen function for h with respect to μ. By Fact 5.6, for each $G < [h]_\mu$, there is a minimal code $\beta \in J_h^\Xi$ so that $\Xi^\beta \in G$. Thus $(J_h^\Xi, \prec_h^\Xi)$ has the same ordertype as $[h]_\mu$. By Fact 5.8, $[h]_\mu$ is a wellordering. Let $\epsilon_h^\Xi \in \mathrm{ON}$ denote the ordertype of $([h]_\mu, <)$ which is equal to the ordertype of $(J_h^\Xi, \prec_h^\Xi)$. Let $\pi_h^\Xi : \epsilon_h^\Xi \to (J_h^\Xi, \prec_h^\Xi)$ be the unique order-preserving isomorphism.

Note that every function $f : \omega_1 \to \omega_1$ is (everywhere) a block function.

THEOREM 5.10. *(Kunen) Assume* ZF + AD. *Every function* $f : \omega_1 \to \omega_1$ *has a Kunen function with respect to the club measure* μ.

PROOF. The Kunen function is derived from the Kunen tree which will be defined below.

Recall that each $x \in {}^\omega\omega$ codes a relation R_x defined in Definition 4.1. Also WF denotes the collection of x so that R_x is a wellfounded relation. Let $\pi_{\text{pair}} : \omega \times \omega \to \omega$ denote a fixed recursive bijection.

Let $S \subseteq \omega \times \omega_1$ be the tree of partial rankings of relations on ω which is defined as follows: $(s, \bar{\alpha}) \in S$ if and only if $|s| = |\bar{\alpha}|$ and for all $i, j < |s|$, if $s(\pi_{\text{pair}}(i,j)) = 1$, then $\bar{\alpha}(i) < \bar{\alpha}(j)$.

If $(x, f) \in [S] = \{(y, g) : (\forall n)(y \restriction n, g \restriction n) \in S)\}$, then $x \in \text{WF}$ since f is a ranking of R_x into ω_1. Conversely, if $x \in \text{WF}$, then R_x has a ranking f with image in ω_1. Thus $(x, f) \in [S]$. It has been shown that $\text{WF} = \pi_1[[S]]$ is the projection of $[S]$ onto the first coordinate.

Let T be a recursive tree on $\omega \times \omega$ so that $\pi_1[[T]] = \{a \in {}^\omega\omega : (\exists b)((a, b) \in [T])\} = {}^\omega\omega \setminus \text{WF}$. The important observation is that ${}^\omega\omega \setminus \text{WF}$ is a Σ^1_1 set which is $\boldsymbol{\Sigma}^1_1$-complete.

Let $\pi_{\text{seq}} : {}^{<\omega}\omega \to \omega$ be a recursive bijection. For each $x \in {}^\omega\omega$, let $\sigma_x : {}^{<\omega}\omega \to \omega$ be a strategy in the usual integer game defined by $\sigma_x(s) = n$ if and $x(\pi_{\text{seq}}(s)) = n$. Suppose $p \in {}^{<\omega}\omega$ and $s \in {}^{<\omega}\omega$, then let $\sigma_p * s$ denote the partial play where Player 1 uses the partial strategy σ_p and Player 2 plays the bits of s turn by turn. The game goes on for as long as p codes a response to the partial play that is produced each turn.

Define a tree K on $\omega \times \omega \times \omega_1 \times \omega \times \omega$ as follows: $(p, s, \bar{\alpha}, t, u) \in K$ if and only if the conjunction of the following holds:
(1) $(s, \bar{\alpha}) \in S$.
(2) $(t, u) \in T$.
(3) $\sigma_p * s$ is a substring of t.

The meaning of K becomes clear if one looks at what a path through K represents: If $(x, y, f, v, w) \in [K]$, then $\sigma_x(y) = v$, $(y, f) \in [S]$, and $(v, w) \in [T]$. Therefore, $y \in \text{WF}$ and $v \in \mathbb{R} \setminus \text{WF}$ since S and T are trees that project onto WF and $\mathbb{R} \setminus \text{WF}$.

The above defines the tree K. One now introduces an arbitrary function $f : \omega_1 \to \omega_1$. Consider the game G_f defined as follows:

	I	$y(0)$		$y(1)$		$y(2)$		$y(3)$			y
G_f											
	II		$v(0)$		$v(1)$		$v(2)$		$v(3)$		v

Player 2 wins if and only if $y \in \text{WO} \Rightarrow (v \in \text{WF} \wedge \text{rk}(T_v) > \sup\{f(\alpha) : \alpha \leq \text{ot}(y)\})$. Here $T_v = \{u \in {}^{<\omega}\omega : (u, v \restriction |u|) \in T\}$. Note that since T projects onto ${}^\omega\omega \setminus \text{WF}$, if $v \in \text{WF}$, then T_v is a wellfounded tree.

Player 2 must have the winning strategy: Suppose ρ is a Player 1 winning strategy. $\rho[{}^\omega\omega] \subseteq \text{WO}$ since otherwise Player 2 can win against ρ. Thus $\rho[{}^\omega\omega]$ is a $\boldsymbol{\Sigma}^1_1$ subset of WO. By the bounding principle, let $\mu < \omega_1$ be such that $\text{ot}(r) < \mu$ for all $r \in \rho[{}^\omega\omega]$. Let $\zeta = \sup\{f(\alpha) : \alpha < \mu\}$. $\zeta < \omega_1$ because ω_1 is regular. Since the projection of T is $\mathbb{R} \setminus \text{WF}$ which is a $\boldsymbol{\Sigma}^1_1$-complete set, one must have that $\sup\{\text{rk}(T_r) : r \in \text{WF}\} = \omega_1$. Choose some $v \in \text{WF}$ so that $\text{rk}(T_v) > \zeta$. If Player 2 plays v against ρ, Player 2 will win. This contradicts the fact that ρ is a Player 1 winning strategy.

Thus there is a winning strategy σ for the game G_f. Let $x \in {}^\omega\omega$ be such that $\sigma_x = \sigma$.

Let K_x be the tree consisting of $(s, \bar{\alpha}, t, u)$ so that there is an $n \in \omega$ such that $|s| = |\bar{\alpha}| = |t| = |u| = n$ and $(x \restriction n, s, \bar{\alpha}, t, u) \in K$. For each $\alpha < \omega_1$, let $K_x \restriction \alpha$ denote the restrict of K_x to ${}^{<\omega}(\omega \times \alpha \times \omega \times \omega)$. Note that $K_x \restriction \alpha$ is wellfounded. To see this, suppose otherwise. This means there is some $y, v, w \in {}^\omega\omega$ and $f \in {}^{<\omega}\alpha$ so that $(x, y, f, v, w) \in [K]$. Thus $\sigma_x(y) = \sigma(y) = v$, $y \in \text{WO}$, and $v \in \mathbb{R} \setminus \text{WF}$. However, σ is a Player 2 winning strategy and $y \in \text{WO}$, so one must have that $v \in \text{WF}$. This is a contradiction.

Let $\alpha \geq \omega$. Suppose $y \in \text{WO}$ with $\text{ot}(y) = \alpha$. Then there is a ranking f of R_y using ordinals below α. Let $v = \sigma_x(y) = \sigma(y)$. Since σ is a Player 2 winning strategy, $\text{rk}(T_v) > f(\text{ot}(y))$. Note that $K_x \restriction \alpha$ has a subtree which is isomorphic to T_v. In particular, this subtree is

$$\hat{T}_v = \{(y \restriction n, f \restriction n, t, u) \in K_x : n \in \omega \wedge |t| = |u| = n \wedge (t, u) \in T_v\}.$$

This implies that $K_x \restriction \alpha$ is a wellfounded tree of rank greater than $\text{rk}(T_v) > f(\alpha)$.

Note that there is a club set of α so that α is closed under the Gödel pairing function. Using this pairing function on α, one can define uniformly a bijection $\pi_\alpha : \alpha \to {}^{<\omega}\alpha$. For all α closed under the Gödel pairing function, define $\Xi(\alpha, \beta)$ to be the rank of $\pi_\alpha(\beta)$ in $K_x \restriction \alpha$ whenever $\beta < \alpha$ and $\pi_\alpha(\beta) \in K_x \restriction \alpha$. (Technically in the definition of a Kunen function, Ξ should be a function on $\omega_1 \times \omega_1$; however, the value of Ξ on (α, β) where $\beta \geq \alpha$ is not relevant in any applications.) If $\pi_\alpha(\beta)$ does not belong to $K_x \restriction \alpha$ then let $\Xi(\alpha, \beta) = 0$.

Ξ is a Kunen function for f. For any $\alpha \geq \omega$ which is closed under the Gödel pairing function, Ξ_α is a surjection onto the $\text{rk}(K_x \restriction \alpha)$. Using the notation of Definition 5.5, $\text{rk}(K_x \restriction \alpha) = \delta^\Xi_\alpha$. Also, it was shown above that $\text{rk}(K_x \restriction \alpha) > f(\alpha)$ and hence $\delta^\Xi_\alpha > f(\alpha)$. This verifies that Ξ is a Kunen function for f with respect to μ. □

REMARK 5.11. The tree K produced in Theorem 5.10 is called the Kunen tree.

There are some simplifications of K that can be made. K is a tree on $\omega \times \omega \times \omega_1 \times \omega \times \omega$. One can merge the last four coordinates on $\omega \times \omega_1 \times \omega \times \omega$ into ω_1 to produce a tree on $\omega \times \omega_1$ with the same features. One can also Kleene-Brouwer order the various trees to produce linear orderings which can be used to produce the desired Kunen functions. See [**12**] for more details on the Kunen tree.

In this survey, one will only use the existence of Kunen functions for functions $f : \omega_1 \to \omega_1$; thus, it is not important here how uniformly these Kunen functions are obtained. However, the proof shows that all Kunen functions are index uniformly by reals. For instance, there is a single tree K, the Kunen tree, so that for any $f : \omega_1 \to \omega_1$, there is a section of this tree by some strategy which can be used to produce the Kunen function for f. The Kunen tree K is also Δ^1_1 in the codes. See [**12**] for the details and the precise meaning of these remarks.

COROLLARY 5.12. *Assume* $\mathsf{ZF} + \mathsf{AD}$. *Let* μ *denote the club measure on* ω_1. $|\prod_{\omega_1} \omega_1/\mu| \leq \omega_2$.

FACT 5.13. *(Martin) Assume* $\mathsf{ZF} + \mathsf{AD}$. *Let* μ *denote the club measure on* ω_1. *Then* $\prod_{\omega_1} \omega_1/\mu = \omega_2$ *and is a regular cardinal.*

PROOF. Note that the club filter on ω_1, μ, is a normal measure on ω_1. Therefore by Fact 5.4, the ultrapower is a regular cardinal. Thus it must be greater than or equal ω_2. Then Corollary 5.12 implies that the ultrapower is ω_2. □

DEFINITION 5.14. Let μ be a normal measure on a cardinal κ. Let $h : \kappa \to \kappa$ be a function so that $h(\alpha) > 0$ μ-almost everywhere. Let $T^h = \{(\alpha, \beta) \in \kappa \times \kappa : \beta < h(\alpha)\}$. Let $\mathcal{T}^h = (T^h, \sqsubset)$ where $\sqsubset$ is the lexicographic ordering. Note that $\text{ot}(\mathcal{T}^h) = \kappa$.

Suppose $F : \mathcal{T}^h \to \kappa$ is an order-preserving function. Let $g \in {}^\kappa\kappa$ be such that $g <_\mu h$. Let $A^g = \{\alpha : g(\alpha) < h(\alpha)\}$. Let $F^g \in {}^\kappa\kappa$ be defined by

$$F^g(\alpha) = \begin{cases} F(\alpha, g(\alpha)) & \alpha \in A^g \\ F(\alpha, 0) & \text{otherwise} \end{cases}$$

Note that if $g_1 <_\mu g_2 <_\mu h$, then $F^{g_1} <_\mu F^{g_2}$.

Fix a Kunen function Ξ for h. Recall that ϵ_h^Ξ is the ordertype of the wellordering $(J_h^\Xi, \prec_h^\Xi)$ and $\pi_h^\Xi : \epsilon_h^\Xi \to (J_h^\Xi, \prec_h^\Xi)$ is the unique order isomorphism. If $\beta \in \epsilon_h^\Xi$, then let $F^{(\beta)} = F^{\Xi^{\pi_h^\Xi(\beta)}}$. Let $\mathsf{funct}(F) : \epsilon_h^\Xi \to \text{ON}$, be defined by $\mathsf{funct}(F)(\alpha) = [F^{(\alpha)}]_\mu$.

$\mathcal{T}^h = (T^h, \sqsubset)$ is order isomorphic to κ. It is merely a reorganization of κ into successive blocks of length $h(\alpha)$. If $F : \mathcal{T}^h \to \kappa$ is order preserving and $\beta < \epsilon_h^\Xi$, then $F^{(\beta)} : \kappa \to \kappa$ is defined by letting $F^{(\beta)}(\alpha)$ be the value of F at $(\alpha, \Xi^{\pi_h^\Xi(\beta)}(\alpha))$, which can be construed to be the $\Xi^{\pi_h^\Xi(\beta)}(\alpha)^{\text{th}}$ element in the αth block of length $h(\alpha)$.

LEMMA 5.15. *(Sliding lemma) Let μ be a normal measure on a cardinal κ. Let $h : \kappa \to \kappa$ be a block function possessing a Kunen function Ξ with respect to μ.*

Let $\langle g_\alpha : \alpha \in \epsilon_h^\Xi \rangle$ be an order preserving sequence elements of ${}^\kappa\kappa$ which are non-constant μ-almost everywhere: order preserving means that if $\alpha < \beta < \epsilon_h^\Xi$, then $g_\alpha <_\mu g_\beta$.

Then there exists an order-preserving $F : \mathcal{T}^h \to \kappa$ so that for all $\beta < \epsilon_h^\Xi$, $F^{(\beta)} =_\mu g_\beta$.

Moreover, if $X \subseteq \kappa$ and for all $\beta < \epsilon_h^\Xi$, $\text{rang}(g_\beta) \subseteq X$, then $\text{rang}(F) \subseteq X$.

PROOF. Fix some $X \subseteq \kappa$ so that $\text{rang}(g_\beta) \subseteq X$ for all $\beta < \epsilon_h^\Xi$, $\text{rang}(g_\beta) \subseteq X$.

Define $F : \mathcal{T}^h \to \kappa$ by recursion as follows:

Let $(\alpha, \beta) \in T^h$. Suppose $F(\alpha', \beta')$ has been defined for all $(\alpha', \beta') \sqsubset (\alpha, \beta)$. (Recall $\mathcal{T}^h = (T^h, \sqsubset)$ where $\sqsubset$ is the lexicographic ordering.)

(Case I) There is a $\gamma < \epsilon_h^\Xi$ so that $\beta = \Xi^{\pi_h^\Xi(\gamma)}(\alpha)$:

Let γ be least with this property. Let $F(\alpha, \beta)$ be the least element in the range g_γ which is greater than or equal to $\sup\{F(\alpha', \beta') : (\alpha', \beta') \sqsubset (\alpha, \beta)\}$.

(Case II) There is no $\gamma < \epsilon_h^\Xi$ so that $\beta = \Xi^{\pi_h^\Xi(\gamma)}(\alpha)$:

Let $F(\alpha, \beta)$ be the ωth-element of X above $\sup\{F(\alpha', \beta') : (\alpha', \beta') \sqsubset (\alpha, \beta)\}$.

Claim: For all $\gamma < \epsilon_h^\Xi$, $F^{(\gamma)} =_\mu g_\gamma$.

This will be proved by induction on ϵ_h^Ξ.

It is clear $B = \{\alpha \in \kappa : \Xi^{\pi_h^\Xi(0)}(\alpha) = 0\} \in \mu$. For $\alpha \in B$, $F^{(0)} = F(\alpha, \Xi^{\pi_h^\Xi(0)}(\alpha)) \in \text{rang}(g_0)$. Thus on B, one must have that $F^{(0)}(\alpha) \geq g_0(\alpha)$. Suppose that $F^{(0)} >_\mu g_0$. Let $C = \{\alpha \in B : F^{(0)}(\alpha) > g_0(\alpha)\} \in \mu$.

Since for $\alpha \in C$, $\Xi^{\pi_h^\Xi(0)}(\alpha) = 0$, there must be some $(\alpha', \beta') \sqsubset (\alpha, 0)$ with $\alpha' < \alpha$ so that $F(\alpha', \beta') \geq g_0(\alpha)$. Define $\Phi : C \to \kappa$ by letting $\Phi(\alpha)$ be the least

$\alpha' < \alpha$ so that there exists some β' with $(\alpha', \beta') \sqsubset (\alpha, \beta)$ and $F(\alpha', \beta') \geq g_0(\alpha)$. Thus Φ is regressive on $C \in \mu$. There is some α' and a $D \in \mu$ so that $\Phi(\alpha) = \alpha'$ for all $\alpha \in D$. Thus for all $\alpha \in D$, $g_0(\alpha) \leq F(\alpha', \beta')$ for some $\beta' < h(\alpha')$. By κ-completeness, there is some $\bar{\beta} < h(\alpha')$ and an $E \subseteq D$ with $E \in \mu$ so that for all $\alpha \in E$, $g_0(\alpha) \leq F(\alpha', \bar{\beta})$. This is impossible since g_0 is not constant μ-almost everywhere. It has been shown that $F^{(0)} = g_0$.

Suppose $\gamma < \epsilon_h^\Xi$ and that it has been shown that $F^{(\gamma')} =_\mu g_{\gamma'}$ for all $\gamma' < \gamma$. One will seek to show that $F^{(\gamma)} = g_\gamma$.

Let $A = \{\alpha : (\forall \gamma' \in \gamma)(\Xi^{\pi_h^\Xi(\gamma)}(\alpha) \neq \Xi^{\pi_h^\Xi(\gamma')}(\alpha))\}$. If $A \notin \mu$, then by κ-completeness, there is some $\gamma' \in \gamma$ so that $\Xi^{\pi_h^\Xi(\gamma')} =_\mu \Xi^{\pi_h^\Xi(\gamma)}$. This is impossible since $\gamma' < \gamma$ implies $\Xi^{\pi_h^\Xi(\gamma')} <_\mu \Xi^{\pi_h^\Xi(\gamma)}$.

By induction, it has been shown that $F^{(0)} = g_0$. Let $B \subseteq A$ with $B \in \mu$ have the property that for all $\alpha \in B$, $\Xi^{\pi_h^\Xi(0)}(\alpha) = 0$, $F^{(0)}(\alpha) = g_0(\alpha)$, and $g_\gamma(\alpha) > g_0(\alpha)$.

Suppose $F^{(\gamma)}$ is not equal to g_γ for μ-almost all α. For $\alpha \in B$, $F^{(\gamma)}(\alpha) \in \operatorname{rang}(g_\gamma)$. Thus one must have that there is a set $C \subseteq B$ with $C \in \mu$ so that for all $\alpha \in C$, $F^{(\gamma)}(\alpha) > g_\gamma(\alpha)$. There must be some $(\alpha', \beta') \sqsubset (\alpha, \Xi^{\pi_h^\Xi(\gamma)}(\alpha))$ so that $F(\alpha', \beta') \geq g_\gamma(\alpha)$. However, B was chosen so that for all $\alpha \in B$, $\Xi^{\pi_h^\Xi(0)}(\alpha) = 0$ and $F^{(0)}(\alpha) = g_0(\alpha) < g_\gamma(\alpha)$. Therefore, for $\alpha \in C$, $F^{(0)}(\alpha) < g_\gamma(\alpha) < F^{(\gamma)}(\alpha)$. Thus the least $(\alpha', \beta') \sqsubset (\alpha, \Xi^{\pi_h^\Xi(\gamma)}(\alpha))$ with $F(\alpha', \beta') \geq g_\gamma(\alpha)$ must have that $\alpha' = \alpha$.

Therefore, for all $\alpha \in C$, there is some $\beta' < \Xi^{\pi_h^\Xi(\gamma)}(\alpha)$ so that $F(\alpha, \beta') \geq g_\gamma(\alpha)$. Let $\Phi(\alpha)$ be this β'. Then $\Phi <_\mu \Xi^{\pi_h^\Xi(\gamma)} < h$. Thus there is some $\bar{\gamma} < \gamma$ so that $\Phi =_\mu \Xi^{\pi_h^\Xi(\bar{\gamma})}$. By induction, $F^{(\bar{\gamma})} =_\mu g_{\bar{\gamma}}$. However by definition of Φ, $F^\Phi \geq_\mu g_\gamma$. Also $F^\Phi =_\mu F^{(\bar{\gamma})} =_\mu g_{\bar{\gamma}}$. Thus $g_{\bar{\gamma}} \geq_\mu g_\gamma$. Since $\langle g_\delta : \delta \in \epsilon_h^\Xi \rangle$ is an increasing sequence and $\bar{\gamma} < \gamma$, one has that $g_{\bar{\gamma}} <_\mu g_\gamma$. This is a contradiction. This shows that $F^\gamma = g_\gamma$.

The lemma has been proved. □

It will be helpful to have the correct type version of the sliding lemma. A sketch of the necessary modification will be given:

LEMMA 5.16. *(Correct-type sliding lemma) Let μ be a normal measure on a cardinal κ. Let $h : \kappa \to \kappa$ be a block function possessing a Kunen function Ξ with respect to μ.*

Let $\langle g_\alpha : \alpha < \epsilon_h^\Xi \rangle$ be an increasing sequence of functions from κ to κ of the correct type which are μ-almost everywhere non-constant. Further, suppose there is a sequence $\langle G_\alpha : \alpha < \epsilon_h^\Xi \rangle$ where each $G_\alpha : \kappa \times \omega \to \kappa$ is a function witnessing that g_α has uniform cofinality ω. Suppose $\langle [g_\alpha]_\mu : \alpha < \epsilon_h^\Xi \rangle$ is discontinuous everywhere.

Then there exists an order preserving function $F : \mathcal{T}^h \to \kappa$ which is of the correct type so that for all $\alpha < \epsilon_h^\Xi$, $F^{(\alpha)} =_\mu g_\alpha$.

Moreover, if $C \subseteq \kappa$ is an ω-club set such that for all $\beta \in \epsilon_h^\Xi$, $\operatorname{rang}(g_\beta) \subseteq C$, then $\operatorname{rang}(F) \subseteq C$.

PROOF. Fix some $C \subseteq \kappa$ be an ω-club so that $\operatorname{rang}(g_\beta) \subseteq C$ for all $\beta < \epsilon_h^\Xi$.

Define $F : \mathcal{T}^h \to \kappa$ by recursion as follows:

Let $(\alpha, \beta) \in T^h$. Suppose $F(\alpha', \beta')$ has been defined for all $(\alpha', \beta') \sqsubseteq (\alpha, \beta)$.

(Case I) There is some $\gamma < \epsilon_h^\Xi$ so that $\beta = \Xi^{\pi_h^\Xi(\gamma)}(\alpha)$:

Let γ be least with this property. Let $F(\alpha, \beta)$ be the least element in the range of g_γ which is greater than $\sup\{F(\alpha', \beta') : (\alpha', \beta') \sqsubset (\alpha, \beta)\}$.

(Case II) There is no $\gamma < \epsilon_h^\Xi$ so that $\beta = \Xi^{\pi_h^\Xi(\gamma)}(\alpha)$:

Find δ least so that $\mathsf{enum}_C(\delta) > \sup\{F(\alpha',\beta') : (\alpha',\beta') \sqsubset (\alpha,\beta)\}$. Let $F(\alpha,\beta) = \mathsf{enum}_C(\delta+\omega)$.

Claim 1: $F : \mathcal{T}^h \to C$ is a function of the correct type.

To prove this: Note that it is clear from the construction that F is discontinuous everywhere.

One will create a function $H : T^h \times \omega \to \kappa$ to witness the uniform cofinality of F.

Let $(\alpha,\beta) \in T^h$.

Suppose Case I had occurred at (α,β) with $\gamma < \epsilon_h^\Xi$ least so that $\beta = \Xi^{\pi_h^\Xi(\gamma)}(\alpha)$. Let $\delta < \kappa$ be least so that $g_\gamma(\delta) > \sup\{F(\alpha',\beta') : (\alpha',\beta') \sqsubset (\alpha,\beta)\}$. Now let n be least so that $G_\gamma(\delta,n) > \sup\{F(\alpha',\beta') : (\alpha',\beta') \sqsubset (\alpha,\beta)\}$. Define $H((\alpha,\beta),m) = G_\gamma(\delta,n+m)$.

Suppose Case II had occurred at (α,β). Let $H((\alpha,\beta),m)$ be the mth element of C above $\sup\{F(\alpha',\beta') : (\alpha',\beta') \sqsubset (\alpha,\beta)\}$.

This function H witnesses that F has uniform cofinality ω. F has the correct type.

Claim 2: For all $\gamma < \epsilon_h^\Xi$, $F^{(\gamma)} =_\mu g_\gamma$.

To show Claim 2: This will be proved by induction on ϵ_h^Ξ. One will indicate some of the necessary modification from the proof of Lemma 5.15.

Suppose that $\gamma < \epsilon_h^\Xi$ and that it has been shown that $F^{(\gamma')} =_\mu g_{\gamma'}$ for all $\gamma' < \gamma$. Let $A = \{\alpha : (\forall\gamma' \in \gamma)(\Xi^{\pi_h^\Xi(\gamma)}(\alpha) \neq \Xi^{\pi_h^\Xi(\gamma')}(\alpha))\}$. As in Lemma 5.15, $A \in \mu$. Also just as before, one can show that $B \in \mu$ where $B \subseteq A$ and has the property that for all $\alpha \in B$, $\Xi^{\pi_h^\Xi(0)}(\alpha) = 0$, $F^{(0)}(\alpha) = g_0(\alpha)$, and $g_\gamma(\alpha) > g_0(\alpha)$.

Now suppose $F^{(\gamma)}$ is not equal to g_γ for μ-almost all α. In the present situation, there are two ways this can happen:

(i) There is a $D \subseteq B$ with $D \in \mu$ so that for all $\alpha \in D$, there exists some $(\alpha',\beta') \sqsubset (\alpha, \Xi^{\pi_h^\Xi(\gamma)}(\alpha))$ so that $F(\alpha',\beta') \geq g_\gamma(\alpha)$. As argued in Lemma 5.15, this can not occur.

(ii) There is a $D \subseteq B$ with $D \in \mu$ so that for all $\alpha \in D$, $g_\gamma(\alpha) = \sup\{F(\alpha',\beta') : (\alpha',\beta') \sqsubset (\alpha, \Xi^{\pi_h^\Xi(\gamma)}(\alpha))\}$.

For each $n \in \omega$, let $\Phi_n : D \to \omega_1$ be defined as follows: Since $\alpha \in D \subseteq B$, one has that $F^{(0)}(\alpha) = g_0(\alpha) < g_\gamma(\alpha)$. For each $n \in \omega$ and $\alpha \in D$, let $\Phi_n(\alpha)$ be the least $\beta' < \Xi^{\pi_h^\Xi(\gamma)}(\alpha)$ so that $F(\alpha,\beta') > G_\gamma(\alpha,n)$.

By Fact 5.6, there is some $\gamma_n < \gamma$ so that $\Phi_n =_\mu \Xi^{\gamma_n}$. By the induction hypothesis, $F^{\Xi^{\pi_h^\Xi(\gamma_n)}} = F^{(\gamma_n)} =_\mu g_{\gamma_n}$. By construction, $\sup\{[g_{\gamma_n}]_\mu : n \in \omega\} = [g_\gamma]_\mu$. However, by assumption, $\langle [g_\alpha]_\mu : \alpha < \epsilon_h^\Xi \rangle$ was discontinuous everywhere.

This complete the proof of the lemma. □

FACT 5.17. *Let μ be a normal measure on a cardinal κ. Let $h : \kappa \to \kappa$ be a block function possessing a Kunen function Ξ with respect to μ. Suppose $F_0, F_1 \in [\omega_1]^{\mathcal{T}^h}$ have the property that $F_0^{(\beta)} =_\mu F_1^{(\beta)}$ for all $\beta < \epsilon_h^\Xi$. Then for μ-almost all α, $F_0(\alpha,\beta) = F_1(\alpha,\beta)$ for all $\beta < h(\alpha)$.*

PROOF. Suppose $A = \{\alpha : (\exists\beta < h(\alpha))(F_0(\alpha,\beta) \neq F_1(\alpha,\beta))\} \in \mu$. Define $g : \omega_1 \to \omega_1$ by $g(\alpha)$ is the least $\beta < h(\alpha)$ so that $F_0(\alpha,\beta) \neq F_1(\alpha,\beta)$. Since $g <_\mu h$, Fact 5.6 implies there is some $\gamma < \epsilon_h^\Xi$ so that $g =_\mu \Xi^{\pi_h^\Xi(\gamma)}$. Then $\neg(F_0^{(\gamma)} =_\mu F_1^{(\gamma)})$. This contradicts the assumptions. □

FACT 5.18. *Assume* ZF+AD. *Let* μ *denote the club measure on* ω_1. *Let* $\alpha < \omega_2$. *Let* $A \subseteq \omega_1$ *with* $|A| = \omega_1$. *Let* $B = [A]^{\omega_1}/\mu$, *which is a set of cardinality* ω_2. *Let* $h : \omega_1 \to \omega_1$ *have the property that* $h(\beta) > 0$ *for all* $\beta < \omega_1$ *and* $[h]_\mu = \alpha$. *Let* Ξ *be a Kunen function for* h *with respect to* μ.

For all $I \in [B]^\alpha$, *there exists some* $F : \mathcal{T}^h \to A$ *which is order preserving and for all* $\gamma < \alpha = \epsilon_h^\Xi$, $[F^{(\gamma)}]_\mu = I(\gamma)$.

PROOF. Since $\omega_2 = {}^{\omega_1}\omega_1/\mu$ is regular by Fact 5.4, $\mathrm{rang}(I) < \omega_2$. Let $k \in [\omega_1]^{\omega_1}$ be such that $\mathrm{rang}(I) < [k]_\mu$. Let Ξ' be a Kunen function for k with respect to μ.

Let $J = \{\beta \in \epsilon_k^{\Xi'} : (\exists \gamma < \alpha)(I(\gamma) = [\Xi'^{\pi_k^{\Xi'}(\beta)}]_\mu)\}$. $J \subseteq \epsilon_k^{\Xi'}$ and $\mathrm{ot}(J) = \alpha = \epsilon_h^\Xi$. Let $\rho : \epsilon_h^\Xi \to J$ be the unique order isomorphism. Define $g'_\gamma = \Xi'^{\pi_k^{\Xi'}(\rho(\gamma))}$. Since $[g'_\gamma]_\mu = I(\gamma)$ and $I(\gamma) \in B = [A]^{\omega_1}/\mu$, the set $B_\gamma = \{\alpha \in \omega_1 : g'_\gamma(\alpha) \in A\} \in \mu$. By Fact 5.1, $g'_\gamma =_\mu g'_\gamma \circ \mathsf{enum}_{B_\gamma}$. Let $g_\gamma = g'_\gamma \circ \mathsf{enum}_{B_\gamma}$. $[g_\gamma]_\mu = I(\gamma)$ and $\mathrm{rang}(g_\gamma) \subseteq A$. The result now follows from Lemma 5.15. ☐

THEOREM 5.19. *(Martin-Paris) Assume* ZF + AD. *Let* μ *be the club measure on* ω_1. *Then for all* $\alpha < \omega_2$, *the partition relation* $\omega_2 \to (\omega_2)^\alpha_2$ *holds. That is,* ω_2 *is a weak partition cardinal.*

PROOF. By Fact 5.13, ω_2 is isomorphic to ${}^{\omega_1}\omega_1/\mu$. One will identify ω_2 with ${}^{\omega_1}\omega_1/\mu$.

Let $\alpha < \omega_2$ and $P : [\omega_2]^\alpha \to 2$ be a partition.

Let $h : \omega_1 \to \omega_1$ be such that $[h]_\mu = \alpha$ and fix a Kunen function Ξ for h. Define a partition $Q : [\omega_1]^{\mathcal{T}^h} \to 2$ by $Q(F) = P(\mathsf{funct}(F)))$, where $\mathsf{funct}(F)$ is an α-sequence of ordinals in ω_2 defined in Definition 5.14 (relative to Ξ).

Since $\mathcal{T}^h$ is order isomorphic to ω_1, $\omega_1 \to (\omega_1)^{\omega_1}_2$ implies that there is a $A \subseteq \omega_1$ with $|A| = \omega_1$ which is homogeneous for Q. Without loss of generality, suppose A is homogeneous for Q taking value 0.

Note that $[A]^{\omega_1}/\mu$ has cardinality ${}^{\omega_1}\omega_1/\mu = \omega_2$. Let B denote $[A]^{\omega_1}/\mu$.

Let $I \in [B]^\alpha$. By Fact 5.18, let $F : \mathcal{T}^h \to A$ be such that $[F^{(\gamma)}]_\mu = I(\gamma)$ for all $\gamma < \alpha$. Since $F \in [A]^{\mathcal{T}^h}$, one has $0 = Q(F) = P(\mathsf{funct}(F)) = P(I)$ since $I = \mathsf{funct}(F)$. B is homogeneous for P taking value 0. The proof is complete. ☐

By Fact 2.6, a suitable form of the ordinary partition property will imply the appropriate correct type version of the partition property which uses a club set as its version of the homogeneous set. In particular, the ordinary weak partition property for ω_2 proved in Theorem 5.19 implies the correct type club version of the weak partition property, that is $\omega_2 \to_* (\omega_2)^\alpha_2$ for all $\alpha < \omega_2$.

COROLLARY 5.20. *Assume* ZF + AD. $W^{\omega_2}_\omega$ *and* $W^{\omega_2}_{\omega_1}$ *are the only two* ω_2*-complete normal ultrafilters on* ω_2.

PROOF. By Fact 2.15. ☐

FACT 5.21. *Assume* ZF + AD. *Let* μ *be the club measure on* ω_1. *If* $C \subseteq \omega_1$ *is a club subset of* ω_1, *then* $[C]^{\omega_1}/\mu$ *is a club subset of* ω_2.

If $D \subseteq \omega_2$ *is club, then there is a club* $C \subseteq \omega_1$ *so that* $[C]^{\omega_1}/\mu \subseteq D$.

PROOF. Let $\epsilon < \omega_2$. Suppose $\langle \nu_\gamma : \gamma < \epsilon \rangle$ is an increasing sequence in $[C]^{\omega_1}/\mu$. Let $h : \omega_1 \to \omega_1$ with $h(\alpha) > 0$ for all $\alpha \in \omega_1$ be such that $\epsilon = [h]_\mu$. Let Ξ be a Kunen function for h with respect to μ.

By Fact 5.18, there is a function $F : \mathcal{T}^h \to C$ which is order preserving so that $[F^{(\gamma)}]_\mu = \nu_\gamma$.

Let $\ell : \omega_1 \to \omega_1$ be defined by $\ell(\alpha) = \sup\{F(\alpha, \beta) : \beta < h(\alpha)\}$. Since $\mathrm{rang}(F) \subseteq C$, one has that $\ell(\alpha) \in C$ for each $\alpha \in \omega_1$.

Suppose $j : \omega_1 \to \omega_1$ is such that $j <_\mu \ell$. Let $B = \{\alpha \in \omega : j(\alpha) < \ell(\alpha)\} \in \mu$. For $\alpha \in B$, let $p(\alpha)$ be the least $\beta < h(\alpha)$ so that $j(\alpha) < F(\alpha, \beta)$. For $\alpha \notin B$, let $p(\alpha) = 0$. Since $p <_\mu h$, Fact 5.6 implies there $\gamma < \alpha$ so that $p =_\mu \Xi^{\pi^{\Xi}_h(\gamma)}$. Then $j <_\mu F^{(\gamma)}$. Thus $[j]_\mu < \nu_\gamma$. This establishes that $[\ell]_\mu$ is the limit of $\langle \nu_\gamma : \gamma < \epsilon \rangle$. Since $\ell \in [C]^{\omega_1}$, this shows the supremum belongs to $[C]^{\omega_1}/\mu$. This shows that $[C]^{\omega_1}/\mu$ is closed. It is easy to see that $[C]^{\omega_1}/\mu$ is unbounded.

Now suppose $D \subseteq \omega_2$ is club. Let $\mathcal{T}^2 = (\omega_1 \times 2, \sqsubset)$ where $\sqsubset$ is the lexicographic ordering. If $F : \mathcal{T}^2 \to \omega_1$ is an increasing function, then let $F_0, F_1 \in [\omega_1]^{\omega_1}$ be defined by $F_i(\alpha) = F(\alpha, i)$. Note that $F_0 <_\mu F_1$.

Define $P : [\omega_1]^{\mathcal{T}^2} \to 2$ by $P(F) = 0$ if and only if $(\exists \alpha \in D)([F_0]_\mu < \alpha < [F_1]_\mu)$. By $\omega_1 \to_* (\omega_1)^{\omega_1}_2$ (the correct type strong partition property), there is a club $C \subseteq \omega_1$ which is homogeneous for P (for all functions $F \in [C]^{\mathcal{T}^2}_*$ of correct type).

Suppose C is homogeneous for P taking value 1. Fix some $\alpha_0 \in [C]^{\omega_1}_*/\mu$. Let $f_0 \in [C]^{\omega_1}_*$ with $[f_0]_\mu = \alpha_0$. Pick $\beta < \omega_2$ with $\alpha_0 < \beta$. Since $[C]^{\omega_1}_*/\mu$ is cofinal through ω_2, pick some $\alpha_1 \in [C]^{\omega_1}_*/\mu$ with $\beta < \alpha_1$. Let $f_1 \in [C]^{\omega_1}_*$ be such that $[f_1]_\mu = \alpha_1$. By an argument as in Fact 5.3 with some additional care to maintain the correct type, there is a function $F \in [C]^{\mathcal{T}^2}_*$ so that $[F_0]_\mu = \alpha_0$ and $[F_1]_\mu = \alpha_1 > \beta$. Since $P(F) = 1$, there is no element of D between α_0 and α_1. In particular, there are no elements of D between α_0 and β. However since β is arbitrary, this implies $D \subseteq \alpha_0$. This contradicts that D is unbounded.

This shows that C is homogeneous for P taking value 0. Let $\tilde{C} = \{\alpha \in C : \mathsf{enum}_C(\alpha) = \alpha\}$ which is a club subset of ω_1. Let $f \in [\tilde{C}]^{\omega_1}$. Suppose $h <_\mu f$. For each $\alpha < \omega_1$, let γ_α be least so $h(\alpha) < \mathsf{enum}_C(\gamma_\alpha)$. Let $f_0(\alpha) = \mathsf{enum}_C(\gamma_\alpha + \omega)$ and $f_1 = \mathsf{enum}_C(\gamma_\alpha + \omega + \omega)$. Since $f(\alpha) \in \tilde{C}$, $f_0(\alpha) < f_1(\alpha) < f(\alpha)$. Let $g_0(\alpha, n) = \mathsf{enum}_C(\gamma_\alpha + n)$ and $g_1(\alpha, n) = \mathsf{enum}_C(\gamma_\alpha + \omega + n)$. Note that $h <_\mu f_0 <_\mu f_1 <_\mu f$, $f_0, f_1 \in [C]^{\omega_1}_*$, and g_0, g_1 witnesses that f_0, f_1 are functions of the correct type. As above, one can find some $F \in [C]^{\mathcal{T}^2}_*$ so that $F_0 =_\mu f_0$ and $F_1 =_\mu f_1$. Since $P(F) = 0$, there is some $\alpha \in D$ so that $[h]_\mu < [f_0]_\mu < \alpha < [f_1]_\mu < [f]_\mu$. Since D is a club and $h <_\mu f$ was arbitrary, this shows that $[f]_\mu \in D$.

It has been shown that for all $D \subseteq \omega_1$, there is some club $\tilde{C} \subseteq \omega_1$ so that $[\tilde{C}]^{\omega_1}/\mu \subseteq D$. □

6. Failure of the Strong Partition Property at ω_2

FACT 6.1. *Assume* ZF+AD. *Let μ be the club measure on ω_1. Suppose $f \in {}^{\omega_1}\omega_1$ is a function of uniform cofinality ω. Then* $\mathrm{cof}([f]_\mu) = \omega$.

PROOF. Suppose f has uniform cofinality ω. Let $g : \omega_1 \times \omega \to \omega_1$ be such that for all $\alpha < \omega_1$, $f(\alpha) = \sup\{g(\alpha, n) : n \in \omega\}$. Let $f_n : \omega_1 \to \omega_1$ be defined by $f_n(\alpha) = g(\alpha, n)$. Then $m < n$ implies $f_m <_\mu f_n$. Suppose $h <_\mu f$. Let $A = \{\alpha : h(\alpha) < f(\alpha)\}$. Define $k(\alpha)$ to be the least $n \in \omega$ so that $h(\alpha) < f_n(\alpha)$ whenever $\alpha \in A$. By the countable additivity of μ, there is some n^* so that $k(\alpha) = n^*$ for μ-almost all $\alpha \in \kappa$. Then $h <_\mu f_{n^*}$. Thus $[f]_\mu = \sup\{[f_n]_\mu : n \in \omega\}$. So $\mathrm{cof}([f]_\mu) = \omega$. □

FACT 6.2. *Assume* $\mathsf{ZF}+\mathsf{AD}$. *Let* μ *be the club measure on* ω_1. *Let* $C \subseteq \omega_1$ *be a club. Then* $[C]^{\omega_1}_*/\mu$ *is an* ω*-club subset of* ω_2. *Moreover, for every* ω*-club* $D \subseteq \omega_2$, *there is a club* $C \subseteq \omega_1$ *so that* $[C]^{\omega_1}_*/\mu \subseteq D$.

PROOF. It is clear that $[C]^{\omega_1}_*/\mu$ is unbounded.

Suppose $\langle \nu_n : n \in \omega\rangle$ is an increasing ω-sequence in $[C]^{\omega_1}_*/\mu$. By $\mathsf{AC}^\omega_\mathbb{R}$, let $\langle g_n : n \in \omega\rangle$ be such that $\nu_n = [g_n]_\mu$ and each g_n is of the correct type. Using $\mathsf{AC}^\mathbb{R}_\omega$, one can also select a sequence $\langle k_n : n \in \omega\rangle$ so that k_n witnesses that g_n has uniform cofinality ω. By Fact 5.1, one may assume $g_n : \omega_1 \to C$. Using Lemma 5.16, let $F : \mathcal{T}^\omega \to C$ be an order preserving map of the correct type so that for all $n \in \omega$, $F^n =_\mu g_n$. Let $g : \omega_1 \to C$ be defined by $g(\alpha) = \sup\{F^n(\alpha) : n \in \omega\}$. Note that g is of the correct type. As before, one can check that $[g]_\mu = \sup\{\nu_n : n \in \omega\}$ and $[g]_\nu \in [C]^{\omega_1}_*/\mu$. This shows that $[C]^{\omega_1}_*/\mu$ is an ω-club subset of ω_2.

Let $D \subseteq \omega_2$ be an ω-club subset of ω_2. Define $P : [\omega_1]^{\omega_1}_* \to 2$ by

$$P(f) = \begin{cases} 0 & [f]_\mu \notin D \\ 1 & [f]_\mu \in D \end{cases}$$

By the correct type partition property $\omega_1 \to_* (\omega_1)^{\omega_1}_*$, there is a club $C \subseteq \omega_1$ which is homogeneous for P (in the correct type sense). Since it was shown that $[C]^{\omega_1}_*/\mu$ is an ω-club, $([C]^{\omega_1}_*/\mu) \cap D \neq \emptyset$. Thus C must be homogeneous for D taking value 1. It has been shown that every ω-club $D \subseteq \omega_2$ contains an ω-club of the form $[C]^{\omega_1}_*/\mu$. □

FACT 6.3. *Assume* $\mathsf{ZF}+\mathsf{AD}$. *Let* μ *denote the club measure on* ω_1. *Let* $C \subseteq \omega_1$ *be club. Let* $B = [C]^{\omega_1}_*/\mu$ *which is an* ω*-club subset of* ω_2.

Let $\epsilon < \omega_2$. *Let* $h : \omega_1 \to \omega_1$ *with* $h(\alpha) > 0$ *for all* $\alpha < \omega_1$ *and* $[h]_\mu = \epsilon$. *Let* Ξ *be a Kunen function for* h.

Let $\mathcal{F} \in [B]^\epsilon_*$ *(be of correct type). Let* $\mathcal{G} : [\omega_2]^\epsilon \times \omega$ *be a function witnessing that* $\mathcal{F}$ *has uniform cofinality* ω *with the property that for all* $\alpha_0 < \alpha_1 < \epsilon$ *and* $m, n \in \omega$, $\mathcal{G}(\alpha_0, m) < \mathcal{G}(\alpha_1, n)$.

Then there is a sequence $\langle G_n : n \in \omega\rangle$ *with each* $G_n : \mathcal{T}^h \to \omega_1$ *so that* $\mathcal{G}(\alpha, n) = [G^{(\alpha)}_n]_\mu$.

There is an $F \in [C]^{\mathcal{T}^h}_*$ *so that for all* $\alpha < \epsilon$, $[F^{(\alpha)}]_\mu = \mathcal{F}(\alpha)$.

PROOF. For each n, let $\mathcal{G}_n : \epsilon \to \omega_1$ be defined by $\mathcal{G}_n(\alpha) = \mathcal{G}(\alpha, n)$. Using Fact 5.18 on h, Ξ, and $\mathcal{G}_n$, one obtains G_n.

Fix $\gamma < \epsilon$. Define $g'_{\gamma,n} : \omega_1 \to \omega_1$ by defined by

$$g'_{\gamma,n}(\alpha) = \begin{cases} G^{(\gamma)}_n(\alpha) & \Xi^{\pi^\Xi_h(\gamma)}(\alpha) < h(\alpha) \\ 0 & \text{otherwise} \end{cases}$$

Let $f'_\gamma : \omega_1 \to \omega_1$ be defined by $f'_\gamma(\alpha) = \sup\{g'_{\gamma,n}(\alpha) : n \in \omega\}$. Note that $\mathcal{F}(\gamma) = [f'_\gamma]_\mu$.

Claim 1: The set D_γ of $\alpha < \omega_1$ so that f'_γ is discontinuous at α belongs to μ.

Suppose $\omega_1 \setminus D_\gamma \in \mu$ and hence contains a club E_γ. Since $\mathcal{F}(\gamma) \in [C]^{\omega_1}_*/\mu$, there is some $\bar{f}_\gamma \in [C]^{\omega_1}_*$ (of the correct type) so that $f'_\gamma =_\mu \bar{f}_\gamma$. Let $J_\gamma = \{\alpha : \bar{f}_\gamma(\alpha) = f'_\gamma(\alpha)\} \in \mu$. Let $K_\gamma \subseteq J_\gamma$ be a club. Then $E_\gamma \cap K_\gamma$ is a club subset of ω_1. Let $\langle \lambda_n : n \in \omega\rangle$ be an increasing sequence in $E_\gamma \cap K_\gamma$. Then $\lambda = \sup\{\lambda_n : n \in \omega\} \in E_\gamma \cap K_\gamma$ since $E_\gamma \cap K_\gamma$ is a club. Since λ_n and λ belong to $K_\gamma \subseteq J_\gamma$, $\bar{f}_\gamma(\lambda_n) = f'_\gamma(\lambda_n)$ and $\bar{f}_\gamma(\lambda) = f'_\gamma(\lambda)$. Since $\lambda \in E_\gamma$, one has that f'_γ is continuous at λ. However, the previous statement implies that $\bar{f}_\gamma$ is also continuous at λ. But $\bar{f}_\gamma$ is discontinuous everywhere since it is of the correct type. This proves the claim.

Since $[f'_\gamma]_\mu = \mathcal{F}(\gamma) \in [C]^{\omega_1}_*/\mu$, there is some $\bar{f}_\gamma \in [C]^{\omega_1}_*$ so that $f'_\gamma =_\mu \bar{f}_\gamma$. Thus $P_\gamma = \{\alpha \in \omega_1 : f'_\gamma(\alpha) \in C\} \in \mu$. Then $Q_\gamma = P_\gamma \cap D_\gamma \in \mu$. Let $f_\gamma = f'_\gamma \circ \mathsf{enum}_{Q_\gamma}$. By Fact 5.1, $f'_\gamma =_\mu f_\gamma$. Thus $[f_\gamma]_\mu = \mathcal{F}(\gamma)$. Note that f_γ is discontinuous everywhere. Define $g_\gamma : \omega_1 \times \omega \to \omega_1$ by $g_\gamma(\alpha, n) = g'_{\gamma,n}(\mathsf{enum}_{Q_\gamma}(\alpha))$. g_γ witnesses that f_γ has uniform cofinality ω. Thus f_γ is a function of the correct type.

By the uniformity of the construction, one has now produced a sequence $\langle f_\gamma : \gamma < \epsilon \rangle$ of functions of the correct type and $\langle g_\gamma : \gamma < \epsilon \rangle$ so that $[f_\gamma]_\mu = \mathcal{F}(\gamma)$ and g_γ witnesses the uniform cofinality of f_γ. Now apply Lemma 5.16 to h, Ξ, $\langle f_\gamma : \gamma < \epsilon \rangle$, and $\langle g_\gamma : \gamma < \epsilon \rangle$ to obtained the desired function $F \in [C]^{\mathcal{T}^h}_*$. □

FACT 6.4. *Assume* $\mathsf{ZF} + \mathsf{AD}$. *Let* μ *be the club measure on* ω_1. *Suppose* f *has uniform cofinality* id. *Then* $[f]_\mu$ *(as an ordinal in* ω_2*) has cofinality* ω_1.

PROOF. Let $T^{\mathsf{id}} = \{(\alpha, \beta) : \beta < \alpha\}$. Let $g : T^{\mathsf{id}} \to \omega_1$ witness that f has uniform cofinality id. For each $\beta < \omega_1$, let

$$f_\beta(\gamma) = \begin{cases} g(\gamma, \beta) & \gamma > \beta \\ 0 & \text{otherwise} \end{cases}$$

One can show that $\langle [f_\beta]_\mu : \beta < \omega_1 \rangle$ is an increasing sequence cofinal below $[f]_\mu$. □

FACT 6.5. *Assume* $\mathsf{ZF} + \mathsf{AD}$. *Let* μ *denote the club measure on* ω_1. *Let* $f : \omega_1 \to \omega_1$ *be a function so that* $f(\alpha)$ *is a limit ordinal for* μ*-almost all* α. *Then either* f *has uniform cofinality* ω *or uniform cofinality* id μ*-almost everywhere but not both.*

PROOF. Let Ξ be a Kunen function for f. Let $A = \{\alpha : f(\alpha) \in \text{Lim}\} \in \mu$. By Definition 5.5, for all $\alpha \in K^\Xi_f$, $f(\alpha) < \delta^\Xi_\alpha = \sup\{\Xi(\alpha, \beta) : \beta < \alpha\}$.

Fix $\alpha \in K^\Xi_f \cap A$, define ϵ^α_0 be the least $\epsilon < \alpha$ so that $\Xi(\alpha, \epsilon) < f(\alpha)$. Suppose $\beta < \alpha$ and $\epsilon^\alpha_{\beta'}$ has been defined for all $\beta' < \beta < \alpha$. If $\sup\{\Xi_\alpha(\epsilon^\alpha_{\beta'}) : \beta' < \beta\} = f(\alpha)$, then let α^* denote this β and declare the construction to have terminated. Otherwise, let ϵ^α_β be the least $\epsilon < \alpha$ so that $\sup\{\Xi_\alpha(\epsilon^\alpha_{\beta'}) : \beta' < \beta\} < \Xi_\alpha(\epsilon) < f(\alpha)$. This defines a sequence $\langle \epsilon^\alpha_\beta : \beta < \alpha^* \rangle$ where $\alpha^* \leq \alpha$. Note that $\langle \epsilon^\alpha_\beta : \beta < \alpha^* \rangle$ is an increasing sequence and $\sup\{\Xi_\alpha(\epsilon^\alpha_\beta) : \beta < \alpha^*\} = f(\alpha)$.

Consider the function $\Phi : K^\Xi_f \cap A \to \omega_1$ by $\Phi(\alpha) = \alpha^*$. Let $B = \{\alpha \in A \cap K^\Xi_f : \Phi(\alpha) = \alpha^* < \alpha\}$.

(Case I) $B \in \mu$.

Then Φ is a regressive function. Since μ is normal, there is a $C \subseteq B$ and $\delta < \omega_1$ so that $C \in \mu$ and $\Phi(\alpha) = \delta$ for all $\alpha \in C$. Let $\phi : \omega \to \delta$ be a cofinal sequence. Let $g : \omega_1 \times \omega \to \omega_1$ be defined by

$$g(\alpha, n) = \begin{cases} n & \alpha \notin C \\ \Xi_\alpha(\epsilon^\alpha_{\phi(n)}) & \alpha \in C \end{cases}$$

The function g witnesses that f is μ-almost everywhere a function of uniform cofinality ω.

(Case II) $B \notin \mu$.

Let $C = (K^{\Xi}_{f} \cap A) \setminus B$. For all $\alpha \in C$, $\Phi(\alpha) = \alpha^* = \alpha$. Let $T^{\mathsf{id}} = \{(\alpha, \beta) : \alpha < \omega_1 \wedge \beta < \alpha\}$. Define $g : T^{\mathsf{id}} \to \omega_1$ by

$$g(\alpha, \beta) = \begin{cases} \beta & \alpha \notin C \\ \Xi_\alpha(\epsilon^\alpha_\beta) & \alpha \in C \end{cases}$$

Then g witnesses that f has uniform cofinality id μ-almost everywhere.

Fact 6.1 implies that if f has uniform cofinality ω, then $[f]_\mu$ has cofinality ω. Fact 6.4 implies that if f has uniform cofinality id, then $[f]_\mu$ has cofinality ω_1. Since ω_1 is regular under AD, f cannot have both uniform cofinality. □

In the previous result, it is shown that f cannot have uniform cofinality ω and id by showing these uniform cofinalities correspond to different cofinalities of $[f]_\mu$. This is however not the case in general. The following argument avoids considering the function in the ultrapower. This argument can be generalized to show functions from $f : [\omega_1]^n \to \omega_1$ can only have one uniform cofinality.

FACT 6.6. *Assume* $\mathsf{ZF} + \mathsf{AD}$. *Every function* $f : \omega_1 \to \omega_1$ *has uniform cofinality* ω *or* id *μ-almost everywhere but not both.*

PROOF. This is already proved in Fact 6.5, so one will just give an argument that no function can have both uniform cofinality which does not involve the ultrapower.

Suppose that $f : \omega_1 \to \omega_1$ μ-almost everywhere has both uniform cofinality ω and id. Let $g_\omega : \omega_1 \times \omega \to \omega_1$ witness that f has uniform cofinality ω μ-almost everywhere. Let $g_{\mathsf{id}} : \omega_1 \times \omega_1 \to \omega_1$ witness that f has uniform cofinality id μ-almost everywhere.

Let $A_\omega \in \mu$ be the set of $\alpha < \omega_1$ so that $f(\alpha) = \sup\{g_\omega(\alpha, n) : n \in \omega\}$. Let $A_{\mathsf{id}} \in \mu$ be the set of $\alpha < \omega_1$ so that $f(\alpha) = \sup\{g_{\mathsf{id}}(\alpha, \beta) : \beta < \alpha\}$. Let $A = A_\omega \cap A_{\mathsf{id}}$ which also belongs to μ.

For each $\beta \in A$ and $\alpha < \beta$, let $n_{\alpha,\beta}$ be least $n \in \omega$ so that $g_\omega(\beta, n) > g_{\mathsf{id}}(\beta, \alpha)$ which exists since $\beta \in A$. Consider the function $\Phi : [A]^2 \to \omega$ defined by $\Phi(\alpha, \beta) = n_{\alpha,\beta}$. By the weak partition property and the countable additivity of μ, one has that there is some $B \subseteq A$ with $B \in \mu$ and $n \in \omega$ so that for all $(\alpha, \beta) \in [B]^2$, $n_{\alpha,\beta} = n$.

Now let $\beta \in B$ be a limit point of B, that is $\sup B \cap \beta = \beta$. For each $\alpha < \beta$ with $\alpha \in B$, one has that $n_{\alpha,\beta} = n$. Then $f(\beta) = \sup\{g_{\mathsf{id}}(\beta, \alpha) : \alpha \in \beta\} \leq g_\omega(\beta, n) < f(\beta)$. Contradiction. □

Next one will consider the ultrapower, $\mathrm{ult}(V, \mu)$, where μ is the club measure on ω_1. One needs some care when working with this structure as one may not have Łoś's theorem without AC. It will be shown later that Łoś's theorem fails for this ultrapower and in fact $\mathrm{ult}(V, \mu)$ is not a model ZF.

FACT 6.7. *Assume* $\mathsf{ZF} + \mathsf{AD}$. *Let μ denote the club measure on ω_1. Suppose $f : \omega_1 \to V$ is such that* $\mathrm{ult}(V, \mu) \models [f]_\mu \subseteq \omega_2$, *then there is an* $f' : \omega_1 \to \mathscr{P}(\omega_1)$ *so that* $[f']_\mu = [f]_\mu$.

PROOF. By Corollary 5.13, one knows that $\omega_2 = \prod_{\omega_1} \omega_1/\mu$. Thus in $\mathrm{ult}(V, \mu)$, each $\zeta < \omega_2$ is represented by some $h : \omega_1 \to \omega_1$.

Now let $f'(\alpha) = f(\alpha) \cap \omega_1$. The claim is that $[f']_\mu = [f]_\mu$.

Suppose $h : \omega_1 \to V$ is a function so that $\mathrm{ult}(V, \mu) \models [h]_\mu \in [f]_\mu$. Since $\mathrm{ult}(V, \mu) \models [f]_\mu \subseteq \omega_2$, one must have that $A = \{\alpha \in \omega_1 : h(\alpha) \in f(\alpha) \cap \omega_1\} \in \mu$.

Let $h' : \omega_1 \to \omega_1$ be defined by $h'(\alpha) = h(\alpha)$ if $\alpha \in A$ and $h'(\alpha) = \emptyset$ otherwise. Note that $[h]_\mu = [h']_\mu$ and $[h']_\mu \in [f']_\mu$. This shows that $[f]_\mu \subseteq [f']_\mu$. It is clear that $[f']_\mu \subseteq [f]_\mu$. Thus $[f]_\mu = [f']_\mu$. $\square$

FACT 6.8. *Assume* $\mathsf{ZF} + \mathsf{AD}$ *and* μ *is the club measure on* ω_1. *Suppose* $h : \omega_1 \to \mathscr{P}(\omega_1)$ *has the property that for all* $\alpha < \omega_1$, $|h(\alpha)| = \omega_1$ *and* $\mathsf{enum}_{h(\alpha)}$ *is a function of correct type. Suppose there is a function* G *so that for each* $\alpha < \omega_1$, $G(\alpha) : \omega_1 \times \omega \to \omega_1$ *is a witness to* $\mathsf{enum}_{h(\alpha)}$ *having uniform cofinality* ω, *then* $\mathsf{enum}_{[h]_\mu} : \omega_2 \to \omega_2$ *is a function of correct type.*

PROOF. For each $\xi < \omega_1$, let $g_\xi : \omega_1 \times \omega \to \omega_1$ be defined by $g_\xi = G(\xi)$. Hence g_ξ is a witnesses to $\mathsf{enum}_{h(\alpha)}$ having the correct type.

For each $\xi < \omega_2$, let $p_\xi : \omega_1 \to \omega_1$ have the property that for all $\alpha < \omega_1$, $p_\xi(\alpha) \in h(\alpha)$ and $[p_\xi]_\mu$ represents the ξth-element of $[h]_\mu$. (Note that one does not have a uniform procedure for finding p_ξ as ξ ranges over ordinals below ω_2.) Define $k_{\xi,n} : \omega_1 \to \omega_1$ by $k_{\xi,n}(\alpha) = g_\xi(\mathsf{enum}_{h(\xi)}^{-1}(p_\xi(\alpha)), n)$. Let $\delta_{\xi,n} = [k_{\xi,n}]_\mu$. Note that if $p'_\xi =_\mu p_\xi$ and $k'_{\xi,n}$ was defined in the same manner as k using p'_ξ instead of p_ξ, then $k'_{\xi,n} =_\mu k_{\xi,n}$. This shows that $\delta_{\xi,n}$ is well defined independent of the choice of p_ξ.

Define $r : \omega_2 \times \omega \to \omega_2$ by $r(\xi, n) = \delta_{\xi,n}$. One can check that r witnesses that $\mathsf{enum}_{[h]_\mu}$ has uniform cofinality ω.

Let $\zeta \in [h]_\mu$. Let $\ell : \omega_1 \to \omega_1$ be such that $[\ell]_\mu = \zeta$. One may assume that for all α, $\ell(\alpha) \in h(\alpha)$. Let $\iota : \omega_1 \to \omega_1$ be defined by $\iota(\alpha) = \sup(h(\alpha) \cap \ell(\alpha))$. Note that $\iota(\alpha) < \ell(\alpha)$ since $\mathsf{enum}_{h(\alpha)}$ was assumed to be discontinuous everywhere. One can check that every element of $[h]_\mu$ which is below $[\ell]_\mu$ is below $[\iota]_\mu < [\ell]_\mu$. Thus $\mathsf{enum}_{[h]_\mu}$ is discontinuous everywhere. $\square$

FACT 6.9. *Suppose* $h : \omega_1 \to \mathscr{P}(\omega_1)$ *with* $|h(\alpha)| = \omega_1$ *for all* $\alpha < \omega_1$. *Let* $\xi < \omega_2$ *and* $\ell : \omega_1 \to \omega_1$ *be such that* $[\ell]_\mu = \xi$. $\mathsf{enum}_{[h]_\mu}(\xi)$ *is represented by the function* $g : \omega_1 \to \omega_1$ *defined by* $g(\alpha) = \mathsf{enum}_{h(\alpha)}(\ell(\alpha))$.

PROOF. For each $\ell : \omega_1 \to \omega_1$, let $g_\ell : \omega_1 \to \omega_1$ be defined by $g_\ell(\alpha) = \mathsf{enum}_{h(\alpha)}(\ell(\alpha))$. Note that if $\ell =_\mu \ell'$, then $g_\ell =_\mu g_{\ell'}$. Note also that $[g_\ell]_\mu \in [h]_\mu$.

For each $\xi < \omega_1$, let $\gamma_\xi \in \omega_2$ be defined by $\gamma_\xi = [g_\ell]_\mu$ where ℓ is any function so that $[\ell]_\mu = \xi$. This is well defined by the previous paragraph. $\langle \gamma_\xi : \xi < \omega_2 \rangle$ is an increasing sequence through $[h]_\mu$.

Let $k : \omega_1 \to \omega_1$ be such that $[k]_\mu \in [h]_\mu$. Note $A = \{\alpha \in \omega_1 : k(\alpha) \in h(\alpha)\} \in \mu$. By modifying k off A, one will assume without loss of generality that $k(\alpha) \in h(\alpha)$ for all $\alpha < \omega_1$. Let $\ell(\alpha) = \mathsf{enum}_{h(\alpha)}^{-1}(k(\alpha))$. Then $k =_\mu g_\ell$.

This shows that $\mathsf{enum}_{[h]_\mu}(\xi) = \gamma_\xi$. This completes the proof. $\square$

FACT 6.10. *Assume* $\mathsf{ZF} + \mathsf{AD}$ *and let* μ *be the club measure on* ω_1. *Suppose* $E \subseteq \omega_2$ *is such that* $|E| = \omega_1$. *Then* $E \in \text{ult}(V, \mu)$.

PROOF. Let $\zeta = \text{ot}(E)$. Let $h : \omega_1 \to \omega_1$ be such that $[h]_\mu = \zeta$ and $h(\alpha) > 0$ for all $\alpha < \omega_1$. Let Ξ be a Kunen function for h. Using Fact 5.18, there is an $F : \mathcal{T}^h \to \omega_1$ which is increasing so that for all $\xi < \zeta$, $[F^{(\xi)}]_\mu = \mathsf{enum}_E(\xi)$.

Let $g : \omega_1 \to \mathscr{P}(\omega_1)$ be defined by $g(\alpha) = \{F(\alpha, \beta) : \beta < h(\alpha)\}$. The claim is that $[g]_\mu = E$.

Suppose $p : \omega_1 \to \omega_1$ is such that $[p]_\mu \in [g]_\mu$. Note $A = \{\alpha : p(\alpha) \in g(\alpha)\} \in \mu$. By modifying p off A, one may assume that for all α, $p(\alpha) \in g(\alpha)$. Let $f(\alpha)$ be the $\beta < h(\alpha)$ so that $F(\alpha,\beta) = p(\alpha)$. Then $p = F^f = F^{(\xi)} = \mathsf{enum}_E(\xi)$ for some $\xi < \zeta$. Then $[g]_\mu \subseteq E$. It is straightforward to see that $E \subseteq [g]_\mu$. □

FACT 6.11. *Assume* $\mathsf{ZF} + \mathsf{AD}$. *Suppose* $D \subseteq \omega_2$ *is such that there is some* $g : \omega_1 \to \mathscr{P}(\omega_1)$ *so that* $[g]_\mu = D$. *Suppose* $E \subseteq \omega_2$ *is such that* $\sup E < \min D$, *then* $D \cup E \in \mathrm{ult}(V,\mu)$.

PROOF. By Fact 6.10, $E \in \mathrm{ult}(V,\mu)$. Thus there is a $g' : \omega_1 \to \mathscr{P}(\omega_1)$ so that $[g']_\mu = E$. Let $g'' : \omega_1 \to \mathscr{P}(\omega_1)$ be defined by $g''(\alpha) = g(\alpha) \cup g'(\alpha)$. One can check that $[g'']_\mu = D \cup E$. □

FACT 6.12. *Assume* $\mathsf{ZF} + \mathsf{AD}$ *and let* μ *be the club measure on* ω_1. *Let* $D = \{\alpha < \omega_2 : \mathrm{cof}(\alpha) = \omega_1\}$. *Then* $D \notin \mathrm{ult}(V,\mu)$.

PROOF. Suppose $D \in \mathrm{ult}(V,\mu)$. By Fact 6.7, there is some $h : \omega_1 \to \mathscr{P}(\omega_1)$ so that $D = [h]_\mu$.

(Case I) $A = \{\alpha < \omega_1 : h(\alpha) \in \mu\} \in \mu$.

By Fact 4.8, there is a sequence $\langle C_\alpha : \alpha \in A\rangle$ of club subsets of ω_1 so that $C_\alpha \subseteq h(\alpha)$ for all $\alpha \in A$.

Define $\ell : \omega_1 \to \omega_1$ by $\ell(\alpha) = \mathsf{enum}_{C_{\mathsf{enum}_A(\alpha)}}(\omega)$. Let $\ell' : \omega_1 \to \omega_1$ be defined by

$$\ell'(\alpha) = \begin{cases} \mathsf{enum}_{C_\alpha}(\omega) & \alpha \in A \\ \min(h(\alpha)) & \alpha \notin A \end{cases}$$

Clearly $[\ell']_\mu \in [h]_\mu$. Note that $\ell = \ell' \circ \mathsf{enum}_A$. By Fact 5.1, $\ell =_\mu \ell'$. Define $g : \omega_1 \times \omega \to \omega$ by $g(\alpha, n) = \mathsf{enum}_{C_{\mathsf{enum}_A(\alpha)}}(n)$. The function g witnesses that ℓ has uniform cofinality ω. Thus $\mathrm{cof}([\ell]_\mu) = \omega$ by Fact 6.1. Since $[\ell]_\mu \in D = \{\alpha < \omega_2 : \mathrm{cof}(\alpha) = \omega_1\}$, one has a contradiction.

(Case II) $A = \{\alpha < \omega_1 : h(\alpha) \in \mu\} \notin \mu$.

Then $B = \omega_1 \setminus A = \{\alpha < \omega_1 : \omega_1 \setminus h(\alpha) \in \mu\} \in \mu$. By Fact 4.8, there is a sequence $\langle C_\alpha : \alpha \in B\rangle$ of club subsets of ω_1 so that $C_\alpha \subseteq \omega_1 \setminus h(\alpha)$. Define $\ell : \omega_1 \to \omega_1$ by $\ell(\alpha) = \mathsf{enum}_{C_{\mathsf{enum}_B(\alpha)}}(\alpha)$. Let $\ell' : \omega_1 \to \omega_1$ be defined by

$$\ell'(\alpha) = \begin{cases} \mathsf{enum}_{C_\alpha}(\alpha) & \alpha \in B \\ \min h(\alpha) & \alpha \notin B \end{cases}$$

Note that $[\ell']_\mu \notin [h]_\mu$. Observe $\ell = \ell \circ \mathsf{enum}_B$. By Fact 5.1, $\ell =_\mu \ell'$. Let $g : \mathcal{T}^{\mathsf{id}} \to \omega_1$ be defined by $g(\alpha,\beta) = \mathsf{enum}_{C_{\mathsf{enum}_B(\alpha)}}(\beta)$ if $\beta < \alpha$. Then g witnesses that ℓ has uniform cofinality id. Therefore, $\mathrm{cof}([\ell]_\mu) = \omega_1$ by Fact 6.4. However, $[\ell]_\mu \notin D = \{\alpha < \omega_2 : \mathrm{cof}(\alpha) = \omega_1\}$ yields a contradiction.

The proof is complete. □

One can now show that μ does not satisfy Łoś's theorem and in fact the ultrapower does not satisfy the ZF axioms.

FACT 6.13. *Assume* $\mathsf{ZF} + \mathsf{AD} + V = L(\mathbb{R})$. *Let* μ *denote the club measure on* ω_1. *Then* $\mathrm{ult}(L(\mathbb{R}),\mu)$ *is not a model of* ZF. *Thus Łoś's theorem fails for* μ.

PROOF. Note that $L(\mathbb{R}) \models \mathsf{AD}$ implies $L(\mathbb{R}) \models \mathsf{DC}_\mathbb{R}$ and hence $L(\mathbb{R}) \models \mathsf{DC}$ by a result of Kechris [**15**]. Thus $\mathrm{ult}(L(\mathbb{R}),\mu)$ may be considered a transitive inner model of $L(\mathbb{R})$. One can check that $\mathbb{R} \subseteq \mathrm{ult}(L(\mathbb{R}),\mu)$. If $\mathrm{ult}(L(\mathbb{R}),\mu)$ is an inner model of ZF containing all the reals of $L(\mathbb{R})$, then one must have that $\mathrm{ult}(L(\mathbb{R}),\mu) = L(\mathbb{R})$. This is impossible since Fact 6.12 asserts that $\mathrm{ult}(L(\mathbb{R}),\mu)$ is missing a subset of ω_2 which belongs to $L(\mathbb{R})$. □

THEOREM 6.14. *(Jackson) Assume* $\mathsf{ZF} + \mathsf{AD}$. *Let* μ *denote the club measure on* ω_1. *Define a partition* $P : [\omega_2]^{\omega_2}_* \to 2$ *by*

$$P(f) = 0 \Leftrightarrow \mathrm{rang}(f) \in \mathrm{ult}(V, \mu).$$

Then there is no club $D \subseteq \omega_2$ *and no* $i \in 2$ *so that* $P(f) = i$ *for all* $f \in [D]^{\omega_2}_*$.

PROOF. (Case I) Suppose there is a club $D \subseteq \omega_2$ so that $P(f) = 1$ for all $f \in [D]^{\omega_2}_*$. By Fact 5.21, there is a club $C \subseteq \omega_1$ so that $[C]^{\omega_1}/\mu \subseteq D$.

Let $A = \{\alpha : (\exists \gamma)(\alpha = \mathsf{enum}_C(\gamma + \omega)\}$. Note that $\mathsf{enum}_A : \omega_1 \to \omega_1$ is a function of the correct type. Let $g : \omega_1 \times \omega \to \omega$ witness that enum_A has uniform cofinality ω. For each $\xi < \omega_1$, let $A_\xi = \{\alpha \in A : \alpha \geq \xi\}$. Let $g_\xi : \omega_1 \times \omega \to \omega$ be defined by $g_\xi(\alpha, n) = g(\mathsf{enum}_A^{-1}(\mathsf{enum}_{A_\xi}(\alpha)), n)$. Then for each $\xi < \omega_1$, g_ξ witnesses that enum_{A_ξ} has uniform cofinality ω. Let $h : \omega_1 \to \mathscr{P}(\omega_1)$ be defined by $h(\xi) = A_\xi$. By Fact 6.8, $\mathsf{enum}_{[h]_\mu} : \omega_2 \to \omega_2$ is a function of the correct type. Let $f = \mathsf{enum}_{[h]_\mu}$. One can check that $\mathrm{rang}(f) = [h]_\mu \subseteq [C]^{\omega_1}/\mu$. Thus $f \in [D]^{\omega_2}_*$. However since $\mathrm{rang}(f) = [h]_\mu \in \mathrm{ult}(V, \mu)$, one must have that $P(f) = 0$. Contradiction.

(Case II) Suppose there is a club $D \subseteq \omega_2$ so that $P(f) = 0$ for all $f \in [D]^{\omega_2}_*$.

Fix any $f \in [D]^{\omega_2}_*$. Since $P(f) = 0$, there is some $h : \omega_1 \to \mathscr{P}(\omega_1)$ so that $\mathrm{rang}(f) = [h]_\mu$ by Fact 6.7.

Now let $A \subseteq \omega_2$ be arbitrary. $\mathsf{enum}_{f[A]} : \omega_2 \to D$ is a function of the correct type since $f : \omega_2 \to D$ is a function of the correct type. Then $P(\mathsf{enum}_{f[A]}) = 0$. Thus $f[A] \in \mathrm{ult}(V, \mu)$. There is some $g : \omega_1 \to \mathscr{P}(\omega_1)$ so that $[g]_\mu = f[A]$. Let $k : \omega_1 \to \mathscr{P}(\omega_1)$ be defined by $k(\alpha) = \mathsf{enum}_{h(\alpha)}^{-1}[g(\alpha)]$.

The claim is that $[k]_\mu = A$.

To show $[k]_\mu \subseteq A$: Suppose $\xi \in [k]_\mu$. There is a representative $\ell : \omega_1 \to \omega_1$ of ξ so that for all μ-almost all α, $\ell(\alpha) \in k(\alpha)$. There is a function $p : \omega_1 \to \omega_1$ so that for μ-almost all α, $p(\alpha) \in g(\alpha)$ and $\ell(\alpha) = \mathsf{enum}_{h(\alpha)}^{-1}(p(\alpha))$. In particular $[p]_\mu \in [g]_\mu = f[A]$. By Fact 6.9, the ξth element of $[h]_\mu$ is represented by the function $q : \omega_1 \to \omega_1$ defined by $q(\alpha) = \mathsf{enum}_{h(\alpha)}(\ell(\alpha))$. However μ-almost everywhere, $q(\alpha) = \mathsf{enum}_{h(\alpha)}(\mathsf{enum}_{h(\alpha)}^{-1}(p(\alpha))) = p(\alpha)$. Thus $q =_\mu p$. It has been shown that the ξth element of $[h]_\mu = \mathrm{rang}(f)$ belongs to $[g]_\mu = f[A]$. Thus $\xi \in A$.

To show that $A \subseteq [k]_\mu$: Let $\xi \in A$. Let $\ell : \omega_1 \to \omega_1$ be such that $[\ell]_\mu = \xi$. By Fact 6.9, the ξth-element of $[h]$ is represented by the function $q(\alpha) = \mathsf{enum}_{h(\alpha)}(\ell(\alpha))$. Thus $[q]_\mu \in f[A] = [g]_\mu$. So μ-almost everywhere, $q(\alpha) = \mathsf{enum}_{h(\alpha)}(\ell(\alpha)) \in g(\alpha)$. Thus for μ-almost all α, $\ell(\alpha) \in \mathsf{enum}_{h(\alpha)}^{-1}[g(\alpha)] = k(\alpha)$. It has been shown that $\xi = [\ell]_\mu \in [k]_\mu$.

The claim has been shown. Since A was arbitrary, this implies that everywhere subset of ω_2 belongs to $\mathrm{ult}(V, \mu)$. This contradicts Fact 6.12.

It has been shown that the partition P has no club set which is homogeneous.

□

COROLLARY 6.15. *(Martin and Paris) Assume* $\mathsf{ZF}+\mathsf{AD}$. *The partition relation* $\omega_2 \to (\omega_2)^{\omega_2}_2$ *does not hold. Thus* ω_2 *is a weak partition cardinal which is not a strong partition cardinal.*

REMARK 6.16. The example of Jackson from Theorem 6.14 gives an explicit example of a partition $P : [\omega_2]^{\omega_2}_* \to 2$ which has no homogeneous club subset.

Martin and Paris original argument roughly shows that if $\omega_2 \to (\omega_2)^{\omega_2}_2$ holds, then ω_3 would satisfy $\omega_3 \to (\omega_3)^{\alpha}_2$ for all $\alpha < \omega_1$. Fact 2.7 implies that ω_3 must be regular. However, it can be shown that ω_3 is a singular cardinal of cofinality ω_2. For more information on the result of Martin and Paris, see [**20**] and specifically Lemma 5.19. Also see Section 13 of [**14**].

COROLLARY 6.17. *Let* $\sigma \in [\omega_2]^{<\omega_2}_*$. *Define* $P_\sigma : [\omega_2 \setminus (\sup(\sigma) + \omega)]^{\omega_2}_* \to 2$ *by* $P(\sigma\hat{}f)$, *where* P *is the partition from Theorem* 6.14. P_σ *also does not have have a club homogeneous set.*

PROOF. Essentially the same argument as Theorem 6.14 with the assistance of Fact 6.11. $\square$

7. $L(\mathbb{R})$ as a Symmetric Collapse Extension of HOD

DEFINITION 7.1. Let $S \subseteq \mathrm{ON}$ be a set of ordinals. Let φ be a formula of set theory. The pair (S, φ) is called an ∞-Borel code. Let $n \in \omega$. Define $\mathfrak{B}^n_{(S,\varphi)} = \{x \in \mathbb{R}^n : L[S, x] \models \varphi(S, x)\}$.

Let $A \subseteq \mathbb{R}^n$. (S, φ) is said to be an ∞-Borel code for A if and only if $A = \mathfrak{B}^n_{(S,\varphi)}$.

If a set $A \subseteq \mathbb{R}^n$ has an ∞-Borel code, then A has a very absolute definition. That is, in ordered to determine membership of x in A, one needs only to ask whether $\varphi(S, x)$ holds in $L[S, x]$, which is the minimal model of ZFC containing x and the code set S.

DEFINITION 7.2. Let P be a set. Recall a set A is ordinal definable from P if and only if there is a formula φ of set theory, a finite tuple of ordinals $\bar{\alpha}$, and a finite tuple $\bar{p}$ of elements of P so that $A = \{x : \varphi(x, \bar{\alpha}, \bar{p})\}$.

Using the reflection theorem, one can show that a set A is ordinal definable if and only if there is some $\xi \in \mathrm{ON}$, tuple of ordinals $\bar{\alpha}$, tuple $\bar{p}$ from P, and formula φ so that all these objects belong to V_ξ and $A = \{x \in V_\xi : V_\xi \models \varphi(x, \bar{\alpha}, \bar{p})\}$.

This shows the collection OD_P of all sets which are ordinal definable from P forms a first order class. If P has an OD_P wellordering, then OD_P has a wellordering which is definable with parameter from ordinals and P. Thus there is a bijection of OD_P with ON.

Let HOD_P denote the subclass of OD_P which is hereditarily OD_P. That is, HOD_P consists of those $x \in \mathrm{OD}_P$ such that $\mathrm{tc}(\{x\}) \subseteq \mathrm{OD}_P$, where tc refers to the transitive closure.

As a matter of convention, if $S \subseteq \mathrm{ON}$ is a set of ordinals, one will often write OD_S and HOD_S for $\mathrm{OD}_{\{S\}}$ and $\mathrm{HOD}_{\{S\}}$.

DEFINITION 7.3. Let $n \in \omega$ and $S \subseteq \mathrm{ON}$ be a set of ordinals. Let ${}_n\mathbb{O}_S$ denote the forcing of nonempty ordinal definable in S subsets of $\mathbb{R}^n$. Let the ordering be $\leq_{{}_n\mathbb{O}_S} = \subseteq$. The largest element is $1_{{}_n\mathbb{O}_S} = \mathbb{R}^n$.

Since there is a definable (in S) bijection of the class OD_S with the ordinal ON, one can identify ${}_n\mathbb{O}_S$ as a set of ordinals in HOD_S. In this way, ${}_n\mathbb{O}_S \in \mathrm{HOD}_S$. ${}_n\mathbb{O}_S$ is called the n-dimensional S-Vopěnka forcing. If $n = 1$, ${}_1\mathbb{O}_S$ will be denoted simply $\mathbb{O}_S$.

DEFINITION 7.4. Let $n \in \omega$ and $S \subseteq \mathrm{ON}$ be a set of ordinals. Let ${}_n\mathbb{A}_S$ denote the forcing of nonempty subsets of $\mathbb{R}^n$ which possess OD_S ∞-Borel codes. ${}_n\mathbb{A}_S$ is ordered by $\leq_{{}_n\mathbb{A}_S} = \subseteq$. It has a largest element $1_{{}_n\mathbb{O}_S} = \mathbb{R}^n$.

Since ${}_n\mathbb{A}_S \subseteq {}_n\mathbb{O}_S$, one can consider ${}_n\mathbb{A}_S$ to be a forcing of HOD_S. ${}_n\mathbb{A}_S$ will be called the n-dimensional OD_S ∞-Borel code forcing.

REMARK 7.5. One can be more specific about how ${}_n\mathbb{A}_S$ is coded as a set of ordinals. One can identify ${}_n\mathbb{A}_S$ with a (set sized) collection of pairs of (S', φ), where S' is a OD_S set of ordinals and φ is a formula. Using the canonical global wellordering of HOD_S, let $\langle (S_\alpha, \varphi_\alpha) : \alpha < \delta \rangle$, for some ordinal δ, be an enumeration of ∞-Borel codes that include at least one code for each element of ${}_n\mathbb{A}_S$. Fix $\langle \phi_n : n \in \omega \rangle$ to be a coding of formulas of set theory by natural numbers. Using the Gödel pairing function, let $K = \{(\alpha, \beta, n) : \beta \in S_\alpha \wedge \phi_n = \varphi_\alpha\}$. One will often identify ${}_n\mathbb{A}_S$ with this set of ordinals K. If this is done, then from ${}_n\mathbb{A}_S$, one can obtain uniformly ∞-Borel codes for each condition in ${}_n\mathbb{A}_S$.

In nearly every regard, the Vopěnka forcing is a more practical forcing than $\mathbb{A}_S$. It will shown that in $\mathsf{ZF} + \mathsf{AD} + V = L(\mathbb{R})$, $\mathbb{O}_S$ and $\mathbb{A}_S$ are identical. To establish this, one will prove a structural theorem about $L(\mathbb{R})$ due to Woodin that involves the forcing $\mathbb{A}_S$. The presentation of Woodin's result that $L(\mathbb{R})$ is a symmetric collapse extension of its HOD follows closely [**21**]. Of particular importance to the study of cardinals and combinatorics in $L(\mathbb{R}) \models \mathsf{AD}$ will be existence of an ultimate ∞-Borel code which follows from the proof.

For simplicity, $S = \emptyset$ in the following results. The result can be appropriately relativized.

The main benefit of $\mathbb{A}$ over $\mathbb{O}$ is the following result:

FACT 7.6. *Let $x \in \mathbb{R}^n$. Then there is a generic filter $G_x \subseteq {}_n\mathbb{A}$ which is ${}_n\mathbb{A}$-generic over* HOD *so that* $\mathrm{HOD}[G_x] = \mathrm{HOD}[x]$, *where* $\mathrm{HOD}[x]$ *refers to the smallest transitive model of* ZF *extending* HOD *and containing x.*

PROOF. For simplicity, let $n = 1$. Let $x \in \mathbb{R}$. Let $G_x = \{p \in \mathbb{A} : x \in p\}$.

Using the convention of Remark 7.5, from $\mathbb{A} \in \mathrm{HOD}$, one can obtain an enumeration, $\langle (S_p, \varphi_p) : p \in \mathbb{A} \rangle$ so that (S_p, φ_p) is an OD ∞-Borel for the condition p.

First to show that G_x is an $\mathbb{A}$-generic filter over HOD: Let $A \subseteq \mathbb{A}$ be a maximal antichain which belongs to HOD and is hence OD. Considering $\mathbb{A}$ as the set K defined in Remark 7.5 and using the fact that A is OD, one can find a formula φ so that $(\mathbb{A}, \varphi)$ is an OD-Borel code for $\bigcup A$. Therefore, $\bigcup A = \mathbb{R}$ since otherwise $\mathbb{R} \setminus \bigcup A$ would be a nonempty set with an OD ∞-Borel code. Then $\mathbb{R} \setminus \bigcup A$ is a condition of $\mathbb{A}$ which is incompatible with every element of A. This contradicts A being a maximal antichain. Thus $x \in \bigcup A$. There is some $a \in A$ so that $x \in a$. Thus $a \in G_x$. It has been shown that $G_x \cap A \neq \emptyset$. G_x is $\mathbb{A}$-generic over HOD.

Thinking of $\mathbb{R} = {}^\omega 2$, let $b_n = \{x \in \mathbb{R} : x(n) = 1\}$. Note that $b_n \neq \emptyset$ and b_n clearly has an OD ∞-Borel code. Thus $b_n \in \mathbb{A}$. Note that $b_n \in G_x$ if and only if $x(n) = 1$. Thus $x \in \mathrm{HOD}[G_x]$.

Note that $p \in G_x$ if and only if $x \in p$ if and only if $L[S_p, x] \models \varphi_p(S_p, x)$. Note that $V \models L[S_p, x] \models \varphi_p(S_p, x)$ if and only if $\mathrm{HOD}[x] \models L[S_p, x] \models \varphi_p(S_p, x)$. (This is an application of the important absoluteness property of ∞-Borel codes.) Thus G_x can be defined in $\mathrm{HOD}[x]$ as the set of p such that $L[S_p, x] \models \varphi_p(S_p, x)$. This shows $G_x \in \mathrm{HOD}[x]$.

It has been shown that $\mathrm{HOD}[x] = \mathrm{HOD}[G_x]$. □

DEFINITION 7.7. Suppose $n \in \omega$. For each $s \in {}^n\omega$, let $b_s = \{x \in \mathbb{R}^n : x(s) = 1\}$.

Let $\dot{x}^n_{\mathrm{gen}} = \{(\check{s}, b_s) : s \in {}^n\omega\}$. In light of the argument of Fact 7.6, if $x \in \mathbb{R}^n$, then $\dot{x}^n_{\mathrm{gen}}[G_x] = x$.

Definition 7.8. Let $\mathbb{P}$ and $\mathbb{Q}$ be two forcings. A surjective map $\pi : \mathbb{Q} \to \mathbb{P}$ is a forcing projection if and only if the following holds:
(1) For all $q_0, q_1 \in \mathbb{Q}$, $q_0 \leq_{\mathbb{Q}} q_1$ implies $\pi(q_0) \leq_{\mathbb{P}} \pi(q_1)$ and $\pi(1_{\mathbb{Q}}) = 1_{\mathbb{P}}$.
(2) For all $q \in \mathbb{Q}$ and $p' \in \mathbb{P}$ such that $p' \leq_{\mathbb{P}} \pi(q)$, there exists some $q' \leq_{\mathbb{Q}} q$ so that $\pi(q') = p'$.

Let $\pi : \mathbb{Q} \to \mathbb{P}$ be a projection. If $D \subseteq \mathbb{P}$ is dense, then $\pi^{-1}[D]$ is dense in $\mathbb{Q}$. This implies that if $G \subseteq \mathbb{Q}$ is a $\mathbb{Q}$-generic over V, then $\pi[G]$ is a $\mathbb{P}$-generic over V.

Let $G \subseteq \mathbb{P}$ is $\mathbb{P}$-generic over V, then let $\mathbb{Q}/G = \{q \in \mathbb{Q} : \pi(q) \in G\}$. Let $\leq_{\mathbb{Q}/G} = \leq_{\mathbb{Q}} \restriction \mathbb{Q}/G$.

Let $\dot{G} \in V^{\mathbb{P}}$ denote the $\mathbb{P}$-name for the generic filter. Let $\mathbb{Q}/\dot{G}$ denote the $\mathbb{P}$-name so that $1_{\mathbb{P}} \Vdash_{\mathbb{P}} \mathbb{Q}/\dot{G} = \{p \in \check{\mathbb{Q}} : \check{\pi}(p) \in \dot{G}\}$.

It can be checked that $\mathbb{Q}$ embeds densely into the iteration $\mathbb{P}*(\mathbb{Q}/\dot{G})$. Therefore, these two forcings are equivalent as forcings. Also if G is $\mathbb{P}$-generic over V then $(\mathbb{Q}/\dot{G})[G] = \mathbb{Q}/G$.

Suppose H is $\mathbb{Q}$-generic over V. Let $G = \pi[H]$ be the associated $\mathbb{P}$-generic filter. One can check that H is $(\mathbb{Q}/G)$-generic over $V[G]$, $G * H$ is $(\mathbb{P} * \mathbb{Q}/\dot{G})$-generic over V and $V[G][H] = V[G * H] = V[H]$.

Definition 7.9. For the moment, consider $\mathbb{R} = {}^{\omega}2 = \mathscr{P}(\omega)$. If $x, y \in \mathbb{R}$, then one writes $x \leq_T y$ if and only if there is a Turing machine (taking oracle input) so that x can be computed from this Turing machine when given y as its oracle. Since Turing programs can be coded by natural numbers, for any $x \in \mathbb{R}$, there are only countably many $y \in \mathbb{R}$ so that $y \leq_T x$. A Turing program is also absolute between models of ZF with the same ω.

Define $x =_T y$ if and only if $x \leq_T y$ and $y \leq_T x$.

Let $\mathcal{D} = \mathscr{P}(\omega)/ =_T$ denote the collection of $=_T$ equivalence classes. An element of $\mathcal{D}$ is called a Turing degree. If $x \in \mathbb{R}$, then one uses the notation $[x]_T$ rather than $[x]_{=_T}$ to denote the Turing degree of x. As observed above, each Turing degree contains only countable many reals.

If $X, Y \in \mathcal{D}$, one defines $X \leq Y$ if and only if there exist $x \in X$ and $y \in Y$ so that $x \leq_T y$. (One can check this is a well defined relation.)

A Turing cone of reals with base x is the collection $C_x = \{y \in \mathbb{R} : x \leq_T y\}$. A Turing cone of degrees with base X is the collection $C_X = \{Y \in \mathcal{D} : X \leq Y\}$.

Let $\mu_{\mathcal{D}} \subseteq \mathscr{P}(\mathcal{D})$ consists of those subsets of $\mathcal{D}$ which contain a Turing cone of degrees. $\mu_{\mathcal{D}}$ is a filter.

Fact 7.10. *(Martin) Assume* $\mathsf{ZF} + \mathsf{AD}$. $\mu_{\mathcal{D}}$ *is a countably complete ultrafilter.*

Proof. Let $K \subseteq \mathcal{D}$. Let $\tilde{K} = \{x \in \mathbb{R} : x \in K\}$. Note that $\tilde{K}$ is a $=_T$-invariant subset of $\mathbb{R}$.

Consider the usual game $G_{\tilde{K}}$

	I	$a(0)$		$a(1)$		$a(2)$		$a(3)$			a
$G_{\tilde{K}}$											
	II		$b(0)$		$b(1)$		$b(2)$		$b(3)$		b

where Player 1 wins if and only if $a \oplus b \in \tilde{K}$, where $a \oplus b$ is defined by $a \oplus b(2n) = a(n)$ and $a \oplus b(2n+1) = b(n)$.

By AD, one of the two players has a winning strategy. Suppose Player 1 has a winning strategy σ.

Let $Z = [\sigma]_T$ be the Turing degree of σ. The claim is that $C_Z \subseteq K$:

Let $Y \in \mathcal{D}$ be such that $Z \leq_T Y$. Pick any $y \in Y$. Thus $\sigma \leq_T y$. Let $\sigma * y$ denote the result of the play where Player 1 uses σ and Player 2 plays the bits of y each turn. Since σ is a Player 1 winnings strategy, $\sigma * y \in \tilde{K}$. Since $\sigma \leq_T y$, $\sigma * y \leq_T y$ and clearly $y \leq_T \sigma * y$. Thus $\sigma * y =_T y$. Since $\tilde{K}$ is $=_T$-invariant, $y \in \tilde{K}$. Thus $[y]_T = Y \in K$.

This shows that if Player 1 has a winning strategy in $G_{\tilde{K}}$, then $K \in \mu_{\mathcal{D}}$. A similar argument shows that if Player 2 has a winning strategy then $\mathcal{D} \setminus K \in \mu_{\mathcal{D}}$. Thus $\mu_{\mathcal{D}}$ is an ultrafilter.

Next to show countable completeness: Suppose $\langle K_n : n \in \omega \rangle$ is a sequence in $\mu_{\mathcal{D}}$. By $\mathsf{AC}^{\mathbb{R}}_{\omega}$, let x_n be such that the cone above $X_n = [x_n]_T$ is contained in K_n. Let $x = \bigoplus x_n = \{(n, m) : m \in x_n\}$ and $X = [x]_T$. Then the cone above X is contained inside of $\bigcap_{n\in\omega} K_n$. Thus $\bigcap_{n\in\omega} K_n \in \mu_{\mathcal{D}}$. $\square$

REMARK 7.11. Let S be a set. Note that if $x =_T y$, then $L[S, x] = L[S, y]$. Therefore, if $X \in \mathcal{D}$, one will often write $L[S, X]$ to denote $L[S, x]$ for any $x \in X$.

The canonical constructibility wellordering is based on the hierarchy $\{L_\alpha[S, x] : \alpha \in \mathrm{ON}\}$. Even if $x =_T y$, the levels $L_\alpha[S, x]$ and $L_\alpha[S, y]$ can differ. Thus, the canonical constructibility wellordering on $L[S, x]$ is not invariant under $=_T$.

If V is a model of ZF (in the language $\dot{\in}$), then the canonical wellordering $<^{\mathrm{HOD}_S^V}$ of HOD_S^V depends only on V. Since $x =_T y$ implies that $L[S, x] = L[S, y]$, one has that $\mathrm{HOD}_S^{L[S,x]} = \mathrm{HOD}_S^{L[S,y]}$ and $<^{\mathrm{HOD}_S^{L[S,x]}} = <^{\mathrm{HOD}_S^{L[S,y]}}$. (Note that although the constructibility hierarchy of $L[S, x]$ is naturally formulated in the language $\{\dot{\in}, \dot{E}_0, \dot{E}_1\}$, where $\dot{E}_0$ and $\dot{E}_1$ are unary predicate symbols meant to interpret S and x, when constructing $\mathrm{HOD}_S^{L[S,x]}$, $L[S, x]$ is considered as merely a $\{\dot{\in}\}$-structure.) Thus if X is a Turing degree, one will often write $\mathrm{HOD}_S^{L[S,X]}$ and $\leq^{\mathrm{HOD}_S^{L[S,X]}}$ to refer to $\mathrm{HOD}_S^{L[S,x]}$ and $<^{\mathrm{HOD}_S^{L[S,x]}}$ for any $x \in X$.

The invariance of the canonical wellordering of HOD allows one to take ultrapowers of local HOD's (models of the form $\mathrm{HOD}_S^{L[S,X]}$) by the Martin's measure. This is a very powerful technique as indicated by the following results.

The existence of the canonical wellorderings of the local $\mathrm{HOD}_S^{L[S,X]}$ shows that the resulting ultraproduct satisfies Łoś's theorem:

FACT 7.12. *Assume* $\mathsf{ZF} + \mathsf{AD}$. *Let* $S \subseteq \mathrm{ON}$ *be a set of ordinals.* $\prod_{X\in\mathcal{D}} \mathrm{HOD}_S^{L[S,X]}/\mu_{\mathcal{D}}$ *satisfies the Łoś's theorem: Let* φ *be a formula. Let* $f_0, ..., f_{n-1}$ *be functions on* $\mathcal{D}$ *with the property that for all* $X \in \mathcal{D}$, $f_i(X) \in \mathrm{HOD}_S^{L[S,X]}$. *Then*

$$\prod_{X\in\mathcal{D}} \mathrm{HOD}_S^{L[S,X]}/\mu_{\mathcal{D}} \models \varphi([f_0]_{\mu_{\mathcal{D}}}, ..., [f_{n-1}]_{\mu_{\mathcal{D}}}) \Leftrightarrow \{X \in \mathcal{D} : \mathrm{HOD}_S^{L[S,X]} \models \varphi(f_0(X), ..., f_{n-1}(X))\} \in \mu_{\mathcal{D}}.$$

PROOF. Only the existential quantification case requires a choice-like principle. One will give a sketch:

Let $\mathcal{M}$ denote this ultraproduct. Let φ be a formula and assume inductively one has already shown the result for φ. Suppose

$$K = \{X \in \mathcal{D} : \mathrm{HOD}_S^{L[S,X]} \models (\exists v)\varphi(v, f_0(X), ..., f_{n-1}(X))\} \in \mu_{\mathcal{D}}.$$

Let g be defined on K by letting $g(X)$ be the $\mathrm{HOD}_S^{L[S,X]}$-least v so that $\mathrm{HOD}_S^{L[S,X]} \models \varphi(v, f_0(X), ..., f_{n-1}(X))$. Then using the induction hypothesis, one can show that $\mathcal{M} \models \varphi([g]_{\mu_{\mathcal{D}}}, [f]_{\mu_{\mathcal{D}}}, ..., [f_{n-1}]_{\mu_{\mathcal{D}}})$. Thus $\mathcal{M} \models (\exists v)\varphi(v, [f]_{\mu_{\mathcal{D}}}, ..., [f_{n-1}]_{\mu_{\mathcal{D}}})$. $\square$

There is no claim that $\prod_{X\in\mathcal{D}} \mathrm{HOD}_S^{L[S,X]}/\mu_{\mathcal{D}}$ is a wellfounded model. This is true assuming DC. It is open whether AD implies this.

FACT 7.13. *(Woodin) Assume* $\mathsf{ZF}+\mathsf{AD}+\mathsf{DC}_{\mathbb{R}}$. *Suppose* $A \subseteq \mathbb{R}^2$ *has an* ∞*-Borel code* (S,φ)*, then* $B(x) = (\exists^{\mathbb{R}} y)A(x,y)$ *has an* ∞*-Borel code which is* OD_S.

PROOF. Note that $L(S,\mathbb{R}) \models \mathsf{ZF}+\mathsf{AD}+\mathsf{DC}$. Work in $L(S,\mathbb{R})$.

Let $\mathcal{M} = \prod_{X\in\mathcal{D}} \mathrm{HOD}_S^{L[S,X]}/\mu_{\mathcal{D}}$. $\mathcal{M}$ is an S-definable class which is wellfounded by DC. Implicitly, it will be assumed that all objects in this ultrapower have been Mostowski collapsed. By Fact 7.12, $\mathcal{M}$ satisfies Łoś's theorem. Define $\Phi_{\mathbb{A}_S^\infty}$ on $\mathcal{D}$ by $\Phi_{\mathbb{A}_S^\infty}(X) = \mathbb{A}_S^{L[S,X]}$. (Recall Remark 7.5 concerning the convention on $\mathbb{A}_S$.) Let $\mathbb{A}_S^\infty = [\Phi_{A_S^\infty}]_{\mu_{\mathcal{D}}}$. By Łoś's theorem, $\mathbb{A}_S^\infty$ is a forcing poset. Let $\lambda_X = |\mathbb{A}_S|^{\mathrm{HOD}_S^{L[S,X]}}$. Let Φ_λ be a function on $\mathcal{D}$ defined by $\Phi_\lambda(X) = \lambda_X$. Let $\lambda = [\Phi_\lambda]_{\mu_{\mathcal{D}}}$. By Łoś's theorem, $\mathcal{M} \models \lambda = |\mathbb{A}_S^\infty|$. Let Φ_{S^∞} be a function of $\mathcal{D}$ defined by $\Phi_{S^\infty}(X) = S$. Let $S^\infty = [\Phi_{S^\infty}]_{\mu_{\mathcal{D}}}$.

Claim: For all $a \in \mathbb{R}$,

$$a \in B \Leftrightarrow L[S^\infty, \mathbb{A}_S^\infty, a] \models 1_{\mathrm{Coll}(\omega,\lambda)} \Vdash (\exists b) L[S^\infty, a, b] \models \varphi(S^\infty, a, b).$$

To prove the claim: First observe that for all $X \in \mathcal{D}$, λ_X is countable in $L(S,\mathbb{R})$ which is a model of AD. To see this: Note that $\mathbb{R}^{L[S,X]}$ is countable since it is a wellorderable collection of reals; thus, there is a bijection of ω with $\mathbb{R}^{L[S,X]}$. $\mathbb{A}_S^{L[S,X]}$ is a collection of subsets of $\mathbb{R}^{L[S,X]}$ in $L[S,X]$. Identifying $\mathbb{R}^{L[S,X]}$ with ω, this collection $\mathbb{A}_S^{L[S,X]}$ can be identified as a wellorderable collection of $\mathbb{R}$, as well. Thus $\lambda_X = |\mathbb{A}_S|^{L[S,X]}$ is countable in $L(S,\mathbb{R})$. By the same reasoning, $(2^{\lambda_X})^{L[S,X]}$ is countable in $L(S,\mathbb{R})$.

($\Leftarrow$) For all $X \in \mathcal{D}$ so that $a \in X$, one can define $\mathrm{HOD}_S^{L[S,X]}[a]$ as in Fact 7.6. Let

$$\mathcal{M}[a] = \prod_{X\in\mathcal{D}} \mathrm{HOD}_S^{L[S,X]}[a]/\mu_{\mathcal{D}}.$$

Assume that

$$V \models L[S^\infty, \mathbb{A}_S^\infty, a] \models 1_{\mathrm{Coll}(\omega,\lambda)} \Vdash (\exists b) L[S^\infty, a, b] \models \varphi(S^\infty, a, b).$$

Thus

$$\mathcal{M}[a] \models L[S^\infty, \mathbb{A}_S^\infty, a] \models 1_{\mathrm{Coll}(\omega,\lambda)} \Vdash (\exists b) L[S^\infty, a, b] \models \varphi(S^\infty, a, b).$$

By the idea of the proof of Łoś's theorem (Fact 7.12), for $\mu_{\mathcal{D}}$-almost all $X \in \mathcal{D}$ with $a \in X$,

$$L[S, \mathbb{A}_S^{L[S,X]}, a] \models 1_{\mathrm{Coll}(\omega,\lambda_X)} \Vdash (\exists b) L[S,a,b] \models \varphi(S,a,b).$$

Fix such an X. Since $(2^{\lambda_X})^{L[S,X]}$ is countable in $L(S,\mathbb{R})$, there is a $g \in L(S,\mathbb{R})$ so that $g \subseteq \mathrm{Coll}(\omega,\lambda_X)$ is $\mathrm{Coll}(\omega,\lambda_X)$-generic over $\mathrm{HOD}_S^{L[S,X]}[a]$. Then g is also generic over $L[S, \mathbb{A}_S^{L[S,X]}, a]$. By the forcing theorem, $L[S, \mathbb{A}_S^{L[S,X]}, a][g] \models (\exists b) L[S,a,b] \models \varphi(S,a,b)$. Pick some $b \in L[S, \mathbb{A}_S^{L[S,X]}, a][g]$ witnessing the existential. Then one has in particular that $L[S,a,b] \models \varphi(S,a,b)$. Since (S,φ) is the ∞-Borel code for A, one has that $(a,b) \in A$. Thus $a \in B$.

($\Rightarrow$) Suppose that $a \in B$. Let $b \in \mathbb{R}$ be such that $(a,b) \in A$. Let X be a Turing degree such that $[a \oplus b] \leq X$. By Fact 7.6, let $G_{a\oplus b}$ be the $\mathbb{A}_S^{L[S,X]}$-generic over $\mathrm{HOD}_S^{L[S,X]}$-filter derived from $a \oplus b$. Note that $G_{a\oplus b}$ is also $\mathbb{A}_S^{L[S,X]}$-generic over

$L[S,\mathbb{A}_S^{L[S,X]}]$. Using the convention from Remark 7.5 and an argument similar to Fact 7.6, one can recover $a \oplus b$ from $G_{a,b}$ in $L[S,\mathbb{A}_S^{L[S,X]}][G_{a,b}]$. Thus

$$L[S,\mathbb{A}_S^{L[S,X]}][G_{a\oplus b}] \models (\exists y)L[S,a,y] \models \varphi(S,a,y).$$

Since $|\mathbb{A}_S^{L[S,X]}|$ is a forcing of size λ_X, one has that $\mathbb{A}_S^{L[S,X]}$ regularly embeds into $\mathrm{Coll}(\omega,\lambda_X)$ by [**13**] Corollary 26.8. There is some $g \subseteq \mathrm{Coll}(\omega,\lambda_X)$ which is generic over $L[S,\mathbb{A}_S^{L[S,X]}]$ so that $G_{a\oplus b} \in L[S,\mathbb{A}_S^{L[S,X]}][g]$. Thus

$$L[S,\mathbb{A}_S^{L[S,X]}][g] \models (\exists y)L[S,a,y] \models \varphi(S,a,y).$$

Note that $a \in L[S,\mathbb{A}_S^{L[S,X]}][g]$. By one of the main properties of the collapse forcing ([**13**] Corollary 26.10), there is some $h \subseteq \mathrm{Coll}(\omega,\lambda_X)$ which is $\mathrm{Coll}(\omega,\lambda_X)$-generic over $L[S,\mathbb{A}_S^{L[S,X]}][a]$ so that $L[S,\mathbb{A}_S^{L[S,X]}][a][h] = L[S,\mathbb{A}_S^{L[S,X]}][g]$. Thus

$$L[S,\mathbb{A}_S^{L[S,X]}][a][h] \models (\exists y)L[S,a,y] \models \varphi(S,a,y).$$

There is some $p \in \mathrm{Coll}(\omega,\lambda_X)$ which forces the inner statement. By the homogeneity of $\mathrm{Coll}(\omega,\lambda_X)$, $1_{\mathrm{Coll}(\omega,\lambda_X)}$ forces this statement:

$$L[S,\mathbb{A}_S^{L[S,X]}][a] \models 1_{\mathrm{Coll}(\omega,\lambda_X)} \Vdash (\exists y)L[S,a,y] \models \varphi(S,a,y).$$

In particular,

$$\mathrm{HOD}_S^{L[S,X]}[a] \models L[S,\mathbb{A}_S^{L[S,X]},a] \models 1_{\mathrm{Coll}(\omega,\lambda_X)} \Vdash (\exists y)L[S,a,y] \models \varphi(S,a,y).$$

By Łoś's theorem (Fact 7.12), one has

$$\mathcal{M}[a] \models L[S^\infty,\mathbb{A}_S^\infty,a] \models 1_{\mathrm{Coll}(\omega,\lambda)} \Vdash (\exists y)L[S^\infty,a,y] \models \varphi(S^\infty,a,y).$$

So in particular,

$$L[S^\infty,\mathbb{A}_S^\infty,a] \models 1_{\mathrm{Coll}(\omega,\lambda)} \Vdash (\exists y)L[S^\infty,a,y] \models \varphi(S^\infty,a,y).$$

This completes the proof of the claim.

Let J be a set of ordinals that codes S^∞ and $\mathbb{A}_S^\infty$ in some fixed way. Let $\psi(J,a)$ be the formula that asserts

$$1_{\mathrm{Coll}(\omega,\lambda)} \Vdash (\exists y)L[S^\infty,a,y] \models \varphi(S^\infty,a,y).$$

(J,ψ) is an ∞-Borel code for B. □

FACT 7.14. *(Woodin) Assume* $\mathsf{ZF}+\mathsf{AD}+\mathsf{DC}$. *Let* $1 \leq m \leq n$. *Let* $\pi_{n,m} : \mathbb{R}^n \to \mathbb{R}^m$ *be defined by* $\pi_{n,m}(s) = s \restriction m$. *Define* $\pi_{n,m} : {}_n\mathbb{A} \to {}_m\mathbb{A}$ *by* $\pi_{n,m}(b) = \pi_{n,m}[b]$, *where the latter* $\pi_{n,m}$ *refers to the earlier function* $\pi_{n,m} : \mathbb{R}^n \to \mathbb{R}^m$. *Then* $\pi_{n,m}$ *is a forcing projection.*

PROOF. By Fact 7.13, one can show that $\pi_{n,m}(p) \in {}_m\mathbb{A}$ for all $p \in {}_n\mathbb{A}$.

Now suppose that $q \in {}_n\mathbb{A}$. Suppose $p' \in {}_m\mathbb{A}$ is such that $p' \leq_{{}_m\mathbb{A}} \pi_{n,m}(q)$. Let $q' = \{x \in \mathbb{R}^n : x \restriction n \in \pi_{n,m}(q)\}$. Since $\pi_{n,m}(q) \in {}_m\mathbb{A}$ (that is, it has an OD ∞-Borel code), q' has an OD ∞-Borel code and hence belongs to ${}_n\mathbb{A}$. Clearly, $\pi_{n,m}(q') = p'$. This establishes that $\pi_{n,m}$ is a forcing projection. □

Now one will work in $L(\mathbb{R}) \models \mathsf{AD}$. Kechris ([**15**]) showed that if AD holds, then $L(\mathbb{R}) \models \mathsf{DC}_\mathbb{R}$. Thus, for the following results, the background theory $\mathsf{ZF}+\mathsf{AD}$ is sufficient.

DEFINITION 7.15. Using the projections from Fact 7.14, let ${}_\omega\mathbb{A}$ be the finite support direct limit of $\langle {}_n\mathbb{A} : n \in \omega \setminus \{0\}\rangle$. That is, ${}_\omega\mathbb{A}$ is the collection of $p : (\omega \setminus \{0\}) \to \bigcup_{n\in\omega} {}_n\mathbb{A}$ so that for all $m \leq n$, $\pi_{n,m}(p(n)) = p(m)$ and there exists a $N \in \omega$ so that for all $k \geq N$, $p(k) = p(N) \times \mathbb{R}^{k-N}$. The least such N is denoted $\dim(p)$, the dimension of p. For $n \in \omega$ and $p \in {}_\omega\mathbb{A}$, let $\pi_{\omega,n}(p) = p(n)$. Each $\pi_{\omega,n} : {}_\omega\mathbb{A} \to {}_n\mathbb{A}$ is a forcing projection.

Since each ${}_n\mathbb{A} \in \mathrm{HOD}$ is identified as a set of ordinals (see Remark 7.5) and the projection maps are in HOD, ${}_\omega\mathbb{A}$ belongs to HOD and may be identified as a set of ordinals having the property expressed in Remark 7.5.

Let $m \leq n$, let τ_m be a ${}_n\mathbb{A}$-name for the last real of the $\dot{x}^m_{\mathrm{gen}}$, which is a name for the generic m-tuple coming from a ${}_n\mathbb{A}$-generic filter. (Technically, τ_m is different for each n, but the projection can be used to interpret it in suitable n's.)

Let $\dot{\mathbb{R}}_{\mathrm{sym}}$ be a ${}_\omega\mathbb{A}$-name so that $1_{{}_\omega\mathbb{A}} \Vdash \dot{\mathbb{R}}_{\mathrm{sym}} = \{\tau_n : n \in \omega \setminus \{0\}\}$.

Observe that from Fact 7.6, every $z \in \mathbb{R}^n$ induces a ${}_n\mathbb{A}$-generic filter G_z over HOD so that $\mathrm{HOD}[z] = \mathrm{HOD}[G_z]$. Note that $\dot{x}_{\mathrm{gen}}[G_z] = z$ and $\tau_m[G_z] = z(m)$ for all $m < n$.

THEOREM 7.16. *(Woodin) Assume* $\mathsf{ZF}+\mathsf{AD}+\mathsf{V} = \mathsf{L}(\mathbb{R})$. *Suppose* $g \subseteq \mathrm{Coll}(\omega, \mathbb{R})$-*generic over* $L(\mathbb{R})$. *From* g, *one can derive a* ${}_\omega\mathbb{A}$-*generic over* $\mathrm{HOD}^{L(\mathbb{R})}$ *filter* G_g.

PROOF. Let $g \subseteq \mathrm{Coll}(\omega, \mathbb{R})$ be a generic over $\mathrm{HOD}^{L(\mathbb{R})}$. Let $G_g \subseteq {}_\omega\mathbb{A}$ be the collection of condition $p \in {}_\omega\mathbb{A}$ so that $g \restriction \dim(p) \in p(\dim(p))$.

The claim is that G_g is ${}_\omega\mathbb{A}$-generic over $\mathrm{HOD}^{L(\mathbb{R})}$.

Suppose $D \subseteq {}_\omega\mathbb{A}$ belongs to $\mathrm{HOD}^{L(\mathbb{R})}$ and is dense. Let $\tilde{D} \subseteq \mathrm{Coll}(\omega, \mathbb{R})$ be the collection of $s \in \mathrm{Coll}(\omega, \mathbb{R})$ so that there is some $p \in {}_\omega\mathbb{A}$ with $\dim(p) = |s|$, $s \in p(|s|)$, and $p \in D$. (Note that s as a condition of $\mathrm{Coll}(\omega, \mathbb{R})$ is a finite tuple of reals.)

One will show that $\tilde{D}$ is dense in $\mathrm{Coll}(\omega, \mathbb{R})$. Let $s \in \mathrm{Coll}(\omega, \mathbb{R})$. Let $n = |s|$. Let

$$E = \{p \in {}_n\mathbb{A} : (\exists q \in {}_\omega\mathbb{A})(\dim(q) \geq n \wedge q \in D \wedge \pi_{\omega,n}(q) = p)\}.$$

First, one will show E is dense in ${}_n\mathbb{A}$. Let $r \in {}_n\mathbb{A}$. Since D is dense in ${}_\omega\mathbb{A}$, there is some $q \in {}_\omega\mathbb{A}$ with $\dim(q) \geq n$ so that $q \leq_{{}_\omega\mathbb{A}} r$ and $q \in D$. Then $p = \pi_{\omega,n}(q)$ belongs to E and $p \leq_{{}_n\mathbb{A}} r$. Thus E is dense in ${}_n\mathbb{A}$.

Let G^n_s be the ${}_n\mathbb{A}$-generic over $\mathrm{HOD}^{L(\mathbb{R})}$ filter derived from s. By genericity, pick some $p \in G^n_s \cap E$. Thus there is some $q \in D$ so that $\pi_{\omega,n}(q) = p$. Let $s' \supseteq s$ be such that $s' \in q$. This means that $s' \leq_{\mathrm{Coll}(\omega,\mathbb{R})} s$ and $s' \in \tilde{D}$. It has been shown that $\tilde{D}$ is dense in $\mathrm{Coll}(\omega, \mathbb{R})$.

Now since $g \subseteq \mathrm{Coll}(\omega, \mathbb{R})$ is $\mathrm{Coll}(\omega, \mathbb{R})$-generic over $L(\mathbb{R})$, $g \cap \tilde{D} \neq \emptyset$. There is some $n \in \omega$ so that $g \restriction n \in \tilde{D}$. By definition, there is some $p \in D$ so that $g \restriction n \in p(n)$. Thus $p \in G_g \cap D$.

This shows that G_g is ${}_\omega\mathbb{A}$-generic over HOD. □

FACT 7.17. *(Woodin) Assume* $\mathsf{ZF} + \mathsf{AD} + \mathsf{V} = \mathsf{L}(\mathbb{R})$. $\mathrm{HOD}^{L(\mathbb{R})} \models 1_{{}_\omega\mathbb{A}} \Vdash$ *"the reals of* $L(\dot{\mathbb{R}}_{\mathrm{sym}})$ *is* $\dot{\mathbb{R}}_{\mathrm{sym}}$*".*

PROOF. Let $p \in {}_\omega\mathbb{A}$ be some condition and let $n = \dim(p)$. Let $s \in \mathbb{R}^n$ with $s \in p(n)$. Consider s as a condition of $\mathrm{Coll}(\omega, \mathbb{R})$. Let $g \subseteq \mathrm{Coll}(\omega)$ be $\mathrm{Coll}(\omega, \mathbb{R})$-generic over $L(\mathbb{R})$ containing s. An easy density argument shows that if g is considered as a function from ω to $\mathbb{R}$, g must be a surjection onto $\mathbb{R}^{L(\mathbb{R})}$. Let

G_g be the ${}_\omega\mathbb{A}$-generic filter over $\mathrm{HOD}^{L(\mathbb{R})}$ derived from g. Then $\mathrm{HOD}^{L(\mathbb{R})}[G_g] \models$ "the reals of $L(\dot{\mathbb{R}}_{\mathrm{sym}}[G_g])$ is $\dot{\mathbb{R}}_{\mathrm{sym}}[G_g]$" since $L(\dot{\mathbb{R}}_{\mathrm{sym}}[G_g]) = L(\mathbb{R})$. Thus there is some $q \leq_{{}_\omega\mathbb{A}} p$ so that $\mathrm{HOD}^{L(\mathbb{R})} \models q \Vdash_{{}_\omega\mathbb{A}}$ "the reals of $L(\dot{\mathbb{R}}_{\mathrm{sym}})$ is $\dot{\mathbb{R}}_{\mathrm{sym}}$". Since p was an arbitrary condition, one has that $1_{{}_\omega\mathbb{A}}$ forces this same statement. □

THEOREM 7.18. *(Woodin) Assume* $\mathsf{ZF} + \mathsf{AD} + \mathsf{V} + \mathsf{L}(\mathbb{R})$. *Let* $s \in \mathbb{R}^n$. *Suppose* G_s^n *is the* ${}_n\mathbb{A}$*-generic filter over* $\mathrm{HOD}^{L(\mathbb{R})}$ *derived from* s. *Let* φ *be a formula. Suppose* $z \in \mathrm{HOD}^{L(\mathbb{R})}[G_s^n]$. *Then*

$$\mathrm{HOD}^{L(\mathbb{R})}[G_s^n] \models 1_{{}_\omega\mathbb{A}/G_s^n} \Vdash_{{}_\omega\mathbb{A}/G_s^n} L(\dot{\mathbb{R}}_{\mathrm{sym}}) \models \varphi(\check{z}, \dot{x}_{\mathrm{gen}}^n)$$

or

$$\mathrm{HOD}^{L(\mathbb{R})}[G_s^n] \models 1_{{}_\omega\mathbb{A}/G_s^n} \Vdash_{{}_\omega\mathbb{A}/G_s^n} L(\dot{\mathbb{R}}_{\mathrm{sym}}) \models \neg\varphi(\check{z}, \dot{x}_{\mathrm{gen}}^n)$$

PROOF. In $L(\mathbb{R})$, one either has $L(\mathbb{R}) \models \varphi(z, s)$ or $L(\mathbb{R}) \models \neg\varphi(z, s)$. The claim is that which ever case occurs, this is the side that is forced over $\mathrm{HOD}^{L(\mathbb{R})}[G_s^n]$.

So without loss generality, suppose $L(\mathbb{R}) \models \varphi(z, s)$.

Take any $q \in {}_\omega\mathbb{A}/G_s^n$ with $\dim(q) \geq n$. Let $m = \dim(q)$. This means that $\pi_{\omega,n}(q) \in G_s^n$. Let $r \in q$ be a $\mathbb{R}^m$ sequence extending s. Let $g \subseteq \mathrm{Coll}(\omega, \mathbb{R})$ be a $\mathrm{Coll}(\omega, \mathbb{R})$-generic over $L(\mathbb{R})$ filter so that $r \subseteq g$. Let G_g be the ${}_\omega\mathbb{A}$-generic over $\mathrm{HOD}^{L(\mathbb{R})}$ filter derived from g. Note that G_g is a ${}_\omega\mathbb{A}/G_s^n$-generic filter over $\mathrm{HOD}^{L(\mathbb{R})}[G_s^n]$ (see the basic properties of projection from Definition 7.8) and that $q \in G_g$. Using Fact 7.17 and the fact that $\mathrm{HOD}[G_s^n][G_g] = \mathrm{HOD}[G_g]$ (also see Definition 7.8),

$$\mathrm{HOD}[G_s^n][G_g] \models L(\dot{\mathbb{R}}_{\mathrm{sym}}[G_g]) \models \varphi(z, \dot{x}_{\mathrm{gen}}^n[G_s^n])$$

since $L(\mathbb{R}) \models \varphi(z, s)$. Hence there is some $q' \leq_{{}_\omega\mathbb{A}/G_s^n} q$ so that

$$\mathrm{HOD}[G_s^n] \models q' \Vdash_{{}_\omega\mathbb{A}/G_s^n} L(\dot{\mathbb{R}}_{\mathrm{sym}}) \models \varphi(\check{z}, \dot{x}_{\mathrm{gen}}^n).$$

Since $q \in {}_\omega\mathbb{A}/G_s^n$ was arbitrary, $1_{{}_\omega\mathbb{A}/G_s^n}$ forces this statement. This completes the proof. □

THEOREM 7.19. *(Woodin) Assume* $\mathsf{ZF} + \mathsf{AD} + \mathsf{V} = \mathsf{L}(\mathbb{R})$. *Let* $s \in \mathbb{R}^n$, $z \in L[{}_\omega\mathbb{A}, s]$, *and* φ *be a formula. Then* $L(\mathbb{R}) \models \varphi(s, z)$ *if and only if*

$$L[{}_\omega\mathbb{A}, s] \models 1_{{}_\omega\mathbb{A}/G_s^n} \Vdash_{{}_\omega\mathbb{A}/G_s^n} L(\dot{\mathbb{R}}_{\mathrm{sym}}) \models \varphi(\dot{x}_{\mathrm{gen}}^n, \check{z})$$

PROOF. Recall ${}_\omega\mathbb{A} \in \mathrm{HOD}^{L(\mathbb{R})}$. Using the convention from Remark 7.5, one may assume that from a $p \in {}_\omega\mathbb{A}$, one can obtain an ∞-Borel code for this condition. From this, one can see that given s and ${}_\omega\mathbb{A}$, one can reconstruct G_s^n. Also the names $\dot{x}_{\mathrm{gen}}^n$ and $\dot{\mathbb{R}}_{\mathrm{sym}}$ belong to $L[{}_\omega\mathbb{A}]$.

($\Rightarrow$) Using the idea from Fact 7.6, one can show that $L[{}_\omega\mathbb{A}, s] = L[{}_\omega\mathbb{A}][G_s^n]$. Take any $p \in {}_\omega\mathbb{A}/G_s^n$. In particular, $p \in {}_\omega\mathbb{A}$. Pick some $r \in \mathrm{Coll}(\omega, \mathbb{R})$ with $r \supseteq s$ and $r \in p(|r|)$. Let $g \subseteq \mathrm{Coll}(\omega, \mathbb{R})$ be $\mathrm{Coll}(\omega, \mathbb{R})$-generic over $L(\mathbb{R})$ and $r \subseteq g$. Using ${}_\omega\mathbb{A}$, one can reconstruct G_g which is a ${}_\omega\mathbb{A}$-generic filter over $\mathrm{HOD}^{L(\mathbb{R})}$. Thus G_g is $({}_\omega\mathbb{A}/G_s^n)$-generic over $\mathrm{HOD}^{L(\mathbb{R})}[G_s^n]$ and hence generic over $L[{}_\omega\mathbb{A}, s]$. Since $L(\mathbb{R}) \models \varphi(s, z)$, $L[{}_\omega\mathbb{A}, s][G_g] \models L(\dot{\mathbb{R}}_{\mathrm{sym}}[G_g]) \models \varphi(s, \check{z})$ because $L(\dot{\mathbb{R}}_{\mathrm{sym}}[G_g]) = L(\mathbb{R})$. Pick a $q \leq_{{}_\omega\mathbb{A}} p$ so that

$$L[{}_\omega\mathbb{A}, s] \models q \Vdash_{{}_\omega\mathbb{A}/G_s^n} L(\dot{\mathbb{R}}_{\mathrm{sym}}) \models \varphi(s, \check{z}).$$

Since p was arbitrary, $1_{{}_\omega\mathbb{A}/G_s^n}$ forces this statement.

($\Leftarrow$) Let $g \supseteq s$ be a $\mathrm{Coll}(\omega, \mathbb{R})$-generic filter over $L(\mathbb{R})$. Let G_g be the ${}_\omega\mathbb{A}$-generic filter of $\mathrm{HOD}^{L(\mathbb{R})}$ derived from g. It is $({}_\omega\mathbb{A}/G^n_s)$-generic over $\mathrm{HOD}^{L(\mathbb{R})}[G^n_s]$ and hence generic over $L[{}_\omega\mathbb{A}, s]$. Then one has

$$L[{}_\omega\mathbb{A}, s][G_g] \models L(\dot{\mathbb{R}}_{\mathrm{sym}}[G_g]) \models \varphi(s, z).$$

However, $L(\dot{\mathbb{R}}_{\mathrm{sym}}[G_g]) = L(\mathbb{R})$. Thus $L(\mathbb{R}) \models \varphi(s, z)$. □

COROLLARY 7.20. *Assume* $\mathsf{ZF} + \mathsf{AD} + \mathsf{V} = \mathsf{L}(\mathbb{R})$. *Every set of reals has an* ∞*-Borel code. Moreover, if* A *is* OD_s *where* s *is a finite tuple of reals, then* A *has an* ∞*-Borel code in* $L[{}_\omega\mathbb{A}, s] \subseteq \mathrm{HOD}_s$.

PROOF. Suppose $A \subseteq \mathbb{R}$. In $L(\mathbb{R})$, every set is definable from ordinals and some reals. Let $s \in \mathbb{R}^n$, $\bar{\alpha}$ be a tuple of ordinals, and φ be a formula so that $r \in A \Leftrightarrow \varphi(s, r, \bar{\alpha})$. By Theorem 7.19, one has that $r \in A$ if and only if

$$L(\mathbb{R}) \models L[{}_\omega\mathbb{A}, s, r] \models 1_{{}_\omega\mathbb{A}/G^{n+1}_{s\hat{}r}} \Vdash L(\dot{\mathbb{R}}_{\mathrm{sym}}) \models \varphi(s, r, \bar{\alpha}).$$

Let S be a set of ordinals coding $({}_\omega\mathbb{A}, s, \bar{\alpha})$. Let ψ be the formula expressing the inner forcing statement. One has that (S, ψ) is an ∞-Borel code for A. □

Note that ${}_\omega\mathbb{A}$ can be considered an ultimate ∞-Borel code in the sense that for any $A \subseteq \mathbb{R}$, there is some tuple of reals, tuple of ordinals, and a formula so that together with ${}_\omega\mathbb{A}$ they form an ∞-Borel code for A. This is very useful in diagonalization arguments.

COROLLARY 7.21. *Assume* $\mathsf{ZF} + \mathsf{AD} + \mathsf{V} = \mathsf{L}(\mathbb{R})$. *Then* $\mathrm{HOD}^{L(\mathbb{R})} = L[{}_\omega\mathbb{A}]$.

PROOF. Since ${}_\omega\mathbb{A} \in \mathrm{HOD}^{L(\mathbb{R})}$, $L[{}_\omega\mathbb{A}] \subseteq \mathrm{HOD}^{L(\mathbb{R})}$. $\mathrm{HOD}^{L(\mathbb{R})} \subseteq L[{}_\omega\mathbb{A}]$ by Theorem 7.19. □

COROLLARY 7.22. *Assume* $\mathsf{ZF}+\mathsf{AD}+\mathsf{V} = \mathsf{L}(\mathbb{R})$. $\mathbb{A} = \mathbb{O}$, *that is the forcing with nonempty subsets of* $\mathbb{R}$ *possessing* OD ∞*-Borel codes is the same as the Vopěnka forcing.*

PROOF. This follows from Corollary 7.20. □

Therefore, in the following, one will use the more practical Vopěnka forcing $\mathbb{O}$ rather than $\mathbb{A}$.

8. The Vopěnka Forcing

FACT 8.1. *([9] Theorem 2.4.) Let* M *be a transitive inner model of* ZF. *Let* $S \in M$ *be a set of ordinals. Suppose* K *is an* OD^M_S *set of ordinals and* φ *is a formula. Let* N *be some transitive inner model with* $\mathrm{HOD}^M_S \subseteq N$. *Suppose* $p = \{x \in \mathbb{R}^M : L[K, x] \models \varphi(K, x)\}$ *is a condition of* $\mathbb{O}^M_S$ *(i.e. is nonempty). Then* $N \models p \Vdash_{\mathbb{O}^M_S} L[\check{K}, \dot{x}_{\mathrm{gen}}] \models \varphi(\check{K}, \dot{x}_{\mathrm{gen}})$.

PROOF. Suppose not. Then there is some $q' \leq_{\mathbb{O}^M_S} p$ so that $N \models q' \Vdash_{\mathbb{O}^M_S} L[\check{K}, \dot{x}_{\mathrm{gen}}] \models \neg\varphi(\check{K}, \dot{x}_{\mathrm{gen}})$. Because every $\mathbb{O}^M_S$-generic filter over N is also generic over HOD^M_S, one can find some $q \leq_{\mathbb{O}^M_S} q'$ so that $\mathrm{HOD}^M_S \models q \Vdash_{\mathbb{O}^M_S} L[\check{K}, \dot{x}_{\mathrm{gen}}] \models \neg\varphi(\check{K}, \dot{x}_{\mathrm{gen}})$. Since $q \neq \emptyset$, let $y \in q$. Let G_y be the $\mathbb{O}^M_S$-generic filter over HOD^M_S derived from y. Note that $q \in G_y$. Thus $\mathrm{HOD}^M_S[G_y] \models L[K, y] \models \neg\varphi(K, y)$. Hence $L[K, y] \models \neg\varphi(K, y)$. This contradicts $q \subseteq p$. □

FACT 8.2. *Let M be an inner model of* ZF. *Let $S \in M$ be a set of ordinals. Let N be an inner model of* ZF *such that $N \supseteq \mathrm{HOD}_S^M$. Let $n \geq 1$ be a natural number. Suppose $(g_0, ..., g_{n-1})$ is an ${}_n\mathbb{O}_S^M$-generic reals over N. Then each g_0,...,g_{n-1} is $\mathbb{O}_S^M$-generic over N.*

PROOF. Here g is a $\mathbb{O}_S^M$-generic real over N if and only if there is a filter G which $\mathbb{O}_S^M$-generic over N so that $g = \dot{x}_{\mathrm{gen}}[G]$. For simplicity assume $n = 2$.

Let $\pi_1 : \mathbb{R}^2 \to \mathbb{R}$ be the projection onto the first coordinate. For each $p \in {}_2\mathbb{O}_S^M$, $\pi_1[p] \in \mathbb{O}_S^M$.

Let (g_0, g_1) be ${}_2\mathbb{O}_S^M$-generic over N. Let $G_{(g_0,g_1)}$ be the ${}_2\mathbb{O}_S^M$-generic filter over N which adds (g_0, g_1). Let $G = \{\pi_1(p) : p \in G_{(g_0,g_1)}\}$. G is a filter on $\mathbb{O}_S^M$.

Suppose $D \subseteq \mathbb{O}_S^M$ belongs to N and is dense open. Let $D' = \{p \in {}_2\mathbb{O}_S^M : \pi_1(p) \in D\}$. Let $r \in {}_2\mathbb{O}_S^M$. Since D is dense, there is some $r' \leq_{\mathbb{O}_S^M} \pi_1(r)$ with $r' \in D$. Let $s = (r' \times \mathbb{R}) \cap r$. Note $s \in {}_2\mathbb{O}_S^M$, $\pi_1(s) = r' \in D$, and $s \leq_{{}_2\mathbb{O}_S^M} r$. Hence $s \in D'$. This shows that D' is dense in ${}_2\mathbb{O}_S^M$.

By genericity, there is some $r \in D' \cap G_{(g_0,g_1)}$. Then $\pi_1(r) \in D \cap G$. This shows that G is $\mathbb{O}_S^M$-generic over N. Since g_0 is the real added by G, g_0 is a $\mathbb{O}_S^M$-generic real over N. □

First, one will give a proof of Woodin's countable section uniformization. This follows a presentation of this result in [**19**]. It is not known whether countable section uniformization is provable in AD alone.

LEMMA 8.3. *Assume* $\mathsf{ZF} + \mathsf{AD}$. *Let $R \subseteq \mathbb{R} \times \mathbb{R}$ be a relation. Suppose there is a set of ordinals S with the property that for all $x \in \mathbb{R}$ so that $R_x = \{y \in \mathbb{R} : R(x, y)\} \neq \emptyset$ (i.e. $x \in \mathrm{dom}(R)$), for $\mu_{\mathcal{D}}$-almost all Turing degrees Z, $R_x \cap \mathrm{HOD}_{S,x}^{L[S,x,Z]} \neq \emptyset$. Then R has a uniformization.*

PROOF. Think of $\mathbb{R} = {}^\omega 2$. Let $x \in \mathrm{dom}(R)$. Let $E^x = \{Z \in \mathcal{D} : R_x \cap \mathrm{HOD}_{S,x}^{L[S,x,Z]} \neq \emptyset\}$. By the assumption, $E^x \in \mu_{\mathcal{D}}$.

For each $n \in \omega$ and $i \in 2$, let $E_{n,i}^x$ be the collection of $Z \in E^x$ so that $z(n) = i$, where z is the $\mathrm{HOD}_{S,x}^{L[S,x,Z]}$-least element of $R_x \cap \mathrm{HOD}_{S,x}^{L[S,x,Z]}$ according to the canonical wellordering of $\mathrm{HOD}_{S,x}^{L[S,x,Z]}$. Note that $E^x = E_{n,0}^x \cup E_{n,1}^x$. Since $\mu_{\mathcal{D}}$ is an ultrafilter and $E^x \in \mu_{\mathcal{D}}$, for each $n \in \omega$, there is a unique $a_n^x \in 2$ so that $E_{n,a_n^x}^x \in \mu_{\mathcal{D}}$.

Define a function $\Phi : \mathrm{dom}(R) \to {}^\omega 2$ by $\Phi(x)(n) = a_n^x$. The claim is that Φ is a uniformization for R.

Suppose $x \in \mathrm{dom}(R)$. Since $\mu_{\mathcal{D}}$ is a countably complete ultrafilter on $\mathcal{D}$, $\bigcap_{n\in\omega} E_{n,a_n^x}^x \in \mu_{\mathcal{D}}$. Pick any $Z \in \bigcap_{n\in\omega} E_{n,a_n^x}^x$. Then $\Phi(x)$ is the $\mathrm{HOD}_{S,x}^{L[S,x,Z]}$-least element of $R_x \cap \mathrm{HOD}_{S,x}^{L[S,x,Z]}$. In particular, $\Phi(x) \in R_x$. Φ is a uniformization. □

THEOREM 8.4. *(Woodin's countable section uniformization) Assume* $\mathsf{ZF} + \mathsf{AD}$ *and that all sets of reals have an ∞-Borel code. Let $R \subseteq \mathbb{R} \times \mathbb{R}$ be such that for all $x \in \mathrm{dom}(R)$, $R_x = \{y \in \mathbb{R} : R(x, y)\}$ is countable. Then R has a uniformization function, that is some function $F : \mathrm{dom}(R) \to \mathbb{R}$ so that for all $x \in \mathrm{dom}(R)$, $R(x, F(x))$.*

In particular, countable section uniformization holds in $L(\mathbb{R}) \models \mathsf{AD}$.

PROOF. Let (S, φ) be an ∞-Borel code for R, that is $\mathfrak{B}_{(S,\varphi)}^2 = R$.

Let $x \in \text{dom}(R)$. Thus there is some y^* such that $R(x, y^*)$. Let $Z \in \mathcal{D}$ be such that $Z \geq [y^*]_T$. In the model $L[S, x, Z]$, define

$$p = \{y \in \mathbb{R} : L[S, x, y] \models \varphi(S, x, y)\}.$$

This condition is clearly $\text{OD}_{S,x}^{L[S,x,Z]}$. Note $[y^*]_T \leq Z$ implies that $y^* \in L[S, x, Z]$. Since in the real world V, $R(x, y^*)$ holds, one has that $V \models L[S, x, y^*] \models \varphi(S, x, y^*)$. Thus $L[S, x, Z] \models L[S, x, y^*] \models \varphi(S, x, y^*)$. So $y^* \in p$. Hence $p \neq \emptyset$. It has been shown that $p \in \mathbb{O}_{S,x}^{L[S,x,Z]}$.

Claim 1: There is a dense below p set of conditions in $\mathbb{O}_{S,x}^{L[S,x,Z]}$ which forces a value for $\dot{x}_{\text{gen}}$.

To prove Claim 1: Suppose otherwise. As argued in the proof of Fact 7.13, since $\text{HOD}_{S,x}^{L[S,x,Z]} \models \mathsf{AC}$, $\mathscr{P}(\mathbb{O}_{S,x}^{L[S,x,Z]})^{\text{HOD}_{S,x}^{L[S,x,Z]}}$ is countable in the real world V. Let $\langle D_n : n \in \omega \rangle$ enumerate (in the real world) all the dense open subsets of $\mathbb{O}_{S,x}^{L[S,x,Z]}$ that belong to $\text{HOD}_{S,x}^{L[S,x,Z]}$.

Let $p_\emptyset$ be any condition below p that meets D_0. Let $m_\emptyset = 0$. Suppose for some $\sigma \in {}^{<\omega}2$, p_σ and m_σ have been defined. First find some $p' \leq p_\sigma$ so that $p' \in D_{|\sigma|+1}$. Since p' can not determine $\dot{x}_{\text{gen}}$, there is some $N \geq m_\sigma$ and two conditions q_0 and q_1 below p' so that
(1) q_0 and q_1 determine $\dot{x}_{\text{gen}} \restriction N + 1$ and
(2) For $i \in 2$, $q_i \Vdash \dot{x}_{\text{gen}}(\check{N}) = \check{i}$. (That is, q_i forces the generic real to take value i at N.)

Let q_0 and q_1 be the $\text{HOD}_{S,x}^{L[S,x,Z]}$-least pair with the above property. Now let $p_{\sigma\hat{}i} = q_i$ and $m_{\sigma\hat{}i} = N + 1$. Observe that $p_{\sigma\hat{}i} \in D_{n+1}$ (as D_{n+1} is dense open). This completes the construction of a sequence $\langle p_\sigma : \sigma \in {}^{<\omega}2 \rangle$ of conditions in $\mathbb{O}_{S,x}^{L[S,x,Z]}$.

For each $f \in {}^\omega 2$, let G_f be the $\leq_{\mathbb{O}_{S,x}^{L[S,x,Z]}}$ upward closure of the set $\{p_{f \restriction n} : n \in \omega\}$. By construction, G_f is $\mathbb{O}_{S,x}^{L[S,x,Z]}$-generic over $\text{HOD}_{S,x}^{L[S,x,Z]}$. Also by construction, if $f \neq g$, then $\dot{x}_{\text{gen}}[G_f] \neq \dot{x}_{\text{gen}}[G_g]$.

Hence $A = \{\dot{x}_{\text{gen}}[G_f] : f \in {}^\omega 2\}$ is an uncountable set of reals.

Subclaim 1.1: $A \subseteq R_x$.

To see Subclaim 1.1: Note that for all $f \in {}^\omega 2$, $p \in G_f$. Note that the condition p takes the form specified in Fact 8.1. Hence the fact asserts that

$$\text{HOD}_{S,x}^{L[S,x,Z]} \models p \Vdash_{\mathbb{O}_{S,x}^{L[S,x,Z]}} L[S, x, \dot{x}_{\text{gen}}] \models \varphi(\check{S}, \check{x}, \dot{x}_{\text{gen}}).$$

Therefore,

$$\text{HOD}_{S,x}^{L[S,x,Z]}[G_f] \models L[S, x, \dot{x}_{\text{gen}}[G_f]] \models \varphi(S, x, \dot{x}_{\text{gen}}[G_f]).$$

So in particular,

$$V \models L[S, x, \dot{x}_{\text{gen}}[G_f]] \models \varphi(S, x, \dot{x}_{\text{gen}}[G_f]).$$

Since $R = \mathfrak{B}^2_{(S,\varphi)}$, $R(x, \dot{x}_{\text{gen}}[G_f])$ for each $f \in {}^\omega 2$. This shows that $A \subseteq R_x$ which completes the proof of the subclaim.

However, A is uncountable and R_x was assumed to be countable. Contradiction. This completes the proof of Claim 1.

Let $D \subseteq \mathbb{O}_{S,x}^{L[S,x,Z]}$ be the dense below p set that witnesses Claim 1. Take any $q \in D$ with $q \leq p$. Define $t \in {}^\omega 2$ by $t(n) = i \Leftrightarrow q \Vdash \dot{x}_{\text{gen}}(\check{n}) = \check{i})$. It is clear that $t \in \mathrm{HOD}_{S,x}^{L[S,x,Z]}$. Again since $\mathscr{P}(\mathbb{O}_{S,x}^{L[S,x,Z]})^{\mathrm{HOD}_{S,x}^{L[S,x,Z]}}$ is countable in V, let G be any $\mathbb{O}_{S,x}^{L[S,x,Z]}$-generic filter over $\mathrm{HOD}_{S,x}^{L[S,x,Z]}$ with $q \in G$ (and hence $p \in G$). By the forcing theorem, $\dot{x}_{\mathrm{gen}}[G] = t$. Using Fact 8.1 as above, one has $\mathrm{HOD}_{S,x}^{L[S,x,Z]}[G] \models L[S,x,t] \models \varphi(S,x,t)$. Hence $V \models L[S,x,t] \models \varphi(S,x,t)$. Thus $R(x,t)$. So $t \in R_x \cap \mathrm{HOD}_{S,x}^{L[S,x,Z]}$.

It has been shown that $R_x \cap \mathrm{HOD}_{S,x}^{L[S,x,Z]} \neq \emptyset$ for any $Z \in \mathcal{D}$ with $Z \geq_T [y^*]_T$. Lemma 8.3 implies that R has a uniformization. $\square$

In ZF: The Silver's dichotomy [**22**] states that for every $\mathbf{\Pi}^1_1$ equivalence E on $\mathbb{R}$, $\mathbb{R}/E$ injects into ω or $\mathbb{R}$ injects into $\mathbb{R}/E$. Burgess [**1**] showed that if E is a $\mathbf{\Sigma}^1_1$ equivalence relation, $\mathbb{R}/E$ injects into ω, $\mathbb{R}/E$ is in bijection with ω_1, or $\mathbb{R}$ inject into $\mathbb{R}/E$. Note that both results imply that either $\mathbb{R}/E$ is wellorderable or $\mathbb{R}$ injects into $\mathbb{R}/E$.

Woodin's perfect set dichotomy is an extension of this property to all equivalence relations on $\mathbb{R}$ assuming AD, all sets of reals have ∞-Borel codes, and the ultrapower, $\prod_{X \in D} \omega_1/\mu_{\mathcal{D}}$, is wellfounded. The wellfoundedness of the ultrapower is certainly a consequence of DC.

The proof of Woodin's perfect set dichotomy presented below follows the outline of Hjorth's [**9**] generalization of the [**8**] E_0-dichotomy in $L(\mathbb{R}) \models \mathsf{AD}$ and Harrington's proof of the Silver's dichotomy. Harrington's proof uses the Gandy-Harrington forcing of nonempty Σ^1_1 definable subsets of $\mathbb{R}$. The Vopěnka forcing is simply the OD version of the Gandy-Harrington forcing.

The argument presented below appears in [**6**] where the uniformity of this proof is needed to make further conclusions about wellordered disjoint unions of quotients of "smooth" equivalence relations with countable classes.

THEOREM 8.5. *(Woodin's perfect set dichotomy) Assume* $\mathsf{ZF} + \mathsf{AD} + \mathsf{V} = \mathsf{L}(\mathbb{R})$. *Let E be an equivalence relation on $\mathbb{R}$. Then either*

(i) $\mathbb{R}/E$ is wellorderable.

(ii) $\mathbb{R}$ inject into $\mathbb{R}/E$.

PROOF. Let E be an equivalence relation on $\mathbb{R}$. An E-component is a nonempty set A so that for all $x, y \in A$, $x \mathrel{E} y$. That is, an E-component is simply a nonempty subset of an E-class.

By Corollary 7.20, every set of reals has an ∞-Borel code. Let (S,φ) be an ∞-Borel code for E; that is, $E = \mathfrak{B}^2_{(S,\varphi)}$. Throughout this argument, E will always be considered as the set defined by the ∞-Borel code (S,φ).

(Case I) For all $X \in \mathcal{D}$, for all $a \in \mathbb{R}^{L[S,X]}$, there is some $\mathrm{OD}_S^{L[S,X]}$ E-component containing a.

In other words, for all degrees $X \in \mathcal{D}$, in the local model $L[S,X]$, every real belongs to an OD_S E-component.

For each $\alpha \in \prod_{X \in \mathcal{D}} \omega_1/\mu_{\mathcal{D}}$, let $f : \mathcal{D} \to \omega_1$ be such that f is a representative for α. Define $A_\alpha \subseteq \mathbb{R}$ as follows: $a \in A_\alpha$ if and only if for $\mu_{\mathcal{D}}$-almost all X, a belongs to the $f(X)^{\text{th}}$ $\mathrm{OD}_S^{L[S,X]}$ E-component according to the canonical wellordering of $\mathrm{HOD}_S^{L[S,X]}$. (If there is no $F(X)^{\text{th}}$ $\mathrm{OD}_S^{L[S,X]}$ E-component, then one just let this set be $\emptyset$.) One can check that A_α is well defined, independent of the choice of f representing α.

A_α is an E-component: To see this, let $a, b \in A_\alpha$. Thus there is a set $K \in \mu_\mathcal{D}$ so that for all $X \in K$, $[a \oplus b]_T \leq X$ and a and b both belong to the $f(X)^{\text{th}}$ OD_S E-component of $\text{HOD}_S^{L[S,X]}$. Thus, $L[S,X] \models a\ E\ b$. Since E is always defined by it ∞-Borel code, one has, $L[S,X] \models L[S,a,b] \models \varphi(S,a,b)$. Thus $V \models L[S,a,b] \models \varphi(S,a,b)$. Therefore, in V, $a\ E\ b$. It has been shown that A_α is an E-component.

For every $a \in \mathbb{R}$, there is some $\alpha \in \prod_{X\in\mathcal{D}} \omega_1/\mu_\mathcal{D}$ such that $a \in A_\alpha$: To see this. Define $f : \mathcal{D} \to \omega_1$ by letting $f(X)$ be the least β so that a belongs to the βth $\text{OD}_S^{L[S,X]}$ E-component of $L[S,X]$ according to the canonical wellordering of $\text{HOD}_S^{L[S,X]}$. This β exists due to the Case I assumption. Let $\alpha = [f]_{\mu_\mathcal{D}}$. Then $a \in A_\alpha$.

Since $L(\mathbb{R}) \models \mathsf{DC}$, $\prod_{X\in\mathcal{D}} \omega_1/\mu_\mathcal{D}$ is a wellordered set. Thus $\langle A_\alpha : \alpha \in \prod_{X\in\mathcal{D}} \omega_1/\mu_\mathcal{D}\rangle$ is a wellordered sequence of E-components with the property that every real belongs to some A_α. One can now wellorder $\mathbb{R}/E$ as follows: For two E-classes $u, v \in \mathbb{R}/E$, $u \prec v$ if and only if the least α so that $A_\alpha \subseteq u$ is less than the least α so that $A_\alpha \subseteq v$. Thus $\prec$ wellorders $\mathbb{R}/E$.

(Case II) There exists an $X \in \mathcal{D}$ and an $a \in \mathbb{R}^{L[S,X]}$ so that there is no $\text{OD}_S^{L[S,X]}$ E-component containing a.

In other word, there is a particular local model $L[S,X]$ so that within this model, there is a real a which does not belong to any OD_S E-component.

Fix such a degree $X \in \mathcal{D}$. One will always work in this local model $L[S,X]$.

In $L[S,X]$, define u be the collection of reals of $L[S,X]$ that do not belong to any $\text{OD}_S^{L[S,X]}$ E-component. Note that $u \neq \emptyset$ by the Case II assumption and u is $\text{OD}_S^{L[S,X]}$. Thus u is condition of $\mathbb{O}_S^{L[S,X]}$.

Let $\dot{x}_{\text{left}}$ and $\dot{x}_{\text{right}}$ be $\mathbb{O}_S^{L[S,X]} \times \mathbb{O}_S^{L[S,X]}$-names so that $\dot{x}_{\text{left}}$ and $\dot{x}_{\text{right}}$ is the evaluation of the $\mathbb{O}_S^{L[S,X]}$-name $\dot{x}_{\text{gen}}$ according to the left and right $\mathbb{O}_S^{L[S,X]}$-generic filter, respectively, coming from an $\mathbb{O}_S^{L[S,X]} \times \mathbb{O}_S^{L[S,X]}$-generic filter.

Claim 1: $\text{HOD}_S^{L[S,X]} \models (u,u) \Vdash_{\mathbb{O}_S^{L[S,X]}\times\mathbb{O}_S^{L[S,X]}} \neg(\dot{x}_{\text{left}}\ E\ \dot{x}_{\text{right}})$.

To see Claim 1: Suppose not. Then there is some $(v,w) \leq_{\mathbb{O}_S^{L[S,X]}\times\mathbb{O}_S^{L[S,X]}} (u,u)$ so that $\text{HOD}_S^{L[S,X]} \models (v,w) \Vdash_{\mathbb{O}_S^{L[S,X]}\times\mathbb{O}_S^{L[S,X]}} \dot{x}_{\text{left}}\ E\ \dot{x}_{\text{right}}$.

Subclaim 1.1: Suppose G_0 and G_1 are $\mathbb{O}_S^{L[S,X]}$-generic filter over $\text{HOD}_S^{L[S,X]}$ which belong to V and contain the condition v, then $\dot{x}_{\text{gen}}[G_0]\ E\ \dot{x}_{\text{gen}}[G_1]$. (Note that G_0 and G_1 are not necessarily mutually generic.)

To see Subclaim 1.1: As before, since $\text{HOD}_S^{L[S,X]} \models \mathsf{AC}$, $\mathscr{P}(\mathbb{O}_S^{L[S,X]})^{\text{HOD}_S^{L[S,X]}}$ is countable in the real world which is a model of AD. Therefore, one can find in the real world V, an $H \subseteq \mathbb{O}_S^{L[S,X]}$ which is $\mathbb{O}_S^{L[S,X]}$-generic over both $\text{HOD}_S^{L[S,X]}[G_0]$ and $\text{HOD}_S^{L[S,X]}[G_1]$ and $w \in H$.
Applying the forcing theorem, one has that $\text{HOD}_S^{L[S,X]}[G_0][H] \models \dot{x}_{\text{gen}}[G_0]\ E\ \dot{x}_{\text{gen}}[H]$ and $\text{HOD}_S^{L[S,X]}[G_1][H] \models \dot{x}_{\text{gen}}[G_1]\ E\ \dot{x}_{\text{gen}}[H]$. As E is defined by its ∞-Borel code (S,φ), one has

$$\text{HOD}_S^{L[S,X]}[G_0][H] \models L[S,\dot{x}_{\text{gen}}[G_0],\dot{x}_{\text{gen}}[H]] \models \varphi(S,\dot{x}_{\text{gen}}[G_0],\dot{x}_{\text{gen}}[H])$$

$$\text{HOD}_S^{L[S,X]}[G_1][H] \models L[S,\dot{x}_{\text{gen}}[G_1],\dot{x}_{\text{gen}}[H]] \models \varphi(S,\dot{x}_{\text{gen}}[G_1],\dot{x}_{\text{gen}}[H])$$

In particular

$$V \models L[S,\dot{x}_{\text{gen}}[G_0],\dot{x}_{\text{gen}}[H]] \models \varphi(S,\dot{x}_{\text{gen}}[G_0],\dot{x}_{\text{gen}}[H])$$

$$V \models L[S, \dot{x}_{\text{gen}}[G_1], \dot{x}_{\text{gen}}[H]] \models \varphi(S, \dot{x}_{\text{gen}}[G_1], \dot{x}_{\text{gen}}[H])$$

Thus in the real world, $\dot{x}_{\text{gen}}[G_0] \ E \ \dot{x}_{\text{gen}}[H]$ and $\dot{x}_{\text{gen}}[G_1] \ E \ \dot{x}_{\text{gen}}[H]$. By the transitivity of the equivalence relation E, $\dot{x}_{\text{gen}}[G_0] \ E \ \dot{x}_{\text{gen}}[G_1]$. This proves Subclaim 1.1.

Observe that there must be some $a, b \in v$ so that $\neg(a \ E \ b)$ since otherwise v would be an $\text{OD}_S^{L[S,X]}$ E-component containing a. This is impossible since $v \leq_{\mathbb{O}_S^{L[S,X]}} u$ implies that $v \subseteq u$ and u consists of those $a \in \mathbb{R}$ which do not belong to any $\text{OD}_S^{L[S,X]}$ E-component. Thus $p = v \times v \setminus E = \{(a,b) \in v \times v : \neg(a \ E \ b)\}$ is a nonempty $\text{OD}_S^{L[S,X]}$ subset of $\mathbb{R}^2$. Thus $p \in {}_2\mathbb{O}_S^{L[S,X]}$.

Let $\dot{x}^2_{\text{gen}}$ denote the generic element of $\mathbb{R}^2$ added by ${}_2\mathbb{O}_S^{L[S,X]}$. Let τ_0 and τ_1 be the ${}_2\mathbb{O}_S^{L[S,X]}$-names for the first and second coordinate of $\dot{x}^2_{\text{gen}}$. Using the ∞-Borel code (S, φ), the condition $q = \{(a,b) : \neg(a \ E \ b)\} = \{(a,b) : L[S,a,b] \models \neg\varphi(S,a,b)\}$ takes the form specified in Fact 8.1. Therefore, Fact 8.1 implies that $q \Vdash_{{}_2\mathbb{O}_S^{L[S,X]}} L[S, \tau_0, \tau_1] \models \neg\varphi(S, \tau_0, \tau_1)$.

As before, in the real world, find some $G \subseteq {}_2\mathbb{O}_S^{L[S,X]}$ containing p which is ${}_2\mathbb{O}_S^{L[S,X]}$-generic over $\text{HOD}_S^{L[S,X]}$. Since $p \leq_{{}_2\mathbb{O}_S^{L[S,X]}} q$, one has that $q \in G$. Hence

$$\text{HOD}_S^{L[S,X]}[G] \models L[S, \tau_0[G], \tau_1[G]] \models \neg\varphi(S, \tau_0[G], \tau_1[G]).$$

In particular,

$$V \models L[S, \tau_0[G], \tau_1[G]] \models \neg\varphi(S, \tau_0[G], \tau_1[G]).$$

Since (S, φ) is the ∞-Borel code for E, one has that $\neg(\tau_0[G] \ E \ \tau_1[G])$. However, $\tau_0[G]$ and $\tau_1[G]$ are just the two coordinates of $\dot{x}^2_{\text{gen}}[G]$, which is the generic element of $\mathbb{R}^2$ added by G. Fact 8.2 implies that $\tau_0[G]$ and $\tau_1[G]$ are individually $\mathbb{O}_S^{L[S,X]}$-generic filters over $\text{HOD}_S^{L[S,X]}$. Then Subclaim 1.1 implies that $\tau_0[G] \ E \ \tau_1[G]$. Contradiction. This shows Claim 1.

As before, $\mathscr{P}(\mathbb{O}_S^{L[S,X]} \times \mathbb{O}_S^{L[S,X]})^{\text{HOD}_S^{L[S,X]}}$ is countable in the real world. Let $\langle D_n : n \in \omega \rangle$ enumerate all the dense open subsets of $\mathbb{O}_S^{L[S,X]} \times \mathbb{O}_S^{L[S,X]}$ which belong to $\text{HOD}_S^{L[S,X]}$. By intersecting, one may assume that $D_{n+1} \subseteq D_n$ for all $n \in \omega$. Now in the usual manner, one will build a perfect tree of conditions whose distinct branches correspond to mutually generic filters. The details follow:

Let $p_\emptyset$ be any condition below u. Suppose for some $n \in \omega$, p_σ has been defined for all $\sigma \in {}^n 2$. For each $\sigma \in {}^n 2$, $(p_\sigma, p_\sigma) \leq_{\mathbb{O}_S^{L[S,X]} \times \mathbb{O}_S^{L[S,X]}} (u,u)$ and Claim 1 implies that there must be some $q_0 \leq_{\mathbb{O}_S^{L[S,X]}} p_\sigma$ and $q_1 \leq_{\mathbb{O}_S^{L[S,X]}} p_\sigma$ which are incompatible conditions. Let $(\rho_{\sigma\hat{}0}, \rho_{\sigma\hat{}1})$ denote the $\text{HOD}_S^{L[S,X]}$-least such pair (q_0, q_1).

For $\sigma \in {}^{n+1}2$, let $\rho_\sigma^{-1} = \rho_\sigma$. Let $\langle \sigma_i : i < 2^{n+1} \rangle$ enumerate ${}^{n+1}2$.

Now suppose that ρ_σ^j has been defined for all $\sigma \in {}^{n+1}2$ and $-1 \leq j < 2^{n+1}$. By considering all possible pairs and extending by density, find a collection $\{\rho_{\sigma_i}^{j+1} : 0 \leq i < 2^{n+1}\}$ with the property that

(1) $\rho_\sigma^{j+1} \leq_{\mathbb{O}_s^{L[S,X]}} \rho_\sigma^j$

(2) $(\rho_{\sigma_{j+1}}^{j+1}, \rho_{\sigma_\ell}^{j+1}) \in D_n$ and $(\rho_{\sigma_\ell}^{j+1}, \rho_{\sigma_{j+1}}^{j+1}) \in D_n$ whenever $\ell \neq j+1$.

For $\sigma \in {}^{n+1}2$, let $p_\sigma = \rho_\sigma^{2^{n+1}-1}$.

This defines a sequence $\langle p_\sigma : \sigma \in {}^{<\omega}2 \rangle$. For each $f \in {}^\omega 2$, let G_f be the $\leq_{\mathbb{O}_S^{L[S,X]}}$-upward closure of $\{p_{f \restriction n} : n \in \omega\}$. By construction, if $f \neq g$, then $G_f \times G_g$

is an $\mathbb{O}_S^{L[S,X]} \times \mathbb{O}_S^{L[S,X]}$-generic filter over $\mathrm{HOD}_S^{L[S,X]}$ containing the condition (u, u). Therefore, by Claim 1, for any $f, g \in {}^\omega 2$ with $f \neq g$, $\neg(\dot{x}_{\mathrm{gen}}[G_f] \; E \; \dot{x}_{\mathrm{gen}}[G_g])$. Thus $A = \{\dot{x}_{\mathrm{gen}}[G_f] : f \in {}^\omega 2\}$ is a perfect set of E-inequivalent reals. Thus $\Phi : \mathbb{R} \to \mathbb{R}/E$ defined by $\Phi(f) = [\dot{x}_{\mathrm{gen}}[G_f]]_E$ is an injection. □

COROLLARY 8.6. *Assume* $\mathsf{ZF} + \mathsf{AD} + \mathsf{V} = \mathsf{L}(\mathbb{R})$. *If X is a surjective image of $\mathbb{R}$ (equivalently $X \in L_\Theta(\mathbb{R})$), then X is either wellorderable or $\mathbb{R}$ injects into X.*

In fact, as observed by [**2**], in $L(\mathbb{R})$, this dichotomy actually holds for all sets X in $L(\mathbb{R})$ not just $X \in L_\Theta(\mathbb{R})$.

Acknowledgements

The author would like to thank Stephen Jackson for the numerous conversations concerning the topics that appear here. The author would also like to thank Thilo Wienert for carefully reading this article and suggesting many comments.

References

[1] John P. Burgess, *Effective enumeration of classes in a Σ^1_1 equivalence relation*, Indiana Univ. Math. J. **28** (1979), no. 3, 353–364, DOI 10.1512/iumj.1979.28.28024. MR529670

[2] Andrés Eduardo Caicedo and Richard Ketchersid, *A trichotomy theorem in natural models of* AD^+, Set theory and its applications, Contemp. Math., vol. 533, Amer. Math. Soc., Providence, RI, 2011, pp. 227–258, DOI 10.1090/conm/533/10510. MR2777751

[3] William Chan and Stephen Jackson, *Applications of infinity-borel codes to definability and definable cardinals*, In Preparation.

[4] William Chan and Stephen Jackson, *Definable combinatorics at the first uncountable cardinal*, In preparation.

[5] William Chan and Stephen Jackson, $\mathbf{L}(\mathbb{R})$ *with determinacy satisfies the Suslin hypothesis*, Adv. Math. **346** (2019), 305–328, DOI 10.1016/j.aim.2019.02.012. MR3910797

[6] William Chan and Stephen Jackson, *Cardinality of Wellordered Disjoint Unions of Quotients of Smooth Equivalence Relations*, `arXiv:1903.03902`.

[7] William Chan, Stephen Jackson, and Nam Trang, *More definable combinatorics around the first and second uncountable cardinal*, In preparation.

[8] L. A. Harrington, A. S. Kechris, and A. Louveau, *A Glimm-Effros dichotomy for Borel equivalence relations*, J. Amer. Math. Soc. **3** (1990), no. 4, 903–928, DOI 10.2307/1990906. MR1057041

[9] Greg Hjorth, *A dichotomy for the definable universe*, J. Symbolic Logic **60** (1995), no. 4, 1199–1207, DOI 10.2307/2275883. MR1367205

[10] Steve Jackson, *A new proof of the strong partition relation on ω_1*, Trans. Amer. Math. Soc. **320** (1990), no. 2, 737–745, DOI 10.2307/2001700. MR972702

[11] ———, *A computation of δ^1_5*, Mem. Amer. Math. Soc. **140** (1999), no. 670, viii+94. MR1606035

[12] ———, *Structural consequences of AD*, Handbook of set theory. Vols. 1, 2, 3, Springer, Dordrecht, 2010, pp. 1753–1876. MR2768700

[13] Thomas Jech, *Set theory*, Springer Monographs in Mathematics, Springer-Verlag, Berlin, 2003. The third millennium edition, revised and expanded. MR1940513

[14] Alexander S. Kechris, AD *and projective ordinals*, Cabal Seminar 76–77 (Proc. Caltech-UCLA Logic Sem., 1976), Lecture Notes in Math., vol. 689, Springer, Berlin, 1978, pp. 91–132. MR526915

[15] Alexander S. Kechris, *The axiom of determinacy implies dependent choices in $L(\mathbf{R})$*, J. Symbolic Logic **49** (1984), no. 1, 161–173, DOI 10.2307/2274099. MR736611

[16] Alexander S. Kechris, *Classical descriptive set theory*, Graduate Texts in Mathematics, vol. 156, Springer-Verlag, New York, 1995. MR1321597

[17] Alexander S. Kechris and Yiannis N. Moschovakis, *Notes on the theory of scales*, Cabal Seminar 76–77 (Proc. Caltech-UCLA Logic Sem., 1976), Lecture Notes in Math., vol. 689, Springer, Berlin-New York, 1978, pp. 1–53. MR526913

[18] Alexander S. Kechris and W. Hugh Woodin, *Generic codes for uncountable ordinals, partition properties, and elementary embeddings*, Games, scales, and Suslin cardinals. The Cabal Seminar. Vol. I, Lect. Notes Log., vol. 31, Assoc. Symbol. Logic, Chicago, IL, 2008, pp. 379–397, DOI 10.1017/CBO9780511546488.019. MR2463619

[19] Richard Ketchersid, Paul Larson, and Jindřich Zapletal, *Ramsey ultrafilters and countable-to-one uniformization*, Topology Appl. **213** (2016), 190–198, DOI 10.1016/j.topol.2016.08.018. MR3563079

[20] Eugene M. Kleinberg, *Infinitary combinatorics and the axiom of determinateness*, Lecture Notes in Mathematics, Vol. 612, Springer-Verlag, Berlin-New York, 1977. MR0479903

[21] Itay Neeman and Jindřich Zapletal, *Proper forcings and absoluteness in* $L(\mathbf{R})$, Comment. Math. Univ. Carolin. **39** (1998), no. 2, 281–301. MR1651950

[22] Jack H. Silver, *Counting the number of equivalence classes of Borel and coanalytic equivalence relations*, Ann. Math. Logic **18** (1980), no. 1, 1–28, DOI 10.1016/0003-4843(80)90002-9. MR568914

[23] Robert M. Solovay, *A nonconstructible* Δ^1_3 *set of integers*, Trans. Amer. Math. Soc. **127** (1967), 50–75, DOI 10.2307/1994631. MR211873

Department of Mathematics, University of North Texas, Denton, Texas 76203
Email address: `William.Chan@unt.edu`

Contemporary Mathematics
Volume **752**, 2020
https://doi.org/10.1090/conm/752/15129

Isomorphism and weak conjugacy of free Bernoulli subflows

John D. Clemens

Abstract. We show that for G a countable group which is not locally finite, the isomorphism of free G-subflows is bi-reducible with with the universal countable Borel equivalence relation E_∞, a result obtained independently by Gao-Jackson-Seward. We also show that the same result holds for the relation of weak conjugacy and for any intermediate Borel equivalence relation.

For a countable group G, the *Bernoulli flow* 2^G consists of all functions from G to 2 together with the *shift action* of G given by $(g \cdot \alpha)(h) = \alpha(g^{-1}h)$. A *$G$-subflow* is a compact shift-invariant subset $X \subseteq 2^G$. Two subflows X and Y are *isomorphic*, $X \cong Y$, if there is a homeomorphism $\varphi : X \to Y$ which commutes with the shift, i.e., $\varphi(g \cdot \alpha) = g \cdot \varphi(\alpha)$ for all $\alpha \in 2^G$ and $g \in G$. We say that Y is a *factor* of X if there is a continuous surjection of X onto Y which commutes with the shift, and X and Y are *weakly conjugate* if each is a factor of the other. Various classes of subflows are of particular interest. A subflow X is *free* if $g \cdot \alpha \neq \alpha$ for all $g \neq e$ in G and $\alpha \in X$. A subflow is *minimal* if it contains no proper subflow, i.e., for all $\alpha \in X$, the *orbit closure* of α, $\overline{\mathcal{O}}_\alpha = \overline{G \cdot \alpha}$, is equal to X.

In [**1**] it was shown that the isomorphism problem for all n-dimensional subshifts (i.e., $\mathbb{Z}^n$-subflows) is bi-reducible with E_∞, the universal countable Borel equivalence relation, for each $n \geq 1$. The construction there made use of periodic points in a fundamental way, producing subshifts which are not free, let alone minimal. The authors of [**2**] asked to determine the exact complexity of the isomorphism relation for several classes of G-subflows for various countable groups G, namely the classes of all subflows, of free subflows, and of minimal subflows. This complexity is to be gauged using the notion of Borel reducibility of equivalence relations, explained below.

Here we prove that the same result holds for the collection of free G-subflows for any G which is not locally finite, a result obtained independently by Gao-Jackson-Seward in [**3**], where G is *locally finite* if for any finite $F \subseteq G$, the subgroup $\langle F \rangle$ generated by F is finite. We also show that the same result holds for the relation of weak conjugacy of G-subflows for any G which is not locally finite, and for any Borel equivalence relation intermediate between isomorphism and weak conjugacy. Although the subflows we produce here are free, they contain isolated points and hence are not minimal. The restriction to non-locally finite G is necessary; in [**3**] it is shown that for G locally finite, the isomorphism of minimal free G-subflows or of all

2010 *Mathematics Subject Classification.* Primary 03E15, 54H20.

G-subflows is bi-reducible with E_0. In [**8**] it is shown that isomorphism of Toeplitz G-subshifts (which are free and minimal) is not hyperfinite if G is residually finite and non-amenable, but the general problem for minimal subflows remains open.

In the next section we briefly review important notions from the theory of Borel equivalence relations and basic facts about subflows. In the following section we prove the main result.

1. Countable Borel equivalence relations and G-subflows

A *Borel equivalence relation* is an equivalence relation E on a Polish space X such that E is Borel as a subset of X^2. A standard method of comparing Borel equivalence relations is via the Borel reducibility relation. We say that an equivalence relation E on a space X is *Borel reducible* to an equivalence relation F on a space Y, $E \leq_B F$, if there is a Borel-measurable function $f : X \to Y$ such that for all $x_1, x_2 \in X$ we have $x_1 \, E \, x_2$ if and only if $f(x_1) \, F \, f(x_2)$. Note that having $E \leq_B F$ amounts to having a definable injection from the quotient space X/E into Y/F, so that when $E \leq_B F$ we may view the relation F as being at least as complicated as E.

We say that an equivalence relation E is *countable* if every equivalence class of E is countable. We say E is a *universal countable Borel equivalence relation* if E is a countable Borel equivalence relation and for every countable Borel equivalence relation F we have $F \leq_B E$. Several such examples are known; the one which we will use here is the *(left) shift relation* of the free group on two generators $\mathbb{F}_2$ on the space $2^{\mathbb{F}_2}$, denoted $E(\mathbb{F}_2, 2)$, so that $xE(\mathbb{F}_2, 2)y$ iff there is $g \in \mathbb{F}_2$ so that for all $h \in \mathbb{F}_2$ we have $y(h) = x(g^{-1}h)$. A good introduction to the theory of countable Borel equivalence relations is [**5**].

We introduce the equivalence relations we will study here.

DEFINITION 1. A *homomorphism* of G-subflows X and Y is a continuous surjection $\varphi : X \to Y$ commuting with the shift action of G. We say that Y is a *factor* of X when there is a homomorphism from X to Y. We say that X and Y are *weakly conjugate* when each is a factor of the other, and we say X and Y are *isomorphic* if there is an injective homomorphism between them.

Isomorphic subflows are weakly conjugate, but isomorphism is a stronger relation. A basic observation (see [**7**]) is:

THEOREM (Curtis-Hedlund-Lyndon). *If $\varphi : X \to Y$ is a homomorphism of G-subflows, then φ is given by a block code, i.e., there is a finite $B \subseteq G$ and a function $\pi : 2^B \to 2$ such that $\varphi(x)(h) = \pi((h^{-1} \cdot x) \restriction B)$.*

The following lemma provides a necessary and sufficient condition for minimality:

LEMMA (Gao-Jackson-Seward). *An element $\alpha \in 2^G$ is minimal (i.e., $\overline{\mathcal{O}}_\alpha$ is a minimal subflow) iff for every finite $A \subseteq G$ there is a finite $T \subseteq G$ such that*

$$\forall g \in G \exists t \in T \forall a \in A(\alpha(gta) = \alpha(a)),$$

i.e., α is uniformly recurrent.

DEFINITION 2. By a *pattern* we mean $u \in 2^{<G}$, a finite partial function from G to 2. We say a pattern u *occurs* in $\alpha \in 2^G$ (or in $v \in 2^{<G}$), $u \prec \alpha$ at h if α extends

$h \cdot u$. G acts on patterns by shift. For a pattern u and $\alpha \in 2^G$ we define $\alpha \circledast u \in 2^G$ by

$$\alpha \circledast u(h) = \begin{cases} u(h) & \text{if } h \in \mathrm{dom}(u) \\ \alpha(h) & \text{otherwise} \end{cases}$$

and similarly for multiple patterns (we will always use patterns with disjoint domains). We say that a pattern p *recurs uniformly* in an element of 2^G, the above condition holds with A equal to the domain of p.

We now explain how to view the isomorphism relation on subflows as a countable Borel equivalence relation.

DEFINITION 3. For a countable group G, we let S_G be the set of closed, S-invariant subsets of 2^G, where S is the left shift action. We view this as a subspace of the compact Polish space $K(2^G)$ equipped with the Vietoris topology (see [**6**]).

It is straightforward to note that S_G is a closed subspace of $K(2^G)$. We can also code subflows via a set of forbidden patterns. We now consider the isomorphism relation on the spaces S_G.

DEFINITION 4. For $X, Y \in S_G$, we set $X \, E \, Y$ if $(X, S) \cong (Y, S)$, i.e., there is a homeomorphism of X and Y which commutes with S. We set $X \, E_w \, Y$ if X and Y are weakly conjugate.

It is again straightforward to verify that E and E_w are countable Borel equivalence relations on each S_G.

2. Isomorphism of free G-subflows

We will establish the following result, which gives that both isomorphism and weak conjugacty of free G-subflows are universal countable Borel equivalence relations when G is not locally finite.

THEOREM 1. *Let G be a countable group which is not locally finite. Let E be any Borel equivalence relation containing the isomorphism relation of free G-subflows and contained in weak conjugacy of free G-subflows. Then E is a universal countable Borel equivalence relation.*

The theorem is an immediate consequence of:

THEOREM 2. *There is a Borel map $\psi : x \mapsto X_x$ from $2^{\mathbb{F}_2}$ to S_G so that if $xE(\mathbb{F}_2, 2)y$ then $X_x \, E \, X_y$, and if $\neg xE(\mathbb{F}_2, 2)y$ then X_x is not a factor of X_y and X_y is not a factor of X_x.*

The rest of the paper will be devoted to the proof of this theorem.

Encoding orbits in G-subflows. We begin with some group-theoretic preliminaries. Fix a countable group G which is not locally finite. We develop a suitable sequence of group elements which will be used to encode information in the subflows we define below. Fix a finite symmetric $F \subseteq G$ such that $H = \langle F \rangle \subseteq G$ is infinite; we can assume that no proper symmetric subset of F generates an infinite subgroup. Let $F = \{f_0, \ldots, f_{N-1}\}$ and define the following:

$$F^n = \text{all products of } n \text{ elements of } F$$
$$F^{<n} = \bigcup_{m<n} F^m$$
$$\widetilde{F}^n = F^n \setminus F^{<n}$$

Thus F^n and $F^{<n}$ are the closed and open balls of radius n in the Cayley graph of H generated by F, and $H = \bigcup_{n=0}^{\infty} F^n$. Since H is infinite we must have $\widetilde{F}^n \neq \emptyset$ for all n.

LEMMA 1. *There is a sequence $\langle i(n) : n \in \mathbb{Z}\rangle$ with $0 \leq i(n) < N$ such that if we let $s_0 = e$ and $s_{n+1} = s_n \cdot f_{i(n)}$ for $n \geq 0$ and $s_{n-1} = s_n \cdot f_{i(n-1)}^{-1}$ for $n \leq 0$ and $S = \{s_n : n \in \mathbb{Z}\}$ then we have the following:*

(1) *$s_n \neq s_m$ for $n \neq m$.*
(2) *For all n and m with $m < n$, $s_m^{-1} \cdot s_n = f_{i(m)} \cdot \cdots \cdot f_{i(n-1)} \notin F^{<(n-m)}$.*

PROOF. If some f_i has infinite order we can simply let $i(n) = i$ for all n, so suppose not. Define a tree T whose nodes are finite sequences of elements of F as follows. Let the root of T be $T_0 = \{\langle e\rangle\}$, and for $n \geq 1$ let

$$T_n = \{\langle a_{-n}, \ldots a_0, a_1, \ldots, a_{n-1}\rangle : a_i \in F \wedge a_{-n} \cdot \cdots \cdot a_0 \cdot a_1 \cdot \cdots \cdot a_{n-1} \notin F^{<2n}\},$$

with T ordered by extension. T is a tree (i.e., it is closed under restrictions) since if $a_{-m} \cdot \cdots \cdot a_0 \cdot a_1 \cdot \cdots \cdot a_{m-1} \in F^{<2m}$ for some $m < n$ then $a_{-n} \cdot \cdots \cdot a_0 \cdot a_1 \cdot \cdots \cdot a_{n-1} \in F^{<2n}$. T is finitely branching since F is finite, and T is infinite since $\widetilde{F}^{2n} \neq 0$ for all n. Hence T contains an infinite branch which defines a bi-infinite sequence $\beta \in F^{\mathbb{Z}}$. Take $i(n) = \beta(n)$ and define s_n accordingly. Now if $m < n$, let $k = \max(|m|, |n|)$. If $s_m^{-1} \cdot s_n \in F^{<(m-n)}$, then $s_{-k}^{-1} \cdot s_k \in F^{<2k}$ contradicting our construction of T. □

LEMMA 2. *For any finite $T \subseteq G$ there is a $k \geq 1$ such that the sets $\{s_{kn} \cdot T : n \in \mathbb{Z}\}$ are pairwise disjoint.*

PROOF. Choose k such that $H \cap TT^{-1} \subseteq F^{<k}$. Then if $s_{km} \cdot T \cap s_{kn} \cdot T$ for $m < n$ we have $t_1, t_2 \in T$ such that $s_{km}t_1 = s_{kn}t_2$ so $s_{km}^{-1} \cdot s_{kn} = t_1 t_2^{-1} \in H \cap TT^{-1}$, contradicting that $s_{km}^{-1} \cdot s_{kn} \notin F^{<(kn-km)}$. □

Next, fix $\mu_0 \in 2^G$ such that the orbit closure of μ_0, $\overline{\mathcal{O}}_{\mu_0}$, is a free minimal G-subflow, the existence of which follows from [**2**] and [**4**]. Let $m_0 \in 2^{\mathbb{Z}}$ generate a free minimal subshift, e.g., we can take m_0 to be the bi-infinite Thue-Morse sequence.

LEMMA 3. *There is a finite $T_0 \subseteq G$ with $F \cup \{e\} \subseteq T_0$ such that neither of the patterns with constant value 0 or constant value 1 on T_0 occurs in μ_0, and for any $g \in G$ with $g \neq e$ we have $g \cdot T_0 \neq T_0$.*

PROOF. We can find a T such that the constant patterns on T do not occur in μ_0 by freeness of $\overline{\mathcal{O}}_{\mu_0}$; we may assume that $F \cup \{e\} \subseteq T$. Now choose h such that $h \notin T^2 \cup TT^{-1}$ and let $T_0 = T \cup \{h\}$. Since $e \in T_0$, if $g \notin T_0$ then $g \cdot T_0 \neq T_0$ since $g \in g \cdot T_0$. We have that $h \cdot T_0$ is disjoint from T_0 which has size at least 2, so $h \cdot T_0 \neq T_0'$. Finally, if $g \in T$ with $g \neq e$ then $h \notin g \cdot T$ (since $h \notin T^2$) and $g \cdot h \neq h$ so $h \notin g \cdot T_0$, whence $g \cdot T_0 \neq T_0$. □

In particular, any translate of T_0 in μ_0 contains both a 0 and a 1. Let $T_0' = T_0^2 \cup T_0T_0^{-1}$ and define the pattern d with domain T_0' with $d \restriction T_0 = 1$ and $d \restriction (T_0' \backslash T_0) = 0$. Let $d_1 = d \restriction T_0$.

LEMMA 4. *Any two occurrences of the pattern d in some element $\alpha \in 2^G$ can not overlap on the T_0-block of 1's, and if we replace some T_0'-block in μ_0 by a copy of d then the only occurrence of d is the one we added.*

PROOF. Suppose first that α is the result of replacing a T_0-block in μ_0 by d, and suppose we added the copy of d at g_1 but d also occurs at some $g_2 \neq g_1$. Then $g_1T_0 \cap g_2T_0 \neq \emptyset$ since d_1 does not occur in μ_0, so we have overlapping T_0-blocks. Now if we have any α with d occurring at g_1 and g_2 with overlapping T_0-blocks (i.e., $g_1 \cdot T_0 \cap g_2 \cdot T_0 \neq \emptyset$), then $g_2 \in g_1T_0T_0^{-1} \subseteq g_1T_0'$. If $g_2 \notin g_1T_0$ then we should have $\alpha(g_2) = 0$ from the occurrence of d at g_1, whereas $\alpha(g_2) = 1$ from the occurrence at g_2 since $e \in T_0$. Hence $g_2 \in g_1T_0$. From the properties of T_0 we must have $g_2 \cdot T_0 \neq g_1 \cdot T_0$; hence there is $h \in g_2 \cdot T_0 \setminus g_1 \cdot T_0$. But now $h \in g_1 \cdot T_0^2 \subseteq g_1 \cdot T_0'$ so we should have $\alpha(h) = 0$ from the occurrence of d at g_1, whereas $\alpha(h) = 1$ from the occurrence of d at g_2, a contradiction. $\square$

Let k_0 from Lemma 2 be such that the sets $\{s_{k_0n} \cdot T_0' : n \in \mathbb{Z}\}$ are pairwise disjoint, and denote $s_{k_0}, \ldots, s_{9k_0}$, respectively, by

$$c_{i(0)}, c_{i(1)}, c_{d(0)}, c_{d(1)}, c_e, c_a, c_{a^{-1}}, c_b, c_{b^{-1}},$$

so that these are elements of G with pairwise disjoint T_0'-translates. Let $T_1 = \bigcup_{n=0}^{9} s_{k_0n} \cdot T_0'$ be the union of T_0' and all of the above T_0'-translates, and define patterns with domain T_1,

$$u_{i(0)}, u_{i(1)}, u_{d(0)}, u_{d(1)}, u_e, u_a, u_{a^{-1}}, u_b, u_{b^{-1}},$$

to be equal to d on T_0' and 0 on all the other listed translates of T_0' except that the corresponding element c should have value 1 (e.g., $u_a(c_a) = 1$). Hence, each such pattern encodes one of $i(0)$, $i(1)$, $d(0)$, $d(1)$, e, a, a^{-1}, b, or b^{-1}. None of these patterns occur in μ_0, and any two occurrences of any of these patterns do not overlap on the T_0-block of all 1's. Also, if we replace disjoint T_1-blocks in μ_0 by copies of these patterns, the T_0-block of 1's in each can occur only at the T_0-block in d.

We now want to extend T_1 so that we can encode "links" to neighboring patterns. Let n_0 be such that $H \cap T_1 \subseteq F^{<n_0}$, and let $T_1^{(n)} = T_1 \cup F^{<n}$ for $n > n_0$. For a pattern u with domain T_1 and $p \subseteq F^{<n} \setminus T_1$ we define the pattern u^p on $T_1^{(n+n_0)}$ to be equal to u_0 on T_1, 1 on p, and 0 elsewhere. Note that since $H \cap T_1 \subseteq F^{<n_0}$, if $d \prec \mu_0 \circledast u^p$ then $d \prec u^p$. Let

$$C(n) = \{p \subseteq F^{<n} \setminus T_1 : d \text{ only occurs in } d^p \text{ at } T_1\}.$$

Note that if $h \in \widetilde{F}^{2n+2n_0}$ then $h \cdot T_1^{(n+n_0)} \cap T_1T_1^{(n+n_0)} = \emptyset$.

We want to choose n large enough so that we can encode two elements of $\widetilde{F}^{2n+2n_0}$ in $T_1^{(n+n_0)}$ using elements of $C(n)$. We claim that $|C(n)| \geq 2^{\frac{1}{2} \cdot |F^{<n} \setminus T_1|}$. To see this, observe that if $p \subseteq \bigcup_{\text{odd } k < n} (\widetilde{F}^k \setminus T_1)$ or $p \subseteq \bigcup_{\text{even } k < n} (\widetilde{F}^k \setminus T_1)$ then d only occurs in d^p at T_1 since $F \subseteq T_0$, and at least one of these two sets has size at least $\frac{1}{2} \cdot |F^{<n} \setminus T_1|$. Also,

$$|\widetilde{F}^{2n+2n_0}|^2 \leq (|F^{2n_0+2}| \cdot |F^{n-1}| \cdot |F^{n-1}|)^2 \leq |F^{2n_0+2}|^2 \cdot |F^{<n}|^4.$$

Hence we just need n such that

$$2^{\frac{1}{2}\cdot(|F^{<n}|-|T_1|)} \geq |F^{2n_0+2}|^2 \cdot |F^{<n}|^4, \text{ i.e.,}$$
$$2^{\frac{1}{2}\cdot|F^{<n}|} \geq 2^{\frac{1}{2}\cdot|T_1|} \cdot |F^{2n_0+2}|^2 \cdot |F^{<n}|^4.$$

This will be true for sufficiently large n since $|F^{<n}| \to \infty$ and the exponential on the left hand side dominates any polynomial.

Choose n_1 with this property and let $\tilde{n} = 2(n_1 + n_0)$. For each pair of elements (h_0, h_1) in $\widetilde{F^{\tilde{n}}}$ we can choose a distinct $p(h_0, h_1) \in C(n_1)$. For each of the 9 patterns u introduced above and $n \geq 1$ let $u^n = u^{p_n}$, where $p_n = p(s_{\tilde{n}(n-1)}^{-1} s_{\tilde{n}n}, s_{\tilde{n}n}^{-1} s_{\tilde{n}(n+1)})$, and let $u^0 = u^{p(e, s_{\tilde{n}})}$. These are patterns with domain $T_2 = T_1 \cup F^{<(n_1+n_0)}$. Let $g_0(n) = s_{\tilde{n}\cdot n}$ for $n \in \mathbb{Z}$, and let $g_1(n) = g_0(2n)$.Then $\{g_0(n) \cdot T_2 : n \in \mathbb{N}\}$ are pairwise disjoint.

Let

$$\mu = \mu_0 \circledast_{n\in\mathbb{N}} \left(g_0(2n) \cdot u_{i(0)}^{2n} \circledast g_0(2n+1) \cdot u_{i(m_0(n))}^{2n+1} \right),$$

i.e., we encode the positive part of the minimal element m_0 into odd g_0-translates of T_1, "clear" even translates, and encode "links" to the previous and next translate. Note that the subflow generated by μ is still free. although not minimal.

Defining the subflow. Let $\{\pi_j : 2^{B_j} \to 2\}$ enumerate all of the block codes, with $B_j \subseteq G$ finite, and let $\varphi_j : 2^G \to 2^G$ be the induced mapping. We can assume that the B_j's are increasing and that each B_j is symmetric and contains e. For a finite $T \subseteq G$ we call $T \cdot B$ the *B-saturation of T*; note that when φ is induced by the block code π, then $\varphi(\alpha) \restriction T$ is completely determined by $\alpha \restriction T \cdot B$.

Choose a sufficiently quickly increasing sequence $\{q_n : n \in \mathbb{N}\}$ so that (with $q_{-1} = 0$)

$$H \cap \bigcup_{j\leq n, 0\leq m\leq 2\tilde{n}q_{n-1}} (T_1 B_j B_j^{-1} T_1^{-1} F^m T_1 B_j B_j^{-1} T_1^{-1}) \subseteq F^{<2\tilde{n}q_n}.$$

For $j \in \mathbb{N}$ and $i \in 2$ define:

$$v_i^j = (\mu \circledast u_{d(i)}^0) \restriction T_2 B_j$$
$$p_i^j = \mu_0 \circledast v_i^j \circledast g_1(q_{2j+i}) \cdot v_i^j$$

and let

$$P = \{p_i^j : j \in \mathbb{N}, i \in 2\}.$$

Note that the two copies of v_i^j are disjoint since $g_1(q_{2j+i}) \in \widetilde{F^{2\tilde{n}q_{2j+i}}}$ which is disjoint from $T_1 B_j B_j^{-1} T_1^{-1}$. The point in choosing the q_n's is that the two copies of v_i^j in p_i^j are sufficiently far apart that any block code using the block B_j (or a smaller block) must act "locally" on each copy.

Let a and b generate $\mathbb{F}_2 = \{w_i : i \in \mathbb{N}\}$. For a reduced word $w = r_0 r_1 \cdots r_n$ in a, a^{-1}, b, b^{-1} (and with $r_0 = e$ for $w = e$) and $j \in \mathbb{N}$ we define the pattern $\rho^j(w)$ by

$$\rho^j(w) = g_1(j) \cdot u_{i(0)}^{2j} \left(\circledast_{k=0}^{n} g_1(j+1+k) \cdot u_{r_k}^{2(j+1+k)} \right) \circledast g_1(j+2+n) \cdot u_{i(0)}^{2(j+2+n)}.$$

We next define a function $k \mapsto n_k$ from $\mathbb{N}$ to $\mathbb{N}$ as follows. Partition $\mathbb{N}$ into disjoint infinite sets N_j for $j \in \mathbb{N}$, and let $e(n) = (n)_0$ be the first coordinate of the pairing function $n = \langle (n)_0, (n)_1 \rangle$, so that $\langle e(n) : n \in \mathbb{N} \rangle$ enumerates each natural number infinitely often. Define $\mu_k = \mu \circledast \rho^{1+k^2}(e)$, and for each j and k let

$$i_k^j = \begin{cases} \text{least } i \text{ such that } \rho^l(w_i) \prec \varphi_j(\mu_k) \text{ for some } l \\ \infty \text{ is no such } i \text{ exists} \end{cases}$$

$$i^j = \begin{cases} \text{least } i \text{ such that } \{k \in N_j : i_k^j = i\} \text{ is infinite, if such exists} \\ \infty \text{ otherwise.} \end{cases}$$

We will construct our subflows so that for φ_j's potentially inducing homomorphisms between them, there will always be a unique i satisfying the conditions for i_k^j, and there will always be only finitely many i_k^j's for a given j (since occurrences of some $\rho(w)$ will have to be induced by the B_j-saturation of an occurrence of $\rho(e)$), so i^j will be finite. Now let $\{k_l^j : l \in \mathbb{N}\}$ be an increasing enumeration of the set

$$\widetilde{N}_j = \{k \in N_j : i_k^j = i^j\}$$

when i^j is finite (and simply enumerate N_j when $i^j = \infty$). We can now define n_k for $k \in \mathbb{N}$ as

$$n_k = \begin{cases} e(l) & \text{if } k = k_l^j \text{ for some } j \text{ and } l \\ k & \text{otherwise.} \end{cases}$$

The point is that for appropriate j, the set $\{n_k : k \in \widetilde{N}_j\}$ will enumerate each element of $\mathbb{N}$ infinitely often. We are now ready to define our subflows.

DEFINITION 5. For $x \in 2^{\mathbb{F}_2}$ we define the G-subflow $\psi(x) = X_x$ as the orbit closure of the set $A_x = P \cup Q_x$, where Q_x is given by

$$Q_x = \{\mu \circledast u^0_{d(w_i \cdot x(w_{n_k}))} \circledast \rho^{1+k^2}(w_i) : i, k \in \mathbb{N}\}.$$

Note that the points of A_x (and shifts thereof) are the isolated points in X_x. The role of the points in P is to force a block code implementing a homomorphism to preserve certain patterns, and the role of Q_x is to encode the $\mathbb{F}_2$-orbit of x. The map $\psi : x \in 2^{\mathbb{F}_2} \to X_x$ can be implemented in a Borel way as in [**1**]. We verify that these are all free G-subflows.

LEMMA 5. *The subflow $\psi(x) = X_x$ is free.*

PROOF. Suppose we have $\alpha \in X_x$ with $g \cdot \alpha = \alpha$ for some $g \neq e$. Since $\overline{\mathcal{O}}_{\mu_0}$ is free, we can assume $\alpha \notin \overline{\mathcal{O}}_{\mu_0}$. Hence there is some pattern $u \prec \alpha$ with $u \not\prec \mu_0$. Let $T = \mathrm{dom}(u)$. We have $\alpha = \lim \alpha_i$ where $\alpha_i = h_i \cdot \beta_i$ with $\beta_i \in A_x$. We can assume that $\alpha_i \restriction T = u$ for all i. Since u does not occur in μ_0, T must intersect some occurrence of one of the u^n_--patterns in each α_i. Since there are only finitely many possibilities for each pattern, by passing to a subsequence we may assume that it is the same pattern in each case, with the same intersection with T. That is, one of the u^n_--patterns (which we will call u_0) occurs at the same position h_0, i.e., $\alpha_i \restriction h_0 \cdot T_1 = h_0 \cdot u_0$ for all i. Multiplying each α_i by h_0^{-1} (and replacing g with $h_0^{-1} g h_0$) we may assume in fact that $\alpha_i \restriction T_1 = u_0$ for all i.

Since the constant pattern of 1's on T_0. d_1, occurs in elements of A_x only at elements $g_0(k) \in S$, we have that for each i there is a $k_i \in \mathbb{N}$ with $h_i \cdot g_0(k_i) \cdot T_0 = T_0$, and since $g \cdot T_0 \neq T_0$ for $g \neq e$ we must have $h_i = g_0(k_i)^{-1}$. Thus, $\alpha = \lim g_0(k_i)^{-1} \cdot \beta_i$ with each $\beta_i \in A_x$. The T_2-block in α_i must then be of the form $u^{k_i}_-$, so if we consider the two $\widetilde{F^{\tilde{n}}}$ links encoded there, they must be $s^{-1}_{\tilde{n}(k_i-1)} s_{\tilde{n}k_i}$

and $s_{\tilde{n}k_i}^{-1} s_{\tilde{n}(k_i+1)}$, and these must be the same for all i. Choosing i large enough so that the corresponding T_2-translates stabilize, the same holds for the links in the translates, so the limit exists in the definition of the following sequence in $g_\infty \in S^{\mathbb{Z}}$:

$$g_\infty(n) = \lim_i g_0(k_i + n)$$

(we may assume that the k_i's are unbounded since otherwise α will have a unique occurrence of a $u_{d(i)}$-block, and hence can not be a fixed point).

Passing to a subsequence we can assume that all of the k_i's are even or all are odd; we assume they are all even (the following sequence may be shifted by 1 if they are odd). We then have that the $g_\infty(2i+1) \cdot T_1$-blocks in α must be either $u_{i(0)}$ or $u_{i(1)}$, so we may define the sequence $\gamma \in 2^{\mathbb{Z}}$ by

$$\gamma(i) = \begin{cases} 0 & \text{if } \alpha \upharpoonright g_\infty(2i+1) \cdot T_1 = u_{i(0)} \\ 1 & \text{if } \alpha \upharpoonright g_\infty(2i+1) \cdot T_1 = u_{i(1)}. \end{cases}$$

Each finite subsequence occurring in γ is induced by a subsequence in some β_i, and hence must be a subsequence of our minimal element m_0 encoded at these positions. Thus, $\gamma \in \overline{\mathcal{O}}_{m_0}$ must be a minimal free element.

We similarly see that since $g \cdot \alpha = \alpha$, by considering i so the that the limit stabilizes at $g \cdot T_0$, we must have that g carries the $g_0(k_i)$-translate of T_0 to some other $g_0(k_i')$-translate, so $g = g_0(k_i')g_0(k_i)^{-1}$ for some $k_i' \in \mathbb{N}$ for each sufficiently large i, and we may assume $k_i' > k_i$ by using g^{-1} if necessary. But now the $g_0(k_i') \cdot T_2$-blocks of β_i must be the same as the $g_0(k_i) \cdot T_2$-blocks, and so (for large enough i), we must have the same links encoded at each such translate. Hence g_∞ is actually periodic with period $p = k_i' - k_i$ for some sufficiently large i. Squaring g, we can take $p = 2q$ to be even, and we see γ must actually be periodic under the shift by q. This contradicts that γ is aperiodic, and hence we could not have had $g \cdot \alpha = \alpha$. □

Forward direction of the reduction. We now check that ψ satisfies the conclusions of Theorem 2. We begin with the forward direction.

LEMMA 6. *If $x, y \in 2^{\mathbb{F}_2}$ satisfy $x \, E(\mathbb{F}_2, 2) \, y$ then $X_x \cong X_y$.*

PROOF. It will be sufficient to show this when $y = a \cdot x$ or $y = b \cdot x$, for a and b generating $\mathbb{F}_2$. Consider the case $y = a \cdot x$ (the case of $y = b \cdot x$ is identical). We have:

$$\begin{aligned} A_x &= P \cup \{\mu \circledast u^0_{d(w_i \cdot x(w_{n_k}))} \circledast \rho^{1+k^2}(w_i) : i, k \in \mathbb{N}\} \\ A_y &= P \cup \{\mu \circledast u^0_{d(w_i \cdot y(w_{n_k}))} \circledast \rho^{1+k^2}(w_i) : i, k \in \mathbb{N}\} \\ &= P \cup \{\mu \circledast u^0_{d(w_i a \cdot x(w_{n_k}))} \circledast \rho^{1+k^2}(w_i) : i, k \in \mathbb{N}\} \\ &= P \cup \{\mu \circledast u^0_{d(w_i \cdot x(w_{n_k}))} \circledast \rho^{1+k^2}(w_i a^{-1}) : i, k \in \mathbb{N}\}. \end{aligned}$$

We can define a block code $\pi : 2^B \to 2$ such the induced map φ_π is an isomorphism of X_x and X_y as follows. The idea is to ensure that each element $\mu \circledast u^0_{d(w_i \cdot x(w_{n_k}))} \circledast \rho^{1+k^2}(w_i)$ is mapped to $\mu \circledast u^0_{d(w_i \cdot x(w_{n_k}))} \circledast \rho^{1+k^2}(w_i a^{-1})$ (and likewise for their shifts) and all other sequences are left unchanged. This will provide a shift-invariant homeomorphism of A_x with A_y and hence an isomorphism of X_x with X_y.

$$
\begin{aligned}
u_{i(0)} \oplus u_e \oplus u_{i(0)} &\mapsto u_{i(0)} \oplus u_{a^{-1}} \oplus u_{i(0)} \\
u_{i(0)} \oplus u_a \oplus g_1 u_{i(0)} &\mapsto u_{i(0)} \oplus u_e \oplus u_{i(0)} \\
u_a \oplus u_a \oplus u_{i(0)} &\mapsto u_a \oplus u_{i(0)} \oplus u_{i(0)} \\
u_b \oplus u_a \oplus u_{i(0)} &\mapsto u_b \oplus u_{i(0)} \oplus u_{i(0)} \\
u_{b^{-1}} \oplus u_a \oplus u_{i(0)} &\mapsto u_{b^{-1}} \oplus u_{i(0)} \oplus u_{i(0)} \\
u_{a^{-1}} \oplus u_{i(0)} \oplus u_{i(0)} &\mapsto u_{a^{-1}} \oplus u_{a^{-1}} \oplus u_{i(0)} \\
u_b \oplus u_{i(0)} \oplus u_{i(0)} &\mapsto u_b \oplus u_{a^{-1}} \oplus u_{i(0)} \\
u_{b^{-1}} \oplus u_{i(0)} \oplus u_{i(0)} &\mapsto u_{b^{-1}} \oplus u_{a^{-1}} \oplus u_{i(0)}
\end{aligned}
$$

FIGURE 1. Patterns to be mapped by the block code

It suffices to map the patterns of the forms shown in Figure 1 (where $u \oplus v \oplus w$ on the left abbreviates a pattern of the form $a_2^{-1}a_1^{-1} \cdot u \circledast s^{p(a_2,-)} \circledast v^{p(a_1,b_1)} \circledast t^{p(-,b_2)} \circledast b_1b_2 \cdot w$ with u, s, v, t, w patterns on T_1, and $u' \oplus v' \oplus w'$ on the right abbreviates $a_2^{-1}a_1^{-1} \cdot u' \circledast s^{p(a_2,-)} \circledast v'^{p(a_1,b_1)} \circledast t^{p(-,b_2)} \circledast b_1b_2 \cdot w'$), and all other patterns should be left unchanged. None of the listed patterns overlaps a shift of another and there are only finitely many, so this can be ensured by choosing the block code π appropriately; taking $B = (F^{2\tilde{n}} \cdot T_2 \cup T_2) \cdot T_2$ (since only the central T_2-block is changed in each of the patterns shown) is sufficient. □

Reverse direction. Finally, suppose x and y are such that X_y is a factor of X_x (the other direction being identical). Let $\varphi = \varphi_{j_0}$ be a homomorphism from X_x to X_y induced by the block code $\pi_{j_0} : 2^{B_{j_0}} \to 2$. We will show that $x\, E(\mathbb{F}_2, 2)\, y$.

Observe that each of the subflows contains two types of sequences: those which differ from an element of the orbit closure of μ_0 on only finitely many coordinates and contain at most two occurrences of the pattern d, which are exactly the elements of the orbit closure of P, and those which contain infinitely many copies of the pattern d and hence differ from μ_0 on infinitely many coordinates, which includes μ as well as the elements of Q_x.

LEMMA 7. *φ can not send an element of the orbit closure of P to a shift of an element of Q_y.*

PROOF. Suppose we have α in the orbit closure of P with $\varphi(\alpha) = \beta = h \cdot \left(\mu \circledast u^0_{d(w_i \cdot y(w_{n_k}))} \circledast \rho^{1+k^2}(w_i)\right)$. Then one of the d-patterns in β must be induced by a $T_1 B_{j_0}$-block in μ_0. This block recurs uniformly in μ_0 and hence in α, and hence the pattern d_1 must recur uniformly in β. Hence there would be a finite T such that $\forall g \exists t \in T \forall t' \in T_0 (\beta(gtt') = d_1(t'))$, i.e., $d_1 \prec \beta \restriction gTT_0$ for all $g \in G$. Since d_1 occurs in β only at $hg_0(k)$ for $k \geq 0$, we have that for all $g \in G$ there is $k \geq 0$ such that $hg_0(k) \in gTT_0$. But we can find some k_0 such that $TT_0 \subseteq F^{<\tilde{n}k_0}$. But now if we let $g = hg_0(-k_0)$ we have that for any $k \geq 0$, $g_0(-k_0)^{-1}g_0(k) \notin F^{<\tilde{n}(k+k_0)}$, so $hg_0(k_0)^{-1}g_0(k) \notin hg_0(k_0)^{-1}F^{<\tilde{n}k_0}$, i.e., for all $k \geq 0$ we have $g_0(k) \notin hg_0^{-k_0}TT_0$, a contradiction. □

LEMMA 8. *φ sends every element of Q_x to a shift of an element of Q_y.*

PROOF. From the previous lemma, for each $q \in Q_y$ there must be an r in the orbit closure of Q_x with $\varphi(r) = q$. Then there are $q_i \to r$ with each q_i the shift of an element of Q_x. Since elements of the orbit closure of P have at most two occurrences of d, we must have $\varphi(q_i)$ in the orbit closure of Q_y for cofinally many i, and since they approach $q \in Q_y$ we must have $\varphi(q_i)$ a shift of an element of Q_y for cofinally many i. In particular, there is some $q_0 \in Q_x$ so that $\varphi(q_0)$ is the shift of an element of Q_y. But now every element of Q_x differs from q_0 on only finitely many coordinates, so the image under φ must differ from the shift of en element of Q_y on only finitely many coordinates, and hence be the shift of an element of Q_y. □

LEMMA 9. *For each $i \in 2$, $\varphi(p_i^{j_0})$ is a shift of $p_i^{j_0}$.*

PROOF. The previous lemma ensures that for each $p_i^j \in P$, there must be some element r of the orbit closure of P with $\varphi(r) = p_i^j$. Note that the v_i^j-blocks in p_i^j cannot be induced by blocks from μ_0 in r or else they would recur uniformly. Since elements of the orbit closure of P which are not shifts of elements in P contain at most one $\mu \circledast u^0_{d(i)}$-block, and the construction of the q_n's ensure that for $j \geq j_0$ the two v_i^j-blocks are sufficiently separated that they must arise from separate $\mu \circledast u^0_{d(i)}$-blocks, we must have that r is a shift of an element $p_{i'}^{j'}$ of P. The construction of the q_n's and p_i^j's further ensures that we must have $2j' + i' = 2j + i$ for $j \geq j_0$, so $j' = j$ and $i' = i$. □

Hence, points in Q_x containing the initial $T_1 B_{j_0}$-block of $\mu \circledast u^0_{d(i)}$ must map to (shifts of) points in Q_y with the same initial T_1-block $u^0_{d(i)}$. Let h_0 and h_1 be the respective shiftings; note that both must be in $T_1 B_{j_0} B_{j_0}^{-1}$. Choose k_1 such that $g_1(1+k^2) \cdot T_1 \cap T_1 B_{j_0} B_{j_0}^{-1} = \emptyset$ for all $k \geq k_1$.

Consider now the points $\alpha_k = \mu \circledast u^0_{d(x(w_{n_k}))} \circledast \rho^{1+k^2}(e)$ in Q_x for $k \geq k_1$. Each must be mapped to the shift by some $h_{x(w_{n_k})}$ of a point $\mu \circledast u^0_{d(w_{i(k)} \cdot y(w_{n_{j(k)}}))} \circledast \rho^{1+j(k)^2}(w_{i(k)})$ in Q_y, with $x(w_{n_k}) = w_{i(k)} \cdot y(w_{n_{j(k)}})$. We claim that $j(k) = k$ for sufficiently large k. To see this, consider α_{k_1}. Note that $\mu \circledast u^0_{d(i)} = \lim_j p_i^j$ (since we took the B_j's to be increasing), so $\varphi(\mu \circledast u^0_{d(i)}) = \lim_j \varphi(p_i^j) = \lim_j h_i \cdot v_{i(j)}^{k(j)}$ (with $k(j) \to \infty$) $= h_i \cdot (\mu \circledast u_{d(i)})$. Hence, $\varphi(\alpha_k) = h_{x(w_{n_k})} \cdot (\mu \circledast u^0_{d(x(w_{n_k}))})$ except on the block $g_1(1+k^2) \cdot T_2 B_{j_0}$. In particular, $g_1(1+j(k)^2) \cdot T_2 \subseteq g_1(1+k^2) \cdot T_2 B_{j_0}$. Since $g_1(j) \cdot T_2$ will be disjoint from $T_2 B_{j_0}$ for all but finitely many j, we must have that $|j(k)^2 - k^2| = (j(k)+k) \cdot |j(k) - k| \geq k \cdot |j(k) - k|$ is bounded, and hence $j(k) = k$ for all but finitely many k. Choose $k_2 \geq k_1$ so that $j(k) = k$ for all $k \geq k_2$. Note also that the above shows that there is a bound on the length of $w_{i(k)}$, so that there are only finitely many possible w_{i_k}'s.

We now have $i(k) = i_k^{j_0}$ from the definition of i_k^j earlier. Since there are only finitely many $i(k)$'s, we have that i^{j_0} is defined. Hence for all but finitely many $k \in \widetilde{N}_{j_0}$ we have $i(k) = i_k^{j_0} = i^{j_0}$, and $\{n_k : k \in \widetilde{N}_{j_0}\}$ enumerates $\mathbb{N}$ infinitely often. Hence the set $\{n_k : k \in N_{j_0} \wedge i(k) = i^{j_0} \wedge j(k) = k\}$ includes all of $\mathbb{N}$, so for all $n \in \mathbb{N}$ we have $x(w_n) = w_{i^{j_0}} \cdot y(w_n)$. Hence $x = w_{i^j} \cdot y$, so $x E(\mathbb{F}, 2) y$ as claimed.

This completes the proof of the theorem.

References

[1] J. D. Clemens, *Isomorphism of subshifts is a universal countable Borel equivalence relation*, Israel J. Math. **170** (2009), 113–123, DOI 10.1007/s11856-009-0022-0. MR2506320

[2] S. Gao, S. Jackson, and B. Seward, *A coloring property for countable groups*, Math. Proc. Cambridge Philos. Soc. **147** (2009), no. 3, 579–592, DOI 10.1017/S0305004109002655. MR2557144

[3] S. Gao, S. Jackson, and B. Seward, *Group colorings and Bernoulli subflows*, Mem. Amer. Math. Soc. **241** (2016), no. 1141, vi+241, DOI 10.1090/memo/1141. MR3478758

[4] E. Glasner and V. V. Uspenskij, *Effective minimal subflows of Bernoulli flows*, Proc. Amer. Math. Soc. **137** (2009), no. 9, 3147–3154, DOI 10.1090/S0002-9939-09-09905-5. MR2506474

[5] S. Jackson, A. S. Kechris, and A. Louveau, *Countable Borel equivalence relations*, J. Math. Log. **2** (2002), no. 1, 1–80, DOI 10.1142/S0219061302000138. MR1900547

[6] A. S. Kechris, *Classical descriptive set theory*, Graduate Texts in Mathematics, vol. 156, Springer-Verlag, New York, 1995. MR1321597

[7] D. Lind and B. Marcus, *An introduction to symbolic dynamics and coding*, Cambridge University Press, Cambridge, 1995. MR1369092

[8] M. Sabok and T. Tsankov, *On the complexity of topological conjugacy of Toeplitz subshifts*, Israel J. Math. **220** (2017), no. 2, 583–603, DOI 10.1007/s11856-017-1541-8. MR3666438

Boise State University, 1910 University Dr., Boise, Idaho 83725
Email address: johnclemens@boisestate,edu

Contemporary Mathematics
Volume **752**, 2020
https://doi.org/10.1090/conm/752/15130

Distance from marker sequences in locally finite Borel graphs

Clinton T. Conley and Andrew S. Marks

1. Introduction

The investigation of structure in marker sequences has been a recurring theme of the study of countable Borel equivalence relations and Borel graphs. Recall that if E is a countable Borel equivalence relation on a standard Borel space X, then we say that $A \subseteq X$ is a *complete section* or *marker set* for E if A meets every E-class. We similarly say that A is a marker set for a locally countable graph on X if A is a marker set for the connectedness relation of G. Finally, a *marker sequence* $\{A_n\}_{n\in\mathbb{N}}$ for a countable Borel equivalence relation or graph is a countable sequence of marker sets.

The study of marker sequences forms the backbone of our understanding of what groups generate hyperfinite equivalence relations [**GJ**] [**ScSe**], and the combinatorics of Borel graphs generated by these group actions [**GJKS**]. More broadly, these ideas underlie many constructions in the study of Borel graph combinatorics [**KM**]. The first result about markers was proved by Slaman and Steel [**SlSt**] who showed that for every countable Borel equivalence relation E there is a decreasing sequence $A_0 \supseteq A_1 \supseteq \ldots$ of markers with empty intersection $\bigcap_n A_n = \emptyset$.

Suppose Γ is a finitely generated group which acts on the space 2^Γ via the left shift action. Let $\mathrm{Free}(2^\Gamma)$ be the set of $x \in 2^\Gamma$ such that for all nonidentity $\gamma \in \Gamma$ we have $\gamma \cdot x \neq x$, and let $G(\Gamma, 2)$ be the graph on $\mathrm{Free}(2^\Gamma)$ where $x, y \in \mathrm{Free}(2^\Gamma)$ are adjacent if there is a generator γ of Γ such that $\gamma \cdot x = y$. Let $d_{G(\Gamma,2)}$ be the graph distance metric for $G(\Gamma, 2)$. A recent result of Gao, Jackson, Krohne, and Seward states the following.

THEOREM 1.1 ([**GJKS**, Theorem 1.1]). *Suppose Γ is a finitely generated infinite group and $f\colon \mathbb{N} \to \mathbb{N}$ tends to infinity. Then for every Borel marker sequence $\{A_n\}_{n\in\mathbb{N}}$ for $G(\Gamma,2)$, there exists an $x \in \mathrm{Free}(2^\Gamma)$ such that for infinitely many n, we have $d_{G(\Gamma,2)}(x, A_n) < f(n)$*

This result led us to ask the following question: what can we say if the function $f\colon \mathbb{N} \to \mathbb{N}$ is allowed to vary depending on the point x? Of course, we cannot possibly draw an analogous conclusion for an arbitrary Borel way of associating some $f_x\colon \mathbb{N} \to \mathbb{N}$ to each point x in our space; given a Borel marker sequence

2010 *Mathematics Subject Classification.* Primary 03E15.
The first author was partially supported by NSF grant DMS-1500906.
The second author was partially supported by NSF grant DMS-1500974.

$\{A_n\}_{n\in\mathbb{N}}$ for a graph G on X, we could define $f_x(n) = d_G(x, A_n)$ for all $x \in X$. Instead, we show the existence of some Borel map $x \mapsto f_x$ for which we can draw a stronger conclusion than that of Theorem 1.1, showing closeness for all n instead of just infinitely many n. This is true even when we generalize to arbitrary locally finite non-smooth graphs.

THEOREM 1.2. *Suppose G is a locally finite non-smooth Borel graph on X. Then there exists a Borel map associating to each $x \in X$ a function $f_x\colon \mathbb{N} \to \mathbb{N}$ such that for every Borel marker sequence $\{A_n\}_{n\in\mathbb{N}}$ for G, there exists an $x \in X$ such that for all n, we have $d_G(x, A_n) < f_x(n)$.*

Now when G is smooth, an easy diagonalization constructs marker sequences that do not satisfy the conclusion of our theorem. Hence, this result provides a novel way of characterizing smoothness; a locally finite Borel graph is smooth if and only if it does not admit Borel marker sequences that are somewhere "far" from every point.

We remark here that in Theorem 1.2, the map $x \mapsto f_x$ may always be chosen so that it is a Borel homomorphism from the equivalence relation graphed by G to tail equivalence on $\mathbb{N}^\mathbb{N}$. We note that standard "sparse" Borel marker sequence constructions show that a map $x \mapsto f_x$ witnessing Theorem 1.2 cannot take a constant value on each connected component of G.

The theorem is proved in two steps. First, we use the fact that all Borel subsets of $[\mathbb{N}]^\mathbb{N}$ are completely Ramsey to give an example of a Borel graph G satisfying the conclusion of Theorem 1.2. Then, we conclude the full result using the Glimm-Effros dichotomy. We show in the last section that this theorem cannot be proven using measure or category arguments.

2. Distance from marker sequences and the Ramsey property

Let $[\mathbb{N}]^\mathbb{N}$ be Ramsey space, the space of infinite subsets of $\mathbb{N}$. Given a finite set $s \subseteq \mathbb{N}$ and an infinite set $x \subseteq \mathbb{N}$ with $\min(x) > \max(s)$, recall the definition $[s,x]^\mathbb{N} = \{y \in [\mathbb{N}]^\mathbb{N} : s \subseteq y \subseteq s \cup x\}$. We can identify $[\mathbb{N}]^\mathbb{N}$ with a subset of $2^\mathbb{N}$ via characteristic functions, and we use the resulting subspace topology on $[\mathbb{N}]^\mathbb{N}$ throughout. A theorem of Galvin and Prikry [**GP**] states that for every $[s,x]^\mathbb{N}$ and every Borel subset $B \subseteq [s,x]^\mathbb{N}$, there exists some $[t,y]^\mathbb{N} \subseteq [s,x]^\mathbb{N}$ such that either $[t,y]^\mathbb{N} \subseteq B$ or $[t,y]^\mathbb{N} \cap B = \emptyset$. From this, it is easy to see the following:

LEMMA 2.1 (Galvin-Prikry [**GP**]). *If $\{B_n\}_{n\in\mathbb{N}}$ is a Borel partition of $[s,x]^\mathbb{N}$, then there exists some $n \in N$ and $[t,y]^\mathbb{N} \subseteq [s,x]^\mathbb{N}$ such that $[t,y]^\mathbb{N} \subseteq B_n$.*

PROOF. Suppose not. Then we may construct a decreasing sequence $[s,x]^\mathbb{N} \supseteq [s_0,x_0]^\mathbb{N} \supseteq [s_1,x_1]^\mathbb{N} \supseteq \ldots$, where $[s_n,x_n]^\mathbb{N} \cap B_n = \emptyset$, and s_n has at least n elements. But then setting $z = \bigcup_n s_n$, we see that $z \in [s,x]$, and $z \notin B_n$ for all n, hence $\{B_n\}_{n\in\mathbb{N}}$ does not partition $[s,x]^\mathbb{N}$. □

The *odometer* $\sigma\colon [\mathbb{N}]^\mathbb{N} \to [\mathbb{N}]^\mathbb{N}$ is defined via the identification of $[\mathbb{N}]^\mathbb{N}$ as a subspace of $2^\mathbb{N}$ by setting $\sigma(x) = 0^n 1y$ if $x = 1^n 0y$, and fixing $\sigma(111\ldots) = 111\ldots$ on the sequence consisting of all ones. Define also $\tau\colon [\mathbb{N}]^\mathbb{N} \to [\mathbb{N}]^\mathbb{N}$ by setting $\tau(x) = \{n-1 : n \in x \wedge n > 0\}$. Let G_t be the graph on $[\mathbb{N}]^\mathbb{N}$ generated by these two functions, where $x, y \in [\mathbb{N}]^\mathbb{N}$ are adjacent if either $\sigma(x) = y$, $\sigma(y) = x$, $\tau(x) = y$, or $\tau(y) = x$. So G_t is graphing of tail equivalence on $[\mathbb{N}]^\mathbb{N}$, and every vertex in G_t has degree ≤ 5.

LEMMA 2.2. *Consider the graph G_t defined on $[\mathbb{N}]^{\mathbb{N}}$ as above, and for each $x \in [\mathbb{N}]^{\mathbb{N}}$, let $f_x(n)$ be equal to the $(n+1)$st element of x. Then for every Borel marker sequence $\{A_n\}_{n\in\mathbb{N}}$ for G_t, there is an $x \in [\mathbb{N}]^{\mathbb{N}}$ such that for all n, we have $d_{G_t}(x, A_n) < f_x(n)$.*

PROOF. We construct x as the intersection of a decreasing sequence $[s_0, x_0]^{\mathbb{N}} \supseteq [s_1, x_1]^{\mathbb{N}} \supseteq \ldots$, where s_n has exactly n elements. Let $s_0 = \emptyset$, and $x_0 = \mathbb{N}$. Now given $[s_n, x_n]^{\mathbb{N}}$, since the sets $\{\{y \in [s_n, x_n]^{\mathbb{N}} : d(y, A_n) = k\}\}_{k\in\mathbb{N}}$ partition $[s_n, x_n]$, we may apply Lemma 2.1 to obtain some $[t, y] \subseteq [s_n, x_n]^{\mathbb{N}}$ and some k such that every element of $[t, y]^{\mathbb{N}}$ is distance exactly k from A_n. Since $s_n \subseteq t$, there is some m such that m applications of the odometer applied to $[s_n, y]^{\mathbb{N}}$ yield $\sigma^m([s_n, y]^{\mathbb{N}}) = [t, y]^{\mathbb{N}}$. Hence by the triangle inequality, we see that there is some $k^* = k + m$ such that all the elements of $[s_n, y]$ are distance $\leq k^*$ from A_n. Now let l be the least element of y that is strictly greater than k^* and $\max(s_n)$, let $s_{n+1} = s_n \cup \{l\}$, and $x_{n+1} = y \setminus \{0, 1, \ldots, l\}$. We have ensured then that every element of $[s_{n+1}, x_{n+1}]^{\mathbb{N}}$ has distance $\leq l$ from A_n. □

We remark here that the above proof works equally well for the usual graphing of E_0 on $[\mathbb{N}]^{\mathbb{N}}$ induced by the odometer. We have used the larger graph G_t because we will need a locally finite graphing of tail equivalence with our desired property in order to finish the proof of Theorem 1.2.

We need one more easy lemma before we complete the theorem. The lemma roughly states that this question of closeness to marker sequences is independent of the particular locally finite Borel graph we choose, and depends only on the equivalence relation we have graphed.

Given a Borel graph G on X, and a Borel map $x \mapsto f_x$ from $X \to \mathbb{N}^{\mathbb{N}}$, say that a marker sequence $\{A_n\}_{n\in\mathbb{N}}$ *satisfies $x \mapsto f_x$ for G* if for all $x \in X$ there exists an n such that $d_G(x, A_n) \geq f_x(n)$.

LEMMA 2.3. *Suppose G and H are locally finite Borel graphs on a standard Borel space X having the same connected components. Then for every Borel map $x \mapsto g_x$ from $X \to \mathbb{N}^{\mathbb{N}}$, there exists a Borel map $x \mapsto h_x$ such that for every marker sequence $\{A_n\}_{n\in\mathbb{N}}$, if $\{A_n\}_{n\in\mathbb{N}}$ satisfies $x \mapsto h_x$ for H, then $\{A_n\}_{n\in\mathbb{N}}$ satisfies $x \mapsto g_x$ for G.*

PROOF. Since the graphs are locally finite, there are only finitely many points a fixed distance from each $x \in X$. Hence, we may define $h_x(n)$ to be the least k such that $d_H(x, y) \leq k$ for all $y \in X$ such that $d_G(x, y) \leq g_x(n)$. □

We now complete the proof of Theorem 1.2.

PROOF OF THEOREM 1.2. Suppose G is not smooth. Let E be the equivalence relation graphed by G, and E_t be the equivalence relation of tail equivalence on $[\mathbb{N}]^{\mathbb{N}}$. By the Glimm-Effros dichotomy, there must be some E-invariant Borel set A such that $E \restriction A \cong_B E_t$. But then $G \restriction A$ and the graph G_t from Lemma 2.2 are two different locally finite Borel graphings of the same equivalence relation. Hence, by Lemma 2.3, we can find a Borel $x \mapsto h_x$ from $A \to \mathbb{N}^{\mathbb{N}}$ so that no Borel marker sequence can satisfy $G \restriction A$ for $x \mapsto h_x$. Hence, any Borel extension of $x \mapsto h_x$ to a function $x \mapsto f_x$ defined on X suffices to prove the theorem. □

3. Measure and category

In this section, we prove the following:

PROPOSITION 3.1. *Suppose G is a locally finite Borel graph on X, and $x \mapsto f_x$ is a Borel map from $X \to \mathbb{N}^{\mathbb{N}}$. Then*

(1) *For every Borel probability measure μ on X, there is a G-invariant μ-conull set B and a Borel marker sequence $\{A_n\}_{n\in\mathbb{N}}$ for $G \upharpoonright B$ such that for every $x \in X$, there is an n such that $d_G(x, A_n) \geq f_x(n)$.*
(2) *For every compatible Polish topology τ on X, there is a G-invariant τ-comeager set B and a Borel marker sequence $\{A_n\}_{n\in\mathbb{N}}$ for $G \upharpoonright B$ such that for every $x \in X$, there is an n such that $d_G(x, A_n) \geq f_x(n)$.*

PROOF. Let $B_0 \supseteq B_1 \supseteq B_2 \dots$ be a decreasing sequence of Borel markers for G with empty intersection. Such a sequence exists by [SlSt]. Let $C_{i,n} = \{x \in X : d_G(x, B_i) < f_x(n)\}$. Note that since $\{B_n\}_{n\in\mathbb{N}}$ is decreasing with empty intersection, for each n, we have $\bigcap_i C_{i,n} = \emptyset$.

For part (1), we may assume as usual that μ is G-quasi-invariant. Observe that for each n, the μ-measure of the sets $C_{i,n}$ goes to 0. Hence, we may find a sequence $i_0, i_1, i_2, \dots$ such that $\mu(C_{i_n,n}) \to 0$. Now choose our marker sequence to be $\{A_n\}_{n\in\mathbb{N}}$ where $A_n = B_{i_n}$. This marker sequence has the required property on the complement of the nullset $\bigcap_i C_{i_n,n}$.

Part (2) follows using a similar argument, since relative to any basic open set, the set $C_{i,n}$ can be comeager for only finitely many i. □

References

[GP] F. Galvin and K. Prikry, *Borel sets and Ramsey's theorem*, J. Symbolic Logic **38** (1973), 193–198, DOI 10.2307/2272055. MR337630

[GJ] S. Gao and S. Jackson, *Countable abelian group actions and hyperfinite equivalence relations*, Invent. Math. **201** (2015), no. 1, 309–383, DOI 10.1007/s00222-015-0603-y. MR3359054

[GJKS] S. Gao, S. Jackson, E. Krohne, and B. Seward *Forcing constructions and countable Borel equivalence relations*, arXiv:1503.07822.

[KM] A.S. Kechris and A.S. Marks, *Descriptive Graph combinatorics*, preprint.

[ScSe] S. Schneider and B. Seward, *Locally nilpotent groups and hyperfinite equivalence relations*, Preprint.

[SlSt] T. A. Slaman and J. R. Steel, *Definable functions on degrees*, Cabal Seminar 81–85, Lecture Notes in Math., vol. 1333, Springer, Berlin, 1988, pp. 37–55, DOI 10.1007/BFb0084969. MR960895

DEPARTMENT OF MATHEMATICAL SCIENCES, CARNEGIE MELLON UNIVERSITY, PITTSBURGH, PENNSYLVANIA 15213

Email address: clintonc@andrew.cmu.edu

DEPARTMENT OF MATHEMATICS, UNIVERSITY OF CALIFORNIA AT LOS ANGELES

Email address: marks@math.ucla.edu

Contemporary Mathematics
Volume **752**, 2020
https://doi.org/10.1090/conm/752/15131

The Ultrapower Axiom

Gabriel Goldberg

1. Introduction

This paper is a survey of some consequences of the Ultrapower Axiom (UA), a principle with origins in the developing theory of canonical inner models for supercompact cardinals. UA roughly states that any two ultrapowers of the universe have a common ultrapower. It is therefore a weak consequence of the comparison lemma of inner model theory, which roughly states that any two canonical inner models at the same large cardinal level have a common iterated ultrapower.

There are some essential differences between UA and other inner model theoretic principles that, taken together, motivate the research directions pursued in this paper.

First, unlike principles like $V = L$ and $\square$, UA is compatible with all large cardinal axioms accessible to contemporary inner model theory. In fact, UA holds in all known canonical inner models. We will argue that the basic methodology of modern inner model theory cannot produce a model in which UA fails.

Second, unlike principles like GCH and $\diamondsuit$, the only known proofs that UA is consistent with large cardinal axioms are the inner model constructions. In fact, even a proof of the consistency of the first nontrivial instance of UA (namely, "UA + there is a measurable cardinal") seems to be out of reach of known forcing techniques.

Third, unlike any other set theoretic principle that has been proposed, UA permits the development of a semi-fine structural theory of large cardinals in the region of supercompact cardinals.

Given these distinguishing features, a key question arises:

QUESTION. Is UA consistent with a supercompact cardinal?

Granting that all canonical inner models satisfy UA and that the theory "UA + there is a supercompact cardinal" cannot be proved consistent except by the techniques of inner model theory, this question is no more than a precise formulation of the well-known Inner Model Problem:

QUESTION. Is there a canonical inner model with a supercompact cardinal?

2010 *Mathematics Subject Classification.* Primary 03E45 (inner models), 03E55 (large cardinals), and 03E65 (other set theoretic axioms).

This suggests a fruitful approach to the Inner Model Problem: develop the theory of UA in the presence of supercompact cardinals.

If an inconsistency arose from this theory, this would amount to an anti-inner model theorem, showing that the basic methodology of inner model theory cannot build a canonical inner model with a supercompact cardinal. So far, however, the theory "UA + there is a supercompact cardinal" has resisted attack. We take this to be strong evidence that the Inner Model Problem can be solved positively. We are also optimistic that the techniques developed here will provide some clues in the search for canonical inner models with supercompact cardinals, and that they may someday become part of the machinery used to analyze these models.

1.1. Outline. We now outline the paper section by section.

In Section 3, we state the Ultrapower Axiom. To support our claim that the current methodology of inner model theory only produces models of UA, we also state Woodin's Weak Comparison axiom, a form of the comparison lemma that can be stated without reference to fine structure. Weak Comparison manifestly follows from the comparison lemma, while the argument for UA is less direct. In fact, we sketch a proof that UA follows from Weak Comparison.

Next, in Section 4, we show that UA implies the linearity of the Mitchell order on normal ultrafilters, generalizing Schlutzenberg's theorem [**15**] for the short extender models. We also state some theorems from [**4**] regarding the generalized Mitchell order on generalized normal ultrafilters.

Analyzing the proof of this theorem leads to the definition under UA of a new order on ultrafilters called the *seed order*, whose antecedents are the Mitchell order on normal ultrafilters and Ketonen's order [**7**] on weakly normal ultrafilters. In Section 5, we define the seed order and show assuming UA that it wellorders the class of uniform countably complete ultrafilters on ordinals.

The seed order is the key to the deeper consequences of UA in the context of a supercompact cardinal. In fact, we immediately derive a somewhat surprising corollary. The inner models typically satisfy $V = \text{HOD}$ or something close. Since UA is preserved by small forcing, one cannot hope to prove $V = \text{HOD}$ from UA supplemented by any large cardinal hypothesis consistent with UA. But using the linearity of the seed order, we easily prove the next best thing:

THEOREM 1.1 (UA). *If there is a supercompact cardinal, then V is a generic extension of* HOD. □

Similarly the usual independence proof shows that CH is independent of UA supplemented by any large cardinal hypothesis consistent with UA. But in Section 6, we sketch a proof that UA implies the next best thing:

THEOREM 1.2 (UA). *If κ is supercompact, then $2^\lambda = \lambda^+$ for all $\lambda \geq \kappa$.* □

Finally, in Section 7, we discuss the results that we think provide the most compelling evidence for the conjecture that UA is consistent with a supercompact cardinal. These results constitute the beginnings of a structure theory for supercompactness and strong compactness under the assumption of UA. A sample result is that one can classify all instances of strong compactness in terms of supercompactness:

THEOREM 1.3 (UA). *Suppose κ is an uncountable cardinal. Then the following are equivalent:*

(1) *κ is strongly compact.*
(2) *κ is supercompact or κ is a measurable limit of supercompact cardinals.* □

The conclusion of Theorem 1.3 was originally proved consistent with the existence of many supercompact cardinals in unpublished work of Kimchi-Magidor (but see [**1**] for a stronger result). This consistency proof uses the method of forcing, but the true nature of Theorem 1.3 seems completely out of reach current forcing techniques: the proof shows that except for a class of trivial counterexamples and the intriguing possibility of a class of anomalous counterexamples, *every countably complete ultrafilter is maximally supercompact*; see Theorem 7.13. If it is provable from UA plus some large cardinal hypothesis that these anomalous counterexamples really do exist, then the Ultrapower Axiom arguably reveals a feature of the large cardinal hierarchy that is obscured in ZFC alone.

Since the paper is a survey, we will often state results without proof. All these results will appear in the author's forthcoming thesis [**5**].

2. Preliminary definitions

To keep this paper self-contained, we very briefly define some standard notation for ultrafilters and ultrapowers.

DEFINITION 2.1. Suppose P is a transitive model of ZFC and U is a P-ultrafilter. We denote the ultrapower of P by U using functions in P by $j^P_U : P \to M^P_U$. If M^P_U is wellfounded, we identify it with its transitive collapse.

Note that we write $j^P_U : P \to M^P_U$ as if we are relativizing the ultrapower construction to P, but this is a bit of an abuse of notation since we may have $U \notin P$.

In the case that $P = V$, we will denote the ultrapower by $j_U : V \to M_U$.

DEFINITION 2.2. Suppose P and Q are transitive models of ZFC and $j : P \to Q$ is a cofinal elementary embedding.

We say that j is an *ultrapower embedding* of P if there is some $a \in Q$ such that $Q = \{j(f)(a) : f \in P\}$. In this case we say Q is an *ultrapower* of P.

We say that j is an *internal ultrapower embedding* of P if j is an ultrapower embedding of P that is definable over P from parameters. In this case we say that Q is an *internal ultrapower* of P.

Notice that we are only concerned with ultrapower embeddings *between transitive models*. In our context, this turns out to amount to restricting to countably complete ultrafilters.

With only one exception, we will be concerned only with the case that P and Q are proper class inner models. The next lemma easily implies that we can talk about internal ultrapower embeddings in a first-order way.

LEMMA 2.3. *An elementary embedding $j : P \to Q$ is an ultrapower embedding if and only if there is a P-ultrafilter U such that $Q = M^P_U$ and $j = j^P_U$. The embedding j is an internal ultrapower embedding of P if and only if U can be chosen to be an element of P.* □

We will be informal with our use of proper classes throughout.

3. The Ultrapower Axiom

Motivated by the comparison lemma of inner model theory, we use the following terminology:

DEFINITION 3.1. If $j_0 : V \to M_0$ and $j_1 : V \to M_1$ are ultrapower embeddings, a *comparison* of (j_0, j_1) with *target model* N is a pair of elementary embeddings (i_0, i_1) such that $i_0 : M_0 \to N$ and $i_1 : M_1 \to N$ are internal ultrapower embeddings of their respective domains and $i_0 \circ j_0 = i_1 \circ j_1$.

ULTRAPOWER AXIOM. Every pair of ultrapower embeddings of V admits a comparison.

We devote this section to the inner model theoretic justification for UA, intended for the reader familiar with inner model theory at the level a Woodin cardinal [**11**] although we will not use any fine structure. The rest of the paper can be understood without reading this section.

The embeddings given by the comparison lemma of inner model theory are *not* typically internal ultrapower embeddings. Moreover, even applying the comparison lemma requires the iterability of the models being compared, and it is not clear that an arbitrary countably complete ultrafilter (which is not assumed to lie on an extender sequence) should have an iterable ultrapower. Therefore it is not obvious that UA should hold in all canonical inner models.

Instead, UA is proved by a hull argument. To present this argument abstractly, we prove UA from a more complicated principle that has the benefit of following essentially immediately from the comparison lemma. This principle is Woodin's axiom *Weak Comparison.*

DEFINITION 3.2. A transitive set P is *finitely generated* if there is some $a \in P$ such that every element of P is definable in $(P, \in)$ from a, and P is *pointwise definable* if every element of P is definable in $(P, \in)$ without parameters.

DEFINITION 3.3. An elementary embedding $j : M \to N$ *factors into* an elementary embedding $j' : M \to N'$ if there is an elementary embedding $k : N \to N'$ such that $k \circ j = j'$.

A cofinal elementary embedding $j : M \to N$ is *close* to M if any ultrapower embedding $\bar{j} : M \to \bar{N}$ that factors into j is an internal ultrapower embedding of M.

Although they are not internal, the embeddings given by the comparison lemma turn out to be close.[1]

WEAK COMPARISON. Suppose M_0 and M_1 are finitely generated transitive models of ZFC that are Σ_2-elementarily embeddable in V. Suppose moreover that $\mathbb{R}^{M_0} = \mathbb{R}^{M_1}$. Then there is a transitive set N admitting elementary embeddings $i_0 : M_0 \to N$ and $i_1 : M_1 \to N$ such that i_0 is close to M_0 and i_1 is close to M_1.

Woodin [**20**] argues that Weak Comparison must hold in any inner model constructed by the current inner model theoretic methodology. Granting this, the following proposition shows that the same is true of UA.

[1]For the reader familiar with the inner model theoretic definition of a close extender, if P is a model of ZFC, an elementary embedding $i : P \to Q$ is close to P if and only if $i = j_E^P$ for some extender E that is close to P. The claim that the branch embeddings obtained from the comparison process are close then follows from [**11**] Lemma 6.1.5 and a lemma of [**20**] stating that the transfinite iteration of close embeddings is close.

PROPOSITION 3.4. *Assume Weak Comparison, $V = \text{HOD}$, and the existence of a transitive model of* ZFC *that is Σ_2-elementarily embeddable in V. Then the Ultrapower Axiom holds.*

SKETCH. Let P be a pointwise definable transitive model of ZFC that is Σ_2-elementarily embeddable in V. This exists by the assumption that $V = \text{HOD}$. It is not hard to check that UA is Π_2, so it suffices to show UA holds in P.

Fix internal ultrapowers $j_0 : P \to M_0$ and $j_1 : P \to M_1$ of P. We will show that relative to P, there is a comparison of (j_0, j_1).

Note that M_0 and M_1 are finitely generated transitive models and $\mathbb{R}^{M_0} = \mathbb{R}^{M_1}$. Moreover, by the standard realizability argument,[2] M_0 and M_1 are Σ_2-elementarily embeddable in V. Therefore by Weak Comparison, there is a transitive set N admitting close embeddings $i_0 : M_0 \to N$ and $i_1 : M_1 \to N$.

Let

$$H = H^N(i_0[M_0] \cup i_1[M_1])$$

and let $k : H \to N$ be the inverse of the transitive collapse. Let $h_0 = k^{-1} \circ i_0$ and $h_1 = k^{-1} \circ i_1$. Then $h_0 : M_0 \to H$ and $h_1 : M_1 \to H$.

We claim (h_0, h_1) is a comparison of (j_0, j_1) relative to P, which completes the proof.

First, since P is pointwise definable, $h_0 \circ j_0 = h_1 \circ j_1$ automatically.

It therefore suffices to show that h_0 and h_1 are internal ultrapower embeddings of M_0 and M_1 respectively.

One first shows that h_0 and h_1 are ultrapower embeddings of M_0 and M_1 respectively. For example, $H = \{h_0(f)(h_1(a)) : f \in M_0\}$ for any a such that $M_1 = \{j_1(g)(a) : g \in M_1\}$.

Since i_0 is close and h_0 factors into i_0 via $k : H \to N$, it follows that h_0 is an internal ultrapower embedding of M_0. Similarly h_1 is an internal ultrapower embedding of M_1. This proves the claim and hence the proposition. □

The assumptions beyond Weak Comparison required for the proposition are really artifacts of the way Weak Comparison is stated, so for example it is quite easy to show that UA holds in all of the known canonical inner models, even though $V = \text{HOD}$ does not.

4. The Mitchell order

Recall that a κ-complete ultrafilter U on a cardinal κ is *normal* if $[\text{id}]_U = \kappa$.

DEFINITION 4.1. The *Mitchell order* is defined on normal ultrafilters U and W by setting $U \lhd W$ if $U \in M_W$.

The linearity of the Mitchell order on normal ultrafilters is one of the staples the canonical inner models. We show here that the linearity of the Mitchell order is a consequence of UA.

THEOREM 4.2 (UA). *The Mitchell order wellorders the class of normal ultrafilters.*

[2]See for example [**14**] Theorem 10.3, which treats the realizability of internal ultrapowers of models that are Σ_{1000}-elementarily embeddable in V.

PROOF. Suppose U and W are normal ultrafilters. We will show that either $U = W$, $U \lhd W$, or $W \lhd U$.

By UA, there is a comparison (k_U, k_W) of (j_U, j_W) to a target model N. That is, $k_U : M_U \to N$ and $k_W : M_W \to N$ are internal ultrapower embeddings of M_U and M_W respectively and $k_U \circ j_U = k_W \circ j_W$.

Let $\kappa = [\mathrm{id}]_U = \mathrm{CRT}(U)$ and $\kappa' = [\mathrm{id}]_W = \mathrm{CRT}(W)$. If $\kappa < \kappa'$ then $U \in V_{\kappa'} \subseteq M_W$, so $U \lhd W$. Similarly if $\kappa' < \kappa$, then $W \lhd U$. So we may assume without loss of generality that $\kappa = \kappa'$.

We now consider three cases:

CASE 1. $k_U(\kappa) = k_W(\kappa)$

CASE 2. $k_U(\kappa) < k_W(\kappa)$

CASE 3. $k_W(\kappa) < k_U(\kappa)$

In Case 1, we will show $U = W$. In Case 2, we will show $U \lhd W$. In Case 3, we will show $W \lhd U$.

The following calculation is useful in the first two cases: for any $X \subseteq \kappa$,

$$\begin{aligned} X \in U &\Leftrightarrow \kappa \in j_U(X) \\ &\Leftrightarrow k_U(\kappa) \in k_U(j_U(X)) \\ &\Leftrightarrow k_U(\kappa) \in k_W(j_W(X)) \end{aligned} \tag{1}$$

Given this we turn to the casework.

PROOF IN CASE 1. Using (1) and the case assumption that $k_U(\kappa) = k_W(\kappa)$, for any $X \subseteq \kappa$,

$$X \in U \Leftrightarrow k_W(\kappa) \in k_W(j_W(X))$$

By the elementarity of k_W, this holds if and only if $\kappa \in j_W(X)$; that is, if and only if $X \in W$. So $X \in U$ if and only if $X \in W$, and hence $U = W$. □

PROOF IN CASE 2. Using (1) and the case assumption that $k_U(\kappa) < k_W(\kappa)$, for any $X \subseteq \kappa$,

$$X \in U \Leftrightarrow k_U(\kappa) \in k_W(j_W(X)) \cap k_W(\kappa)$$

By the elementarity of k_W, this holds if and only if $k_U(\kappa) \in k_W(j_W(X) \cap \kappa)$; but since $\kappa = \mathrm{CRT}(W)$, this holds if and only if $k_U(\kappa) \in k_W(X)$. So $X \in U$ if and only if $k_U(\kappa) \in k_W(X)$. Since $k_W : M_W \to N$ is an internal embedding of M_W, it follows that U is definable over M_W. Hence $U \in M_W$, so $U \lhd W$. □

PROOF IN CASE 3. Similar to the proof in Case 2. □

This proves the theorem. □

An immediate followup question is whether this result can be extended to a broader class of countably complete ultrafilters than the normal ones. We describe one extension briefly and then turn in the next section to a more important extension.

The Mitchell order generalizes in an obvious way to all countably complete ultrafilters:

DEFINITION 4.3. The *generalized Mitchell order* is defined on countably complete ultrafilters U and W by setting $U \lhd W$ if and only if $U \in M_W$.

Under this definition the generalized Mitchell order is not Rudin-Keisler invariant or even a partial order, but this is irrelevant in the cases we consider.

Solovay introduced a well-known and very important generalization of normality:

DEFINITION 4.4. Suppose A is a set and $\mathcal{U}$ is an ultrafilter on $X \subseteq P(A)$.

(1) $\mathcal{U}$ is *fine* if for all $a \in A$, for $\mathcal{U}$-almost all $\sigma \in X$, $a \in \sigma$.
(2) $\mathcal{U}$ is *normal* if for all $f : X \to A$, if $f(\sigma) \in \sigma$ for $\mathcal{U}$-almost all $\sigma \in X$, then for some $a \in A$, $f(\sigma) = a$ for $\mathcal{U}$-almost all $\sigma \in X$.

Theorem 4.2 generalizes to all normal fine ultrafilters on $P(\lambda)$ for any cardinal λ. In order to state the theorem in its most general form, we consider a modified version of the Mitchell order. The issue is that typically $P(P(A)) \notin M_{\mathcal{W}}$, and in this case for any $\mathcal{U}$ on $P(A)$, $\mathcal{U} \not\lhd \mathcal{W}$. But one might still hope that there is a set $X \in \mathcal{U}$ such that $\mathcal{U} \cap P(X)$, the restriction of $\mathcal{U}$ to X, lies below $\mathcal{W}$ in the Mitchell order.

DEFINITION 4.5. Suppose A is a set and $\mathcal{U}$ and $\mathcal{W}$ are countably complete ultrafilters on $P(A)$. We set $\mathcal{U} \lhd^* \mathcal{W}$ if there is some $X \in \mathcal{U}$ such that $\mathcal{U} \cap P(X) \lhd \mathcal{W}$.

THEOREM 4.6 (UA). *For any cardinal λ, $\lhd^*$ wellorders the set of normal fine ultrafilters on $P(\lambda)$.* □

A proof of this theorem under the additional assumption $2^{<\lambda} = \lambda$ appears in [**4**]. Removing this assumption seems to require a much more involved argument that uses the techniques of Section 6 and Section 7. A proof will appear in the author's thesis [**5**].

A somewhat simpler but less general fact can be stated in terms of the usual Mitchell order:

THEOREM 4.7 (UA). *If $\kappa \leq \lambda$ are cardinals, then the generalized Mitchell order wellorders the set of normal fine κ-complete ultrafilters on $P_\kappa(\lambda)$.* □

5. The seed order

The most important extension of Theorem 4.2 does not involve the Mitchell order at all. Instead, the proof of Theorem 4.2 suggests a new order on ultrafilters which under UA wellorders *all* uniform countably complete ultrafilters.

DEFINITION 5.1. We establish some basic notation for ultrafilters:

(1) An ultrafilter U on an ordinal α is *uniform* if $\alpha \setminus \xi \in U$ for all $\xi < \alpha$.
(2) If U is a uniform ultrafilter, the *space* of U, denoted $\text{SP}(U)$, is the unique ordinal α such that $\alpha \in U$.
(3) We let Un denote the *class of uniform countably complete ultrafilters.*
(4) For any ordinal α,

$$\text{Un}_\alpha = \{U \in \text{Un} : \text{SP}(U) = \alpha\}$$
$$\text{Un}_{<\alpha} = \{U \in \text{Un} : \text{SP}(U) < \alpha\}$$
$$\text{Un}_{\leq\alpha} = \{U \in \text{Un} : \text{SP}(U) \leq \alpha\}$$

For contrast (and since it will come up later), we define a variant of uniformity that does not always coincide with the notion from Definition 5.1.

DEFINITION 5.2. An ultrafilter U on a cardinal λ is *strongly uniform* if for all $A \in U$, $|A| = \lambda$.

If λ is a regular cardinal, then an ultrafilter on λ is strongly uniform if and only if it is uniform. On singular ordinals these concepts diverge. For example, the principal ultrafilter P_α concentrated at an ordinal α is a uniform ultrafilter on $\alpha+1$, but it is not strongly uniform if $\alpha > 0$. The pushforward of a nonprincipal ultrafilter on ω via the function $n \mapsto \aleph_n$ yields a uniform ultrafilter on $\aleph_\omega$ that is not strongly uniform.

In the proof of Theorem 4.2, the question that split the proof into cases was whether $k_U([\text{id}]_U) < k_W([\text{id}]_W)$. The seed order is defined by asking the same question in a more general context:

DEFINITION 5.3 (UA). The *seed order* is defined on $U, W \in \text{Un}$ by setting $U <_S W$ if and only if there is a comparison (k_U, k_W) of (j_U, j_W) such that $k_U([\text{id}]_U) < k_W([\text{id}]_W)$.

The goal of this section is to prove one of the key consequences of UA:

THEOREM 5.4 (UA). *The seed order is a set-like wellorder of the class of uniform countably complete ultrafilters.*

We begin with set-likeness, which does not require UA at all. Recall that a (class) relation $R \subseteq V \times V$ is *set-like* if for all $x \in V$ the class $\{y \in V : y \mathrel{R} x\}$ forms a set.

LEMMA 5.5. *Suppose $U, W \in \text{Un}$ and $U <_S W$. Then $\text{SP}(U) \leq \text{SP}(W)$.*

PROOF. By definition, $\text{SP}(U)$ is the least ordinal α such that $[\text{id}]_U < j_U(\alpha)$. Let $\delta = \text{SP}(W)$; then $[\text{id}]_W < j_W(\delta)$. To show $\text{SP}(U) \leq \delta$, it suffices to show $[\text{id}]_U < j_U(\delta)$.

Let (k_U, k_W) be a comparison of (j_U, j_W) witnessing that $U <_S W$. Then $k_U([\text{id}]_U) < k_W([\text{id}]_W) < k_W(j_W(\delta))$, using here that $[\text{id}]_W \leq j_W(\delta)$. It follows that $k_U([\text{id}]_U) < k_U(j_U(\delta))$, and therefore $[\text{id}]_U < j_U(\delta)$, as desired. □

COROLLARY 5.6. *The seed order is set-like.* □

There are several ways to prove the wellfoundedness and transitivity of the seed order. One way is to check these properties directly, which does not pose great difficulties. Another way is to form the direct limit M_∞ of all ultrapowers and embed the seed order into Ord^{M_∞}. We take the latter course here.

We begin by showing that the collection of all ultrapowers forms a directed system. This comes down to the uniqueness of ultrapower embeddings:

PROPOSITION 5.7 (UA). *Suppose $j, j' : V \to M$ are ultrapower embeddings to a common inner model M. Then $j = j'$.*

The proof uses the following folklore scheme.

LEMMA 5.8. *Given two elementary embeddings $k, k' : V \to N$ that are definable over V from parameters, $k \restriction \text{Ord} = k' \restriction \text{Ord}$.* □

We will only need to use the instance of Lemma 5.8 in which k and k' are ultrapower embeddings. We sketch the proof in this case although it is really no different from the general case. Suppose towards a contradiction that there exist two ultrapower embeddings from the universe into a common model that disagree on the ordinals. Let α be the least ordinal such that there exist two ultrapower embeddings from the universe into a common model that disagree at α. Note that α is definable without parameters, and therefore if $k, k' : V \to N$ are elementary embeddings to a common model, $k(\alpha) = k'(\alpha)$. This contradicts the definition of α.

PROOF OF PROPOSITION 5.7. Fix a comparison (k, k') of (j, j') with target N. Applying Lemma 5.8 in M, $k \restriction \mathrm{Ord} = k' \restriction \mathrm{Ord}$.

We claim that j and j' agree on all sets of ordinals. Fix a set of ordinals X. For any ordinal α,

$$\begin{aligned}\alpha \in j(X) &\Leftrightarrow k(\alpha) \in k(j(X))\\ &\Leftrightarrow k(\alpha) \in k'(j'(X))\\ &\Leftrightarrow k'(\alpha) \in k'(j'(X))\\ &\Leftrightarrow \alpha \in j'(X)\end{aligned}$$

Since j and j' agree on all sets of ordinals, they agree on all sets by the Wellordering Theorem. □

One can therefore divide the Ultrapower Axiom into two separate statements.

THEOREM 5.9. UA *is equivalent to the conjunction of the following statements:*

(1) *Any two ultrapowers of V have a common internal ultrapower.*
(2) *There is a unique ultrapower embedding from V into any of its ultrapowers.*

The uniqueness of ultrapower embeddings gives rise to a directed system:

DEFINITION 5.10. Let Up denote the collection of ultrapowers of V by countably complete ultrafilters.

The notation $\leq_{\mathrm{RF}}$ below stands for Rudin-Frolík (see Definition 7.8).

DEFINITION 5.11 (UA). For $P, Q \in \mathrm{Up}$, we write $P \leq_{\mathrm{RF}} Q$, if $Q \in \mathrm{Up}^P$. If $P \leq_{\mathrm{RF}} Q$, we let $j_{P,Q} : P \to Q$ be the unique internal ultrapower embedding. Let $\mathcal{D} = \langle P, j_{P,Q} : P \leq_{\mathrm{RF}} Q \in \mathrm{Up}\rangle$.

LEMMA 5.12 (UA). *$\mathcal{D}$ is a directed system.*

DEFINITION 5.13 (UA). Let $(M_\infty, E) = \lim \mathcal{D}$, and for each $P \in \mathrm{Up}$, let $j_{P,\infty} : P \to M_\infty$ be the direct limit embedding.

To prove Theorem 5.4, we embed the seed order $(\mathrm{Un}, <_S)$ into the ordinals of M_∞.

DEFINITION 5.14. For any $U \in \mathrm{Un}$, let $|U|_\infty = j_{M_U,\infty}([\mathrm{id}]_U)$.

PROOF OF THEOREM 5.4. We first show that (M_∞, E) is wellfounded in the sense that there are no infinite E-descending sequences. For this, it suffices to check that any linear subsystem of $\mathcal{D}$ of ordertype ω has a wellfounded direct limit.

In other words, we need to check that if $\langle P_n : n < \omega\rangle$ is a sequence of ultrapowers such that for all $n \leq m < \omega$, $P_n \leq_{\mathrm{RF}} P_{n+1}$, then letting $\mathcal{L} = \langle P_n, j_{P_n,P_m} : n \leq m < \omega\rangle$ be the associated linearly directed system, $\lim \mathcal{L}$ is wellfounded. This is a well-known fact due to Mitchell.

It follows that (M_∞, E) is wellfounded. In particular, $(\text{Ord}^{M_\infty}, E)$ is a wellorder, though it may not be set-like.

We first show that any $U \in \text{Un}$ can be recovered from $|U|_\infty$. In fact, U is the uniform ultrafilter derived from $j_{V,\infty}$ using $|U|_\infty$:

$$U = \{X \subseteq \delta : |U|_\infty \in j_{V,\infty}(X)\}$$

where δ is the least ordinal such that $|U|_\infty < j_{V,\infty}(\delta)$. To see this, note that $X \in U$ if and only if $[\text{id}]_U \in j_U(X)$ if and only if $j_{M_U,\infty}([\text{id}]_U) \in j_{M_U,\infty}(j_U(X))$ if and only if $|U|_\infty \in j_{V,\infty}(X)$.

It follows that the map $|-|_\infty$ is an injection on Un. We finish by showing that it is an order embedding $(\text{Un}, <_S) \to (\text{Ord}^{M_\infty}, E)$.

Suppose $U, W \in \text{Un}$. Let (i_U, i_W) be a comparison of (j_U, j_W) with target model N. Then

$$\begin{aligned} i_U([\text{id}]_U) < i_W([\text{id}]_W) &\Leftrightarrow j_{N,\infty} \circ i_U([\text{id}]_U) < j_{N,\infty} \circ i_W([\text{id}]_W) \\ &\Leftrightarrow j_{M_U,\infty}([\text{id}]_U) < j_{M_W,\infty}([\text{id}]_W) \\ &\Leftrightarrow |U|_\infty < |W|_\infty \end{aligned}$$

Thus $U <_S W$ if and only if $|U|_\infty < |W|_\infty$.

Therefore there is an injective embedding from $(\text{Un}, <_S)$ into the wellorder $(\text{Ord}^{M_\infty}, E)$. It follows that $(\text{Un}, <_S)$ is a wellorder. □

The question of the set-likeness of M_∞ naturally arises in the proof. Naively, given the set-likeness of the seed order, one might conjecture that M_∞ is also set-like. It seems likely that this is not true. In connection to this, we state a theorem from the analysis of supercompactness under UA.

THEOREM 5.15 (UA). *The following are equivalent:*

(1) *M_∞ is not set-like.*
(2) *There is a supercompact cardinal.* □

5.1. Ordinal definable ultrafilters. As an application of Theorem 5.4, we prove Theorem 1.1. In fact we just need a coarse corollary of Theorem 5.4:

COROLLARY 5.16 (UA). *Any countably complete ultrafilter on an ordinal is ordinal definable.* □

THEOREM 5.17 (UA). *Suppose κ is supercompact and $X \subseteq \kappa$. If $V_\kappa \subseteq \text{HOD}_X$ then $V = \text{HOD}_X$.*

PROOF. We first claim that for any $U \in \text{Un}$ and for any set of ordinals A, $j_U(\text{HOD}_A) \subseteq \text{HOD}_A$. Easily $j_U(\text{OD}_A) = \text{OD}^{M_U}_{j_U(A)} \subseteq \text{OD}_A$ since U is OD. By definition, then, the hereditarily $j_U(\text{OD}_A)$ sets are hereditarily OD_A. In other words, $j_U(\text{HOD}_A) \subseteq \text{HOD}_A$, as desired.

Now assume $X \subseteq \kappa$ is such that $V_\kappa \subseteq \text{HOD}_X$. Fix $\lambda \in \text{Ord}$, and we will show that $V_\lambda \subseteq \text{HOD}_X$. Since κ is supercompact, there is some $U \in \text{Un}$ such that $j_U(\kappa) > \lambda$ and $V_\lambda \subseteq M_U$. Thus $V_\lambda \subseteq j_U(V_\kappa) \subseteq j_U(\text{HOD}_X) \subseteq \text{HOD}_X$, as desired. □

A less trivial argument shows in ZFC that if every κ-complete ultrafilter is OD and κ is strongly compact, then $V = \text{HOD}_X$ for any $X \subseteq \kappa$ coding V_κ. This argument will appear in the author's thesis [**5**].

We now prove Theorem 1.1, which we restate here for the reader's convenience.

THEOREM 5.18 (UA). *If there is a supercompact cardinal, then V is a set-generic extension of* HOD.

PROOF. This follows from Theorem 5.17 and Vopenka's theorem that HOD_X is a generic extension of HOD for any set of ordinals X. □

By a closer inspection of the proof of Vopenka's theorem, using the results of Section 6, one can compute that the forcing of Theorem 1.1 has cardinality at most κ^{++} in HOD.

QUESTION 5.19 (UA). Assume κ is supercompact. Must V be a generic extension of HOD for a forcing of cardinality κ^+ in HOD?

6. GCH from UA

6.1. The internal relation. The following slight variant of the generalized Mitchell order holds the key to the connection between the Ultrapower Axiom, cardinal arithmetic, and supercompactness.

DEFINITION 6.1. The internal relation $\sqsubset$ is defined on $U, W \in \text{Un}$ by setting $U \sqsubset W$ if $j_U \restriction M_W$ is an internal ultrapower embedding of M_W.

The idea behind the internal relation is to define a variant of the generalized Mitchell order in the "language of comparison": $U \sqsubset W$ if and only if $(j_U(j_W), j_U \restriction M_W)$ is a comparison of (j_U, j_W). It is not surprising that UA, an axiom asserting the existence of comparisons, proves a lot about the internal relation.

If $U \sqsubset W$ then $U \cap M_W \in M_W$ since

$$U \cap M_W = \{X \in P^{M_W}(\text{SP}(U)) : [\text{id}]_U \in (j_U \restriction M_W)(X)\}$$

Therefore if $U \sqsubset W$ and $P(\text{SP}(U)) \subseteq M_W$, then $U \lhd W$.

We remark that if there is a supercompact cardinal κ then there is a nonprincipal κ-complete ultrafilter U on a cardinal λ such that $U \cap M_U \in M_U$. (This example will appear in the author's thesis [**5**].) The internal relation, on the other hand, is irreflexive on nonprincipal ultrafilters. We caution, however, that the internal relation has 2-cycles.[3] It follows that the internal relation is not transitive, even on nonprincipal ultrafilters.

In the details we provide, the only fact about the internal relation we will use explicitly is the following result from [**5**]:

PROPOSITION 6.3 (UA). *Suppose δ is a regular cardinal and U is the $<_S$-least ultrafilter in Un_δ. Then for all $D \in \text{Un}_{<\delta}$, $D \sqsubset U$.* □

6.2. A proof of GCH. There are several different ways to prove GCH above a supercompact from UA, and here we exposit what we think is the easiest one. We then quote the somewhat stronger results one can obtain from the other proofs.

The idea of the proof here is to reduce the problem to a problem about the Mitchell order, since the Mitchell order is quite tractable under UA. It turns out that failures of Weak GCH above a supercompact lead to wild behavior in the Mitchell order:

---

[3]These come from Kunen's commuting ultrapowers lemma:

THEOREM 6.2 (Kunen). *Suppose U is a κ-complete ultrafilter and W is an ultrafilter in V_κ. Then $j_U(j_W) = j_W \restriction M_U$ and $j_W(j_U) = j_U \restriction M_W$.*

LEMMA 6.4. *Suppose κ is supercompact, $\delta \geq \kappa$ is regular, and for some $\lambda < \delta$, $2^\lambda = 2^\delta$. Then for any ultrafilter U on δ, there is a normal fine κ-complete ultrafilter $\mathcal{D}$ on $P_\kappa(\lambda)$ such that $U \lhd \mathcal{D}$.*

PROOF. Fix $A \subseteq 2^\delta$ such that $U \in L[A]$. This is possible because $U \subseteq P(\delta)$. By a theorem of Solovay [**17**], for any $B \subseteq 2^\lambda$, there is a normal fine κ-complete ultrafilter $\mathcal{D}$ on $P_\kappa(\lambda)$ with $B \in M_\mathcal{D}$. Since $2^\lambda = 2^\delta$, we can therefore find $\mathcal{D}$ such that $A \in M_\mathcal{D}$. But then $U \in L[A] \subseteq M_\mathcal{D}$, so $U \lhd \mathcal{D}$, as desired. □

As a consequence, we obtain Weak GCH. This is the key point of the proof:

PROPOSITION 6.5 (UA). *Suppose κ is supercompact and λ is a cardinal satisfying* $\text{cf}(\lambda) \geq \kappa$. *Then $2^\lambda < 2^{(\lambda^+)}$.*

PROOF. Suppose not. Let U be the $<_S$-least ultrafilter in Un_{λ^+} and let $\mathcal{D}$ be a countably complete ultrafilter on $P_\kappa(\lambda)$ such that $U \lhd \mathcal{D}$. Since $\text{cf}(\lambda) \geq \kappa$, Solovay's theorem [**16**] implies $|P_\kappa(\lambda)| = \lambda$. Therefore $\mathcal{D}$ is equivalent to an ultrafilter on λ, so it follows from Proposition 6.3 that $\mathcal{D} \sqsubset U$.

But applying Proposition 6.3 at $j_\mathcal{D}(\lambda^+)$ in $M_\mathcal{D}$, we obtain $U \sqsubset^{M_\mathcal{D}} j_\mathcal{D}(U)$, since $U \in \text{Un}_{\lambda^+}^{M_\mathcal{D}}$ and $\lambda^+ < j_\mathcal{D}(\kappa) < j_\mathcal{D}(\lambda^+)$. In particular, $j_U^{M_\mathcal{D}} \restriction \alpha \in M^{M_\mathcal{D}}_{j_\mathcal{D}(U)}$ for any ordinal α. Since ${}^{\lambda^+}(\lambda^+) \subseteq M_\mathcal{D}$, we have $j_U^{M_\mathcal{D}} \restriction \lambda^+ = j_U \restriction \lambda^+$, and hence

$$j_U \restriction \lambda^+ \in M^{M_\mathcal{D}}_{j_\mathcal{D}(U)} = j_\mathcal{D}(M_U) \subseteq M_U$$

The final inclusion follows from the fact that $\mathcal{D} \sqsubset U$.

But then U is λ^+-supercompact! In particular $P(\lambda^+) \subseteq M_U$ and so since $\mathcal{D} \sqsubset U$, we can conclude that $\mathcal{D} \lhd U$ (see the comments following Definition 6.1). Now $\mathcal{D} \lhd U$ and $U \lhd \mathcal{D}$, and this contradicts that the Mitchell order is asymmetric. □

Using the linearity of the Mitchell order, one can get from Weak GCH to GCH. This uses an observation of Solovay:

LEMMA 6.6. *Suppose κ is supercompact and λ is a cardinal with* $\text{cf}(\lambda) \geq \kappa$. *Assume that the Mitchell order is linear on normal fine κ-complete ultrafilters on $P_\kappa(\lambda)$. Then $2^{2^\lambda} = (2^\lambda)^+$.*

PROOF. By a theorem of Solovay [**17**], there are 2^{2^λ} normal fine κ-complete ultrafilters on $P_\kappa(\lambda)$. Any normal fine κ-complete ultrafilter $\mathcal{D}$ on $P_\kappa(\lambda)$ has at most 2^λ predecessors in the Mitchell order: this is an easily proved, well-known fact. Thus the ordertype of the Mitchell order on normal fine κ-complete ultrafilters on $P_\kappa(\lambda)$ is at most $(2^\lambda)^+$. It follows that $2^{2^\lambda} = (2^\lambda)^+$. □

THEOREM 6.7 (UA). *Suppose κ is supercompact and $\lambda \geq \kappa$. Then $2^\lambda = \lambda^+$.*

PROOF. By induction we may assume $2^{<\lambda} = \lambda$. Therefore if λ is singular, we are done by Solovay's theorem [**16**]. So assume λ is regular.

By Theorem 4.7, the Mitchell order is linear on normal fine κ-complete ultrafilters on $P_\kappa(\lambda)$. (In order to avoid apparent circularity, we should really cite the main theorem of [**4**], which requires the assumption $2^{<\lambda} = \lambda$, which we get by our induction hypothesis.) Therefore by Lemma 6.6, $2^{2^\lambda} = (2^\lambda)^+$.

Applying Proposition 6.5, $2^\lambda < 2^{(\lambda^+)} \leq 2^{2^\lambda}$. But $2^{2^\lambda} = (2^\lambda)^+$ by Lemma 6.6, using Theorem 4.7 to conclude that the Mitchell order is linear. It follows that $2^\lambda < 2^{(\lambda^+)} \leq (2^\lambda)^+$, and therefore $2^{(\lambda^+)} = (2^\lambda)^+ = 2^{2^\lambda}$.

On the other hand applying Proposition 6.5 again, $2^{(\lambda^+)} < 2^{(\lambda^{++})}$ and therefore $2^{2^\lambda} < 2^{(\lambda^{++})}$. It follows that $2^\lambda < \lambda^{++}$, and therefore $2^\lambda = \lambda^+$. □

The proof we have given here leaves open the question of whether it is possible (assuming UA) that κ is 2^κ-supercompact yet 2^κ is weakly inaccessible, even if one localizes the proof in the obvious way. We now discuss some theorems from [**5**] that show that this cannot occur:

THEOREM 6.8 (UA). *Suppose λ is a cardinal and λ^+ carries a uniform countably complete ultrafilter. Then $2^{<\lambda} = \lambda$.* □

COROLLARY 6.9 (UA). *Suppose γ is a cardinal and γ^{++} carries a uniform countably complete ultrafilter. Then $2^\gamma = \gamma^+$.*

PROOF. Take $\lambda = \gamma^+$ in Theorem 6.8. □

COROLLARY 6.10 (UA). *Suppose κ is 2^κ-strongly compact. Then $2^\kappa = \kappa^+$.*

PROOF. Assume towards a contradiction that $2^\kappa > \kappa^+$. Then κ^{++} carries a uniform countably complete ultrafilter. Therefore by Corollary 6.9, $2^\kappa = \kappa^+$, a contradiction. □

There is no reason to believe that one cannot vastly improve our best local results:

QUESTION 6.11 (UA). Suppose κ is measurable. Must $2^\kappa = \kappa^+$?

We would guess that this is independent of ZFC + UA. But maybe one can show $2^\kappa = \kappa^+$ assuming UA plus κ is κ^+-strongly compact. (It would of course suffice to handle the case that κ is κ^+-supercompact.)

One can sometimes derive stronger principles than GCH from UA:

THEOREM 6.12 (UA). *Suppose δ is regular and δ^{++} carries a uniform countably complete ultrafilter. Then $\diamondsuit(S^{\delta^{++}}_{\delta^+})$.* □

This raises an interesting combinatorial question:

QUESTION 6.13 (UA). Suppose κ is supercompact. Does $\diamondsuit(S^{\kappa^+}_\kappa)$ hold?

7. The identity crisis

The results sketched and stated in this section will appear in their complete form in the author's thesis [**5**].

Magidor's classic theorem [**9**] on the "identity crisis" for strongly compact cardinals motivates the results of this section.

THEOREM 7.1 (Magidor). *If κ is strongly compact, then there is a forcing extension in which κ remains strongly compact but becomes the least measurable cardinal.* □

Therefore if it is consistent that there is a strongly compact cardinal, it is consistent that the least measurable cardinal is strongly compact.

On the other hand, an unpublished theorem of Kimchi-Magidor shows that it is consistent for strong compactness and supercompactness to coincide:

THEOREM 7.2 (Kimchi-Magidor). *There is a class forcing extension preserving all supercompact cardinals in which for all uncountable cardinals κ, the following are equivalent:*

(1) *κ is strongly compact.*
(2) *κ is supercompact or κ is a measurable limit of supercompact cardinals.*

One must include the exceptional case in (2) by the following results of Menas **[10]**:

THEOREM 7.3 (Menas). *A measurable limit of strongly compact cardinals is strongly compact.* □

Therefore if there is a measurable limit of strongly compact cardinals, there is a strongly compact cardinal that is not supercompact:

COROLLARY 7.4 (Menas). *Suppose κ is the least measurable limit of strongly compact cardinals. Then κ is strongly compact but the set of measurable cardinals less than κ is nonstationary in κ. In particular, κ is not supercompact.*

PROOF. The set of measurable cardinals less than κ is disjoint from the closed unbounded set of limits of strongly compact cardinals below κ. □

Assuming UA, the Kimchi-Magidor consistency result becomes a theorem. The proof proceeds in two steps: one first carries out an analysis of the *least supercompact cardinal* before handling all higher instances of supercompactness by a separate argument.

It is a remarkable coincidence that the structure of this proof mirrors Woodin's conjectured solution to the inner model problem, in which one first builds a canonical inner model for a single supercompact cardinal before proving by a separate argument that this model inherits all higher instances of supercompactness.

Thus one begins by showing:

THEOREM 7.5 (UA). *The least strongly compact cardinal is supercompact.* □

This is proved by an analysis of the $<_S$-least strongly uniform ultrafilter on a regular cardinal.[4]

THEOREM 7.6 (UA). *Suppose δ is a successor cardinal. The $<_S$-least countably complete uniform ultrafilter on δ is δ-supercompact.* □

This is an application of the theory of the internal relation and some infinitary combinatorics involving independent families of sets. One would expect the theorem to be true for arbitrary regular cardinals, but we are only able to prove the following:

THEOREM 7.7 (UA). *Suppose δ is an inaccessible cardinal. The $<_S$-least countably complete uniform ultrafilter U on δ is $<\delta$-supercompact. Moreover if $A \subseteq M_U$ and $|A| = \delta$, then there is some $A' \in M_U$ with $|A'| = \delta$ and $A \subseteq A'$.* □

We now turn to the argument that the other strongly compact cardinals are supercompact. Corollary 7.4 implies that we cannot hope to prove that *all* strongly compact cardinals are supercompact, and the proof must somehow reflect this.

[4]Ketonen **[7]** actually considered such ultrafilters long before our time; see **[7]** Theorem 5.12, which turns out to be an important part of the proof of Theorem 7.11.

DEFINITION 7.8. Suppose D and U are countably complete ultrafilters. The *Rudin-Frolík order* is defined by setting $D \leq_{\text{RF}} U$ if there is an internal ultrapower embedding $k : M_D \to M_U$ such that $k \circ j_D = j_U$. We say U is *irreducible* if it is minimal in the Rudin-Frolík order; i.e., for any ultrafilter D, if $D \leq_{\text{RF}} U$ then either D is principal or else $j_D = j_U$.

The Rudin-Frolík order was introduced by Frolík [**2**] whose focus was on countably incomplete ultrafilters, and in particular ultrafilters on ω. Kanamori [**6**] studies a version of the Rudin-Frolík order on κ-complete ultrafilters on a measurable cardinal κ, but we warn that what he calls the Rudin-Frolík order is a stronger order that has significantly different properties from the order considered here. The Rudin-Frolík order is of course closely related to the Ultrapower Axiom, since UA is equivalent to the statement that the Rudin-Frolík order is directed.

The key observation is that despite Corollary 7.4, all the pathological countably complete ultrafilters proved to exist from large cardinal hypotheses have been reducible ultrafilters. For example, it is unknown whether one can prove from large cardinal hypotheses that there is an irreducible strongly uniform ultrafilter on a cardinal λ that does not witness λ-supercompactness. Theorem 7.13 below shows that such a proof would constitute an anti-inner model theorem except perhaps if λ is inaccessible.

The main fact about irreducible ultrafilters under UA is also a key component in the proof of the equivalence of strong compactness and supercompactness.

THEOREM 7.9 (UA). *Every countably complete ultrafilter is equivalent to a finite linear iteration of irreducible ultrafilters.* □

At first glance, the theorem looks almost obvious, but in fact it cannot be proved without appealing to UA, essentially by results of Gitik [**3**].

By a comparison argument appealing to Theorem 7.6, one proves the following fundamental fact about irreducible ultrafilters:

PROPOSITION 7.10 (UA). *Suppose δ is a successor cardinal. Then any uniform irreducible ultrafilter W on δ is δ-supercompact.* □

The proof can be seen as an attempt to simulate the phalanx argument of fine structural inner model theory in the abstract context of UA.[5] We give a very vague outline of the proof (glossing over all the difficulties).

By Theorem 7.6, the $<_S$-least ultrafilter U on δ is supercompact. We compare U and W, yielding internal ultrapower embeddings $k_U : M_U \to N$ and $k_W : M_W \to N$. The bulk of the proof is spent showing that one can arrange that $\text{CRT}(k_U) > \delta$. Given this, it follows that $\text{Ord}^\delta = \text{Ord}^\delta \cap M_U \subseteq N \subseteq M_W$, and hence W is δ-supercompact.

A minor issue is that one can easily find a comparison (k'_U, k'_W) of (j_U, j_W) such that $\text{CRT}(k'_U) < \delta$ by taking $F \in \text{Un}^N$ with critical point below λ and setting $k'_U = j^N_F \circ k_U$ and $k'_W = j^N_F \circ k_W$. To ensure that the comparison (k_U, k_W) we started out with was not already of this form, we assume without loss of generality that (k_U, k_W) is a *canonical comparison*, which essentially means that it involves no ultrapowers except for those required to compare (j_U, j_W). (More formally,

[5]More precisely, we are referring to comparison arguments involving double-rooted iteration trees; see [**18**] Section 5.

we assume N is the least upper bound of M_U and M_W in the directed system of ultrapowers, which under UA turns out to be a lattice.)

We then factor k_U over M_U as $h \circ i$ where i is the ultrapower of M_U by an ultrafilter $D \in \mathrm{Un}^{M_U}_{<\delta}$ and h is an internal ultrapower embedding of $(M_D)^{M_U}$ with $\mathrm{CRT}(h) > i(\delta)$. This turns out to be quite complicated, but the underlying reason that it is possible is that by the $<_S$-minimality of U, δ carries no uniform countably complete ultrafilters in M_U.

Since U is δ-supercompact by Theorem 7.6, D is also a countably complete ultrafilter *in* V. We can therefore come back and apply D to V. Comparing M_D, M_U, and M_W simultaneously, and using that (k_U, k_W) is a canonical comparison, one shows that $D \leq_{\mathrm{RF}} W$. Since W is irreducible, D is principal. But then i is the identity, so $k_U = h \circ i = h$, so $\mathrm{CRT}(k_U) > \delta$ as desired.

Given these facts, we prove the equivalence of strong compactness and supercompactness Theorem 1.3, restated here for the reader's convenience:

THEOREM 7.11 (UA). *Suppose κ is an uncountable cardinal. The following are equivalent:*

(1) *κ is strongly compact.*
(2) *κ is supercompact or κ is a measurable limit of supercompact cardinals.*

SKETCH. By Menas's theorem, we only need to show (1) implies (2). Suppose κ is strongly compact. Let $\delta > \kappa$ be a successor cardinal. We claim that either κ is δ-supercompact or κ is a limit of δ-supercompact cardinals.

Let W be the $<_S$-least κ-complete ultrafilter in Un_δ. It follows from Theorem 7.9 that $j_W = j^{M_D}_Z \circ j_D$ for some $D \in \mathrm{Un}_{<\delta}$ and some irreducible ultrafilter $Z \in \mathrm{Un}^{M_D}_{j_D(\delta)}$. By Proposition 7.10, Z is $j_D(\delta)$-supercompact in M_D.

If D is principal, then $W = Z$ and therefore W is δ-supercompact. It follows from a theorem of Ketonen ([**7**] Theorem 5.12) that $\mathrm{CRT}(W) = \kappa$, and therefore W witnesses that κ is δ-supercompact.

In the case that D is nonprincipal, one uses the following lemma:

LEMMA 7.12 (UA). *Suppose D is a nonprincipal divisor of a countably complete ultrafilter W. Let Z be the $<^{M_D}_S$-least element of Un^{M_D} with $j_W = j^{M_D}_Z \circ j_D$. Then $Z <^{M_D}_S j_D(W)$.* □

Thus without loss of generality we may assume $Z <^{M_D}_S j_D(W)$. Since $j_D(W)$ is the $<^{M_D}_S$-least κ-complete ultrafilter in $\mathrm{Un}^{M_D}_{j_D(\delta)}$, it follows that $\mathrm{CRT}(Z) \in [\kappa, j_D(\kappa))$. But $\mathrm{CRT}(Z)$ is $j_D(\delta)$-supercompact in M_D, so by the usual reflection argument, κ is a limit of δ-supercompact cardinals. □

The proof of Proposition 7.10 shows:

THEOREM 7.13 (UA). *Suppose λ is a strong limit singular cardinal or a successor cardinal. Then any irreducible strongly uniform ultrafilter on λ is λ-supercompact.* □

By the proof of Theorem 7.11, one therefore has a version of Apter-Shelah's level-by-level equivalence of strong compactness and supercompactness [**1**] at successor cardinals:

THEOREM 7.14 (UA). *Suppose $\kappa < \delta$ and δ is a successor cardinal. If κ is δ-strongly compact then either κ is δ-supercompact or κ is a limit of δ-supercompact cardinals.* □

If there is a supercompact cardinal, there exist cardinals $\kappa < \lambda$ such that κ is λ-strongly compact but not λ-supercompact; in the known examples, λ is singular of cofinality at least κ. The following theorem extends level-by-level equivalence to singular cardinals.

THEOREM 7.15 (UA). *Suppose $\kappa < \lambda$ and λ is singular. The following are equivalent:*

(1) *There is a κ-complete ultrafilter on $P_\kappa(\lambda)$ extending the club filter.*
(2) *κ is λ-supercompact or a limit of λ-supercompact cardinals.* □

We do not know whether the conclusion of Theorem 7.15 can be proved consistent with large cardinals by the techniques of Apter-Shelah.

These results suggest the main open questions in this topic:

QUESTION 7.16 (UA). Suppose δ is strongly inaccessible and κ is δ-strongly compact. Is κ either δ-supercompact or else a limit of δ-supercompact cardinals? What if one assumes there is a κ-complete ultrafilter on $P_\kappa(\delta)$ extending the club filter?

We do not expect a positive answer to the first question in general.

The strong compact/supercompact phenomenon under UA is an example where UA converts a coarse large cardinal into its fine-structural form. It is natural to wonder whether there other substantial instances of this.

DEFINITION 7.17. A cardinal κ is *κ-compact* if every κ-complete filter on κ extends to a κ-complete ultrafilter on κ.

QUESTION 7.18 (UA). Suppose there is a cardinal κ that is κ-compact. Must there be a superstrong cardinal?

One should actually be able to get a Π_1^2-subcompact cardinal from this hypothesis (see [**12**]), but we do not even know how to get a 2-strong cardinal.

8. Conclusion

We mentioned in the introduction that producing a model of UA with large cardinals by forcing looks like a very hard problem. In fact, it looks extremely difficult even to *preserve* the theory "UA + there is a supercompact cardinal" with a forcing that has any non-trivial effect above the least supercompact cardinal. There is therefore a possibility that UA decides all set theoretic questions "above the least supercompact cardinal," a vague conjecture that is supported to some extent by the results of this paper.

On the other hand, since UA is invariant under various types of forcing below the least supercompact cardinal, it seems to yield no insight into set theoretic questions about small sets. In this conclusion, we discuss the possibility that by simply removing this invariance one arrives at a principle that decides *all* set theoretic questions.

We start by recalling the definition of the *Ground Axiom*, due to Hamkins-Reitz [**13**].

DEFINITION 8.1. An inner model M is a *ground* if $V = M[G]$ for some set G that is generic over M. The *mantle* $\mathbb{M}$ is the intersection of all grounds of V.

The definition is first-order by the uniform version of the ground model definability theorem of Woodin and Laver [**8**].

A theorem of Usuba [**19**] states that the mantle is an inner model of ZFC. Hamkins-Reitz's Ground Axiom is the assertion that $V = \mathbb{M}$:

GROUND AXIOM. The universe has no proper grounds.

A second theorem of Usuba states that the Ground Axiom is almost provable from large cardinals.

THEOREM 8.2 (Usuba). *Suppose there is an extendible cardinal. Then V is a generic extension of $\mathbb{M}$ and so $\mathbb{M}$ satisfies the Ground Axiom.* □

Essentially by definition, the Ground Axiom cannot be forced to hold by nontrivial set forcing. One might therefore conjecture that the Ground Axiom resolves many of the set theoretic propositions proved independent by forcing. But even when supplemented by very strong large cardinal axioms, the Ground Axiom turns out to have very few consequences since it can be forced to hold by various class forcings.

What looks impossible, however, is proving independence results by class forcing to produce a model of the Ground Axiom, large cardinal axioms, and the Ultrapower Axiom. This is because the methodology for forcing the Ground Axiom plus a supercompact cardinal requires nontrivial forcing above the supercompact, which as we have said seems to be incompatible with producing a model of UA. Therefore by combining the Ground Axiom and the Ultrapower Axiom one has a chance of obtaining a theory that decides all set theoretic questions assuming sufficient large cardinal axioms. We now make a precise conjecture to this effect.

Woodin has proposed the axiom $V = \text{Ultimate } L$ (see [**20**]) which, supplemented by large cardinal axioms, is expected to axiomatize the structure of the canonical inner model with a supercompact cardinal *if it exists*. Therefore, supplemented with large cardinal axioms, the axiom $V = \text{Ultimate } L$ should decide all questions of set theory. Woodin has shown that this axiom implies the Ground Axiom and the Continuum Hypothesis and $V = \text{HOD}$. It is natural to expect that it also implies the Ultrapower Axiom, but this is an open problem. The following conjecture, if true, would amount to a recovery result for the axiom $V = \text{Ultimate } L$ from some of the simplest features of canonical inner models:

CONJECTURE 8.3. *Assume the Ultrapower Axiom, the Ground Axiom, and the existence of an extendible cardinal. Then $V = \text{Ultimate } L$.*

It is not at all clear that the theory implicit in Conjecture 8.3 can even resolve CH. This alone would be a startling implication.

References

[1] A. W. Apter and S. Shelah, *On the strong equality between supercompactness and strong compactness*, Trans. Amer. Math. Soc. **349** (1997), no. 1, 103–128, DOI 10.1090/S0002-9947-97-01531-6. MR1333385

[2] Z. Frolík, *Sums of ultrafilters*, Bull. Amer. Math. Soc. **73** (1967), 87–91, DOI 10.1090/S0002-9904-1967-11653-7. MR203676

[3] M. Gitik, *Some constructions of ultrafilters over a measurable cardinal* (To appear).

[4] G. Goldberg, *The linearity of the Mitchell order*, J. Math. Log. **18** (2018), no. 1, 1850005, 17, DOI 10.1142/S0219061318500058. MR3809587

[5] G. Goldberg, *The Ultrapower Axiom*, Forthcoming. Thesis (Ph.D.)–Harvard University.

[6] A. Kanamori, *Ultrafilters over a measurable cardinal*, Ann. Math. Logic **10** (1976), no. 3-4, 315–356, DOI 10.1016/0003-4843(76)90012-7. MR0491186

[7] J. Ketonen, *Strong compactness and other cardinal sins*, Ann. Math. Logic **5** (1972/73), 47–76, DOI 10.1016/0003-4843(72)90018-6. MR0469768

[8] R. Laver, *Certain very large cardinals are not created in small forcing extensions*, Ann. Pure Appl. Logic **149** (2007), no. 1-3, 1–6, DOI 10.1016/j.apal.2007.07.002. MR2364192

[9] M. Magidor, *How large is the first strongly compact cardinal? or A study on identity crises*, Ann. Math. Logic **10** (1976), no. 1, 33–57, DOI 10.1016/0003-4843(76)90024-3. MR0429566

[10] T. K. Menas, *On strong compactness and supercompactness*, Ann. Math. Logic **7** (1974/75), 327–359, DOI 10.1016/0003-4843(75)90009-1. MR0357121

[11] W. J. Mitchell and J. R. Steel, *Fine structure and iteration trees*, Lecture Notes in Logic, vol. 3, Springer-Verlag, Berlin, 1994. MR1300637

[12] I. Neeman and J. Steel, *Equiconsistencies at subcompact cardinals*, Arch. Math. Logic **55** (2016), no. 1-2, 207–238, DOI 10.1007/s00153-015-0465-4. MR3453585

[13] J. Reitz, *The ground axiom*, J. Symbolic Logic **72** (2007), no. 4, 1299–1317, DOI 10.2178/jsl/1203350787. MR2371206

[14] R. Schindler, *Set theory*, Universitext, Springer, Cham, 2014. Exploring independence and truth. MR3243739

[15] F. S. Schlutzenberg, *Measures in mice*, ProQuest LLC, Ann Arbor, MI, 2007. Thesis (Ph.D.)–University of California, Berkeley. MR2711515

[16] R. M. Solovay, *Strongly compact cardinals and the GCH*, Proceedings of the Tarski Symposium (Proc. Sympos. Pure Math., Vol. XXV, Univ. California, Berkeley, Calif., 1971), Amer. Math. Soc., Providence, R.I., 1974, pp. 365–372. MR0379200

[17] R. M. Solovay, W. N. Reinhardt, and A. Kanamori, *Strong axioms of infinity and elementary embeddings*, Ann. Math. Logic **13** (1978), no. 1, 73–116, DOI 10.1016/0003-4843(78)90031-1. MR482431

[18] J. R. Steel, *An outline of inner model theory*, Handbook of set theory. Vols. 1, 2, 3, Springer, Dordrecht, 2010, pp. 1595–1684, DOI 10.1007/978-1-4020-5764-9_20. MR2768698

[19] T. Usuba, *The downward directed grounds hypothesis and very large cardinals*, J. Math. Log. **17** (2017), no. 2, 1750009, 24, DOI 10.1142/S021906131750009X. MR3730565

[20] W. H. Woodin, *In search of Ultimate-L: the 19th Midrasha Mathematicae Lectures*, Bull. Symb. Log. **23** (2017), no. 1, 1–109, DOI 10.1017/bsl.2016.34. MR3632568

Email address: `goldberg@college.harvard.edu`

Contemporary Mathematics
Volume **752**, 2020
https://doi.org/10.1090/conm/752/15132

Quasi-invariant measures for continuous group actions

Alexander S. Kechris

Dedicated to Simon Thomas on his 60th birthday

Abstract. The class of ergodic, invariant probability Borel measure for the shift action of a countable group is a G_δ set in the compact, metrizable space of probability Borel measures. We study in this paper the descriptive complexity of the class of ergodic, quasi-invariant probability Borel measures and show that for any infinite countable group Γ it is $\mathbf{\Pi}^0_3$-hard, for the group $\mathbb{Z}$ it is $\mathbf{\Pi}^0_3$-complete, while for the free group $\mathbb{F}_\infty$ with infinite, countably many generators it is $\mathbf{\Pi}^0_\alpha$-complete, for some ordinal α with $3 \leq \alpha \leq \omega + 2$. The exact value of this ordinal is unknown.

1. Introduction

For any Polish space X, let $P(X)$ be the Polish space of probability Borel measures on X with the usual topology (see, e.g., [**K**, 17.E]). It is compact, metrizable, if X is compact, metrizable. Any $f\colon X \to Y$ induces the map $f_*\colon P(X) \to P(Y)$, defined by $f_*(\mu)(B) = \mu(f^{-1}(B))$.

If E is an equivalence relation on X, a measure $\mu \in P(X)$ is **ergodic** for E if for any Borel E-invariant set $A \subseteq X$, $\mu(A) \in \{0, 1\}$. We denote by ERG_E the set of such measures. Similarly if $a\colon \Gamma \times X \to X$ is an action of a group Γ on X, a measure μ is ergodic for a if for any invariant under a Borel set A, we have $\mu(A) \in \{0, 1\}$. We denote again by ERG_a the set of such measures. Clearly $\mathrm{ERG}_a = \mathrm{ERG}_{E_a}$, where E_a is the equivalence relation induced by (the orbits of) the action a.

Consider now a continuous action a of a countable (discrete) group Γ on a compact, metrizable space K. If a is understood from the context, we write $\gamma \cdot x$ instead of $a(\gamma, x)$. We also let $\gamma^a(x) = a(\gamma, x)$. It is a standard fact that the set INV_a of invariant measures for a is closed in $P(K)$ and the set EINV_a of invariant, ergodic measures for a is G_δ in $P(K)$ (see, e.g., [**G**, Theorem 4.2]).

Recall that $\mu \in P(K)$ is called **quasi-invariant** for the action a if for any $\gamma \in \Gamma$, $\gamma \cdot \mu \sim \mu$, where $\sim$ denotes measure equivalence and $\gamma \cdot \mu = (\gamma^a)_*(\mu)$. Denote by QINV_a the set of quasi-invariant measures for a and by EQINV_a the subset of ergodic, quasi-invariant measures for a. Since the relation $\sim$ of measure equivalence is $\mathbf{\Pi}^0_3$ in $P(K)^2$, it follows that QINV_a is $\mathbf{\Pi}^0_3$ in $P(K)$. From a (more general)

2010 *Mathematics Subject Classification.* Primary 03C13, Secondary 03C15, 05D10, 37B05, 37A15, 54H20.

Key words and phrases. group actions, invariant measures, quasi-invariant measures, ergodic measures.

The author was partially supported by NSF grant DMS-1464475.

result of Ditzen in [**D**], it follows that ERG_a is Borel and, again from a (more general) result in Louveau-Mokobodzki [**LM**, page 4823], this can be improved to $\mathrm{ERG}_a \in \mathbf{\Pi}^0_{\omega+2}$. Thus $\mathrm{EQINV}_a = \mathrm{ERG}_a \cap \mathrm{QINV}_a$ is also $\mathbf{\Pi}^0_{\omega+2}$ in $P(K)$.

In this paper we are interested in the Borel complexity of the sets QINV_a and EQINV_a. To avoid technical complications involving the topology of K, we will consider here the case where K is 0-dimensional and thus can be viewed as a closed subspace of the Cantor space $\mathcal{C} = 2^{\mathbb{N}}$. Under these circumstances, the action a of Γ on K can be topologically embedded, via the map $f(x) = (\gamma^{-1} \cdot x)_\gamma$, into the shift action s_Γ of Γ on $\mathcal{C}^\Gamma$. Therefore QINV_a and EQINV_a are Wadge reducible, via the continuous map $\mu \mapsto f_*(\mu)$, to QINV_{s_Γ} and $\mathrm{EQINV}_{s_\Gamma}$, resp. Recall that if $A \subseteq X, B \subseteq Y$, then A is **Wadge reducible** to B if there is a continuous function $f\colon X \to Y$ such that $A = f^{-1}(B)$. In this case we put $A \leq_W B$. We will thus focus our attention on the study of the Borel complexity of the quasi-invariant and ergodic, quasi-invariant measures for the shift action. For convenience we write

$$\mathrm{QINV}_\Gamma = \mathrm{QINV}_{s_\Gamma}, \mathrm{ERG}_\Gamma = \mathrm{ERG}_{s_\Gamma}, \mathrm{EQINV}_\Gamma = \mathrm{EQINV}_{s_\Gamma}.$$

We prove below the following results, where for a class Φ of sets in Polish spaces, a set $A \subseteq X$, X a Polish space, is called **Φ-hard** if for any $B \subseteq Y$, Y a 0-dimensional Polish space, with $B \in \Phi$, we have $B \leq_W A$. If in addition $A \in \Phi$, then A is called **Φ-complete**.

THEOREM 1. *For any infinite, countable group Γ, QINV_Γ is $\mathbf{\Pi}^0_3$-complete and $\mathrm{ERG}_\Gamma, \mathrm{EQINV}_\Gamma$ are $\mathbf{\Pi}^0_3$-hard.*

THEOREM 2. *The set $\mathrm{EQINV}_\mathbb{Z}$ is $\mathbf{\Pi}^0_3$-complete.*

THEOREM 3. *Let $\mathbb{F}_\infty$ be the group with infinite, countably many generators. Then there is a countable ordinal $\alpha_\infty \geq 3$ such that the set $\mathrm{EQINV}_{\mathbb{F}_\infty}$ is $\mathbf{\Pi}^0_{\alpha_\infty}$-complete.*

Thus $3 \leq \alpha_\infty \leq \omega + 2$.

PROBLEM 4. Calculate α_∞.

We note that from Theorem 3 it follows that $\mathrm{EQINV}_\Gamma \in \mathbf{\Pi}^0_{\alpha_\infty}$, for any countable group Γ.

REMARK 5. The proof of Theorem 3 in Section 4 below also shows that for any countable group Γ that can be mapped onto the direct sum of infinite, countably many copies of itself, there is a countable ordinal α_Γ (thus $3 \leq \alpha_\Gamma \leq \omega + 2$) such that EQINV_Γ is $\mathbf{\Pi}^0_{\alpha_\Gamma}$-complete.

2. Proof of Theorem 1

We first note the following standard fact.

LEMMA 2.1. *For any continuous action a of a countable group Γ on a compact, metrizable space K, $\mathrm{ERG}_a \leq_W \mathrm{EQINV}_a$.*

PROOF. Let $\Gamma = \{\gamma_n \colon n \in \mathbb{N}\}$. The map $\mu \in P(K) \mapsto \sum_n 2^{-(n+1)}\gamma_n \cdot \mu \in P(K)$ is a continuous reduction of ERG_a to EQINV_a. □

Thus to complete the proof of Theorem 1, it is enough to show that QINV_Γ is $\mathbf{\Pi}^0_3$-complete and that ERG_Γ is $\mathbf{\Pi}^0_3$-hard.

(A) QINV_Γ is $\mathbf{\Pi}^0_3$-complete.

Let X be a perfect Polish space and Γ an infinite, countable group, which acts freely and continuously on X. Put

$$S = \{(x_n) \in X^{\mathbb{N}}\colon \{x_n\colon n \in \mathbb{N}\} \text{ is } \Gamma\text{-invariant})$$
$$= \{(x_n) \in X^{\mathbb{N}}\colon \forall n \forall \gamma \exists m (\gamma \cdot x_n = x_m)\}$$

PROPOSITION 2.2. *S is not G_δ.*

PROOF. First notice that S is dense: Given $U_0, \ldots, U_{k-1}$ non-$\emptyset$ open in X consider $U_0 \times \cdots \times U_{k-1} \times X^{\mathbb{N}}$. We will show that it intersects S. Pick $x_i^0 \in V_i, i < k$. Then clearly there are $x_k^0, x_{k+1}^0, \ldots$ such that $(x_n^0) \in S$.

So if S is G_δ, it is comeager. We will show that there is a dense G_δ set G such that $G \cap S = \emptyset$, a contradiction.

Let $\gamma \neq 1, \gamma \in \Gamma$ and put

$$G = \{(x_n)\colon \forall m (\gamma \cdot x_0 \neq x_m)\}.$$

Clearly $G \cap S = \emptyset$. Now

$$G = \bigcap_m G_m, \text{ where}$$
$$G_m = \{(x_n)\colon \gamma \cdot x_0 \neq x_m\}.$$

Clearly G_m is dense, open, so G is comeager. □

Let now K be perfect, compact, metrizable and let a be a free, continuous action of Γ on K.

PROPOSITION 2.3. *QINV_a is not G_δ in $P(K)$.*

PROOF. It is enough to find a continuous function

$$F\colon K^{\mathbb{N}} \to P(K)$$

such that $F^{-1}(\mathrm{QINV}_a) = S$, where S is as above for (K, Γ).

Put

$$F((x_n)) = \sum_{n=0}^{\infty} \frac{1}{2^{n+1}} \delta_{x_n},$$

where δ_x is the Dirac measure at $x \in K$. □

CLAIM. F is continuous.

PROOF. We need to check that if $f \in C(K)$, and $(x_n^i) \to (x_n)$ in $K^{\mathbb{N}}$, then $F((x_n^i))(f) = \sum_{n=0}^{\infty} \frac{1}{2^{n+1}} f(x_n^i) \to F((x_n))(f) = \sum_{n=0}^{\infty} \frac{1}{2^{n+1}} f(x_n)$, which is clear as $f(x_n^i) \to f(x_n), \forall n$. □

CLAIM. $F^{-1}(\mathrm{QINV}_\Gamma) = S$.

PROOF. If $(x_n) \in S$, then clearly $\gamma \cdot (F(x_n)) \sim F((x_n)), \forall \gamma \in \Gamma$. Conversely assume $(x_n) \notin S$. Let then n, γ be such that $\forall m (\gamma \cdot x_n \neq x_m)$. Then $\gamma \cdot F((x_n)) \not\sim F((x_n))$. □

Thus we have shown:

PROPOSITION 2.4. *Let a be a continuous and free action of an infinite countable group Γ on the perfect, compact metrizable space K. Then QINV_a is not G_δ in $P(K)$.*

Let now $Q = \{x \in \mathcal{C}\colon x(n) = 0$ for all but finitely many $n\}$. Then Q is F_σ in the Cantor space $\mathcal{C}$ and for any Polish space X and Borel set $A \subseteq X$, if A is not G_δ, then $Q \leq_W A$ (see [**K**, 24.20 and 22.13]). Thus we have:

COROLLARY 2.5. *Let a be a continuous and free action of an infinite, countable group Γ on the perfect, compact metrizable space K Then there is a continuous function $f\colon \mathcal{C} \to P(K)$ with $f^{-1}(\mathrm{QINV}_a) = Q$.*

Consider now the set $Q^{\mathbb{N}} \subseteq \mathcal{C}^{\mathbb{N}}$. It is known that $Q^{\mathbb{N}}$ is a $\mathbf{\Pi}^0_3$-complete set (see [**K**, page 179] where this set is denoted by P_3). Let now $K = \mathcal{C}^\Gamma$ with the shift action. By a result of Gao-Jackson-Seward [**GJS**, 3.7], there are infinitely many, pairwise disjoint, invariant compact subsets K_n of K on which Γ acts minimally and freely. Note that each K_n is perfect. By the preceding corollary, there is a continuous function $f_n\colon \mathcal{C} \to P(K_n)$ such that $f_n^{-1}(\mathrm{QINV}_{a_n}) = Q$, where a_n is the restriction of the shift action to K_n. Define now $f\colon \mathcal{C}^{\mathbb{N}} \to P(K)$ by $f((x_n)) = \sum_n \frac{1}{2^{n+1}} f_n(x_n)$. Then f is continuous and $f^{-1}(\mathrm{QINV}_\Gamma) = Q^{\mathbb{N}}$, so QINV_Γ is $\mathbf{\Pi}^0_3$-complete.

(B) ERG_Γ is $\mathbf{\Pi}^0_3$-hard.

This follows from the following more general result, where a Borel equivalence relation E on a Polish space X is **smooth** if there is a Borel map $f\colon X \to Y$, Y a Polish space, such that $xEy \Leftrightarrow f(x) = f(y)$.

THEOREM 2.6. *Let E be a non-smooth, Borel equivalence relation on a Polish space X. Then ERG_E is $\mathbf{\Pi}^0_3$-hard.*

PROOF. Let E_0^k be the equivalence relation of $k^{\mathbb{N}}$ given by

$$(x_n) E_0^k (y_n) \Leftrightarrow \exists n \forall m \geq n (x_m = y_m).$$

Then by [**HKL**], E_0^3 can be continuously embedded, say by the function $f\colon 3^{\mathbb{N}} \to X$, into E. The function f_* from $P(3^{\mathbb{N}})$ to $P(X)$ is continuous and μ is ergodic for E_0^3 iff $f_*(\mu)$ is ergodic for E. It is thus enough to prove this result for $E = E_0^3$.

Consider the $\mathbf{\Pi}^0_3$-complete set $P_3 = Q^{\mathbb{N}} \subseteq \mathcal{C}^{\mathbb{N}}$, as in the paragraph following Corollary 2.5. We will define a continuous function $f\colon \mathcal{C}^{\mathbb{N}} \times \mathcal{C} \to 3^{\mathbb{N}}$ as follows:

Fix a bijection $\langle\cdot,\cdot\rangle\colon \mathbb{N}^2 \to \mathbb{N}$. Define first a function $\bar{f}$ by:

$$\bar{f}((a_k), x)(\langle n, m\rangle) = x(n+m), \text{if } a_n(m) = 0; x|[n, n+m], \text{if } a_n(m) = 1.$$

Let then $f((a_k), x) = y$, where letting $y_n(m) = y(\langle n, m\rangle)$, y_n is equal to:

$$2\ \bar{f}((a_k), x)(\langle n, 0\rangle)\ 2\ \bar{f}((a_k), x)(\langle n, 1\rangle)\ 2\ \cdots 2\ \bar{f}((a_k), x)(\langle n, m\rangle)\ 2\ \cdots,$$

which is the concatenation of 2 followed by $\bar{f}((a_k), x)(\langle n, 0\rangle)$ followed by 2 followed by $\bar{f}((a_k), x)(\langle n, 1\rangle) \dots$

Since $y(\langle n, m\rangle)$ depends only on $a_n(l)$, for $l \leq m$, and $x(n), \dots, x(n+m)$, it is clear that f is continuous.

It is also clear that for each fixed (a_k), the section $f_{(a_k)}(x) = f((a_k), x)$ is injective. For each $(a_k) \in \mathcal{C}^{\mathbb{N}}$ let now

$$\mu((a_k)) = (f_{(a_k)})_*(\lambda),$$

where λ is the usual product measure on $\mathcal{C}$. The function $\mu\colon \mathcal{C}^{\mathbb{N}} \to P(3^{\mathbb{N}})$ is continuous, so its is enough to show that

$$(a_k) \in P_3 \Leftrightarrow \mu((a_k)) \in \mathrm{ERG}_{E_0^3}.$$

(A) Let $(a_k) \in P_3$. We claim then that $xE_0^2y \implies f_{(a_k)}(x)E_0^3f_{(a_k)}(y)$. Indeed, if xE_0y, say $x(k) = y(k)$ for all $k \geq n_0$, then for $n \geq n_0$, clearly $f_{(a_k)}(x)(\langle n, m\rangle) = f_{(a_k)}(y)(\langle n, m\rangle), \forall m$. Let also m be large enough so that $a_n(m) = 0$, for all $n < n_0$ and all $m \geq m_0$. Then for some k_0 and all $m \geq k_0, n < n_0$, we have $f_{(a_k)}(x)(\langle n, m\rangle) = f_{(a_k)}(y)(\langle n, m\rangle)$, so $f_{(a_k)}(x)E_0^3f_{(a_k)}(y)$.

Thus if $A \subseteq 3^{\mathbb{N}}$ is Borel, E_0^3-invariant, then $f_{(a_k)}^{-1}(A)$ is Borel E_0^2-invariant, so, since λ is ergodic for E_0^2, it has λ-measure 0 or 1, and thus A has $\mu((a_k))$-measure 0 or 1. So $\mu((a_k)) \in \mathrm{ERG}_{E_0^3}$.

(B) Let $(a_k) \notin P_3$. Fix then n_0 and $1 < m_0 < m_1 < m_2 \ldots$ be such that $a_{n_0}(m_i) = 1, \forall i$. Fix also a tree $T \subseteq 2^{<\mathbb{N}}$ such that $0 < \lambda([T]) < 1$. Put

$$B = \bigcup_{s \in 2^{n_0}} N_s \star [T],$$

where for $s \in 2^{n_0}$:

$$N_s \star [T] = \{a \in \mathcal{C} \colon s \subseteq a \ \& \ (a_{n_0}, a_{n_0+1}, \ldots) \in [T]\}.$$

Then $\lambda(B) = \lambda([T]) \in (0,1)$. Put $f_{(a_k)}(B) = C$ and $A = [C]_{E_0^3}$. Then A is Borel, E_0^3-invariant and we will show that $f_{(a_k)}^{-1}(A) = B$, so that $\mu((a_k))(A) \in (0,1)$, and thus $\mu((a_k))$ is not ergodic for E_0^3, completing the proof.

Let $f_{(a_k)}(x) \in A$ and choose $y \in B$ such that $f_{(a_k)}(x)E_0^3f_{(a_k)}(y)$. Then, in particular, if $f_{(a_k)}(x) = (x_n), f_{(a_k)}(y) = y_n$, we have $x_{n_0}E_0^3y_{n_0}$. Now $x_{n_0} = 2 \ s_0 \ 2 \ s_1 \ldots$, $y_{n_0} = 2 \ t_0 \ 2 \ t_1 \ldots$, where for each i, s_i, t_i are binary sequences of the same length. Let then k be such that for all $i \geq k$, $s_i = t_i$. If $m_j \geq k$, then $t_{m_j} = (y_{n_0}, \ldots, y_{n_0+m_j})$ and so $s_{m_j} = t_{m_j} \in T$. Since also $s_{m_j} = (x_{n_0}, \ldots, x_{n_0+m_j})$, we have that $(x_{n_0}, x_{n_0+1}, \ldots) \in [T]$, i.e., $x \in B$. □

3. Proof of Theorem 2

Ditzen [**D**, page 47] shows that $\mathrm{EQINV}_{\mathbb{Z}}$ is $\mathbf{\Pi}_3^0$ and thus by Theorem 1 it is $\mathbf{\Pi}_3^0$-complete.

4. Proof of Theorem 3

Theorem 3 will follow from the next proposition:

PROPOSITION 4.1. *Let X be a Polish space and let $A \subseteq X$. If $A \leq_W \mathrm{ERG}_{\mathbb{F}_\infty}$, then $A^{\mathbb{N}}(\subseteq X^{\mathbb{N}}) \leq_W \mathrm{ERG}_{\mathbb{F}_\infty}$.*

PROOF. Recall that for any countable Borel equivalence relation E, we denote by ERG_E the set of probability Borel measures that are ergodic for E.

LEMMA 4.2. *Let E_n be a countable Borel equivalence relation in the Polish space X_n and let μ_n be a probability Borel measure on X_n. Let E_∞ be the following equivalence relation on $X^{\mathbb{N}}$:*

$$(x_n)E_\infty(y_n) \Leftrightarrow \forall n(x_nE_ny_n) \ \& \ \exists m\forall n \geq m(x_n = y_n).$$

Then

$$\prod_n \mu_n \in \mathrm{ERG}_{E_\infty} \Leftrightarrow \forall n(\mu_n \in \mathrm{ERG}_{E_n}).$$

PROOF. $\Longrightarrow$: Put $\mu = \prod_n \mu_n$. Let $A \subseteq X_n$ be Borel and E_n-invariant. Let $B = X_0 \times \cdots X_{n-1} \times A \times X_{n+1} \times \cdots$. Then B is Borel and E_∞-invariant, so $\mu_n(A) = \mu(B) \in \{0,1\}$.

$\Longleftarrow$: Assume that each μ_n is ergodic for E_n. Let $A \subseteq \prod_n X_n$ be Borel and E_∞-invariant. For each Borel set $B \subseteq \prod_n X_n$, let $\nu(B) = \mu(A \cap B)$. If we can show that for each Borel cylinder $B \subseteq \prod_n X_n$, $\nu(B) = \mu(A)\mu(B)$, then since the class of all Borel sets B with the property that $\nu(B) = \mu(A)\mu(B)$ is closed under complements and countable disjoint unions, by the $\pi - \lambda$ Theorem (see, e.g., [**K**, 10.1, iii)]) it contains all Borel sets, and in particular A, so $\nu(A) = \mu(A) = \mu(A)^2$, thus $\mu(A) \in \{0,1\}$.

Let then $B = D \times \prod_{i \geq n} X_i$ be a Borel cylinder, where $D \subseteq \prod_{i<n} X_i$. For $y \in \prod_{i \geq n} X_i$, let $A^y = \{(x_i)_{i<n} \in \prod_{i<n} X_i \colon ((x_i)_{i<n}, y)) \in A\}$. Then A^y is $\prod_{i<n} E_i$-invariant. $\square$

CLAIM. $\prod_{i<n} \mu_i \in \mathrm{ERG}_{\prod_{i<n} E_i}$.

PROOF. It is enough to consider the case $n = 2$, so let $A \subseteq X_0 \times X_1$ be Borel and $(E_0 \times E_1)$-invariant. Note that for $x_0 \in X_0$ the section $A_{x_0} \subseteq X_1$ is E_1 invariant, so $\mu_1(A_{x_0}) \in \{0,1\}$. Let $P_i = \{x_0 \colon \mu_1(A_{x_0}) = i\}$, for $i \in \{0,1\}$. Then each P_i is E_0-invariant. If $\mu_0(P_0) = 0$, then $\mu_0(P_1) = 1$, so $(\mu_0 \times \mu_1)(A) = 1$. If $\mu_0(P_0) = 1$, then $(\mu_0 \times \mu_1)(A) = 1$.

Thus A^y has $\prod_{i<n} \mu_i$-measure 0 or 1. Let

$$C = \{y \in \prod_{i \geq n} X_i \colon (\prod_{i<n} \mu_i)(A^y) = 1\}.$$

Then $\mu(A) = (\prod_{i \geq n} \mu_i)(C)$. Now for $y \in C$, $(\prod_{i<n} \mu_i)(A^y \cap D) = (\prod_{i<n} \mu_i)(D)$ and for $y \notin C$, $(\prod_{i<n} \mu_i)(A_y \cap D) = 0$, so

$$\begin{aligned}
\mu(A \cap B) &= \mu(A \cap (D \times \prod_{i \geq n} X_i)) \\
&= \int (\prod_{i<n} \mu_i)(A^y \cap D) d(\prod_{i \geq n} \mu_i)(y) \\
&= (\prod_{i<n} \mu_i)(D) \cdot (\prod_{i \geq n} \mu_i)(C) \\
&= \mu(B)\mu(A).
\end{aligned}$$

$\square$

Let now E be the equivalence relation on $\mathcal{C}^{\mathbb{F}_\infty}$ induced by the shift action of $\mathbb{F}_\infty$. We have to show that if $A \leq_W \mathrm{ERG}_E$, then $A^{\mathbb{N}} \leq_W \mathrm{ERG}_E$. Let $f \colon X \to P(\mathcal{C}^{\mathbb{F}_\infty})$ be a continuous function witnessing that $A \leq_W \mathrm{ERG}_E$. Define $f_\infty \colon X^{\mathbb{N}} \to P((\mathcal{C}^{\mathbb{F}_\infty})^{\mathbb{N}})$ by $f_\infty((x_n)) = \prod_n f(x_n)$. Then f_∞ is continuous and if E_∞ is as in Lemma 4.2 with $E_n = E$ for each n, then

$$f_\infty((x_n)) \in \mathrm{ERG}_{E_\infty} \Leftrightarrow \forall n (f(x_n) \in \mathrm{ERG}_E) \Leftrightarrow (x_n) \in A^{\mathbb{N}}.$$

So $A^{\mathbb{N}} \leq_W \mathrm{ERG}_{E_\infty}$.

Now consider the continuous action of $\bigoplus_n \mathbb{F}_\infty$ on $(\mathcal{C}^{\mathbb{F}_\infty})^{\mathbb{N}}$ given by $(\gamma_n) \cdot (x_n) = (\gamma_n \cdot x_n)$. The equivalence relation it induces is exactly E_∞. Mapping $\mathbb{F}_\infty$ onto $\bigoplus_n \mathbb{F}_\infty$, this gives a continuous action a of $\mathbb{F}_\infty$ on $(\mathcal{C}^{\mathbb{F}_\infty})^{\mathbb{N}}$ for which $\mathrm{ERG}_a =$

ERG_{E_∞} and thus $A^{\mathbb{N}} \leq_W \mathrm{ERG}_a$. Noting that $(\mathcal{C}^{\mathbb{F}_\infty})^{\mathbb{N}}$ is homeomorphic to $\mathcal{C}$, we can embed this action to the shift action of $\mathbb{F}_\infty$ on $\mathcal{C}^{\mathbb{F}_\infty}$ and thus $A^{\mathbb{N}} \leq_W \mathrm{ERG}_{\mathbb{F}_\infty}$. □

Using Proposition 4.1, we now complete the proof of Theorem 3 as follows. Let α be least such that $\mathrm{ERG}_{\mathbb{F}_\infty} \in \mathbf{\Pi}^0_\alpha$. By Theorem 1, $\alpha \geq 3$.

CLAIM. $\mathrm{ERG}_{\mathbb{F}_\infty}$ is $\mathbf{\Pi}^0_\alpha$-complete.

PROOF. Let $\mathcal{A} = \{B \subseteq Y \colon Y \text{ Polish, 0-dimensional}, B \leq_W \mathrm{ERG}_{\mathbb{F}_\infty}\}$. Then $\mathcal{A}$ is closed under countable intersections, since if $B_n \in \mathcal{A}, B_n \subseteq Y$, there is a continuous function $f_n \colon Y \to P(\mathcal{C}^{\mathbb{F}_\infty})$ such that $B_n = f_n^{-1}(\mathrm{ERG}_{\mathbb{F}_\infty})$. Put $X = P(\mathcal{C}^{\mathbb{F}_\infty}), A = \mathrm{ERG}_{\mathbb{F}_\infty}$ and let $f \colon Y \to X^{\mathbb{N}}$ be given by $f(y)_n = f_n(y)$. Then f witnesses that $\bigcap_n B_n \leq_W A^{\mathbb{N}} \leq_W A = \mathrm{ERG}_{\mathbb{F}_\infty}$, so $\bigcap_n B_n \in \mathcal{A}$.

Let now $B \in \mathbf{\Pi}^0_\alpha$, $B \subseteq Y$, Y Polish and 0-dimensional. Then $B = \bigcap_n B_n$, where $B_n \in \mathbf{\Sigma}^0_{\alpha_n}$, for some $\alpha_n < \alpha$. By a result of Saint-Raymond (see [**K**, 24.20 and 22.13]) $B_n \leq_W \mathrm{ERG}_{\mathbb{F}_\infty}$, so $B \leq_W \mathrm{ERG}_{\mathbb{F}_\infty}$.

Now, as $\alpha \geq 3$, $\mathrm{EQINV}_{\mathbb{F}_\infty}$ is in $\mathbf{\Pi}^0_\alpha$. Also by Lemma 2.1, if $B \in \mathbf{\Pi}^0_\alpha$, then $B \leq_W \mathrm{ERG}_{\mathbb{F}_\infty} \leq_W \mathrm{EQINV}_{\mathbb{F}_\infty}$, so $\mathrm{EQINV}_{\mathbb{F}_\infty}$ is $\mathbf{\Pi}^0_\alpha$-complete. □

REMARK 4.3. Note that the only property of $\mathbb{F}_\infty$ that we used in the preceding proof is that it can be mapped onto the direct sum of countably many copies of itself. It follows that Theorem 3 is valid as well for any countable group Γ that has this property.

References

[D] A. Ditzen, *Definable equivalence relations on Polish spaces*, ProQuest LLC, Ann Arbor, MI, 1992. Thesis (Ph.D.)–California Institute of Technology. MR2687873

[GJS] S. Gao, S. Jackson, and B. Seward, *A coloring property for countable groups*, Math. Proc. Cambridge Philos. Soc. **147** (2009), no. 3, 579–592, DOI 10.1017/S0305004109002655. MR2557144

[G] E. Glasner, *Ergodic theory via joinings*, Mathematical Surveys and Monographs, vol. 101, American Mathematical Society, Providence, RI, 2003. MR1958753

[HKL] L. A. Harrington, A. S. Kechris, and A. Louveau, *A Glimm-Effros dichotomy for Borel equivalence relations*, J. Amer. Math. Soc. **3** (1990), no. 4, 903–928, DOI 10.2307/1990906. MR1057041

[K] A. S. Kechris, *Classical descriptive set theory*, Graduate Texts in Mathematics, vol. 156, Springer-Verlag, New York, 1995. MR1321597

[LM] A. Louveau and G. Mokobodzki, *On measures ergodic with respect to an analytic equivalence relation*, Trans. Amer. Math. Soc. **349** (1997), no. 12, 4815–4823, DOI 10.1090/S0002-9947-97-02070-9. MR1451609

DEPARTMENT OF MATHEMATICS, CALIFORNIA INSTITUTE OF TECHNOLOGY, MAIL CODE 253-37, PASADENA, CALIFORNIA 91125

Email address: `kechris@caltech.edu`

Contemporary Mathematics
Volume **752**, 2020
https://doi.org/10.1090/conm/752/15133

An example of a new simple theory

M. Malliaris and S. Shelah

For Simon Thomas on the occasion of his birthday

Abstract. We construct a countable simple theory which, in Keisler's order, is strictly above the random graph and also in some sense orthogonal to the building blocks of the recently discovered infinite descending chain. As a result we prove in ZFC that there are incomparable classes in Keisler's order.

Recent work on the structure of Keisler's order is changing our understanding of the so-called simple unstable theories, a class which includes the random graph and pseudofinite fields. Indeed, when we discovered recently [**15**] that Keisler's order has infinitely many classes – overturning a long-standing idea that it had five or six – these infinitely many classes were within the simple theories. We are starting to see that the simple unstable theories may have very interesting layers of complexity above that of the random graph, arising from the interaction of randomness with underlying constraints, with no obvious analogue in the stable case.

Basic questions remain open about the structure of Keisler's order on the simple theories, in part because of a lack of examples. For instance, among the unstable theories, the Keisler-minimum class is the class of the random graph. The infinitely many classes arise from disjoint unions of certain random hypergraphs with forbidden substructures (higher analogues of the triangle-free random graph [**7**]) forming a descending chain above the random graph. It was not known whether there were theories strictly between the random graph and this infinite descending chain.

In the present paper, we build a new simple theory, really a family of theories, illustrating how the randomness may interact with an underlying structure with some forbidden configurations, even with only a graph edge (no hyperedges required). We analyze its saturation and non-saturation in regular ultrapowers, concluding it is strictly above the random graph but in a precise sense orthogonal to the higher analogues of the triangle-free random graph. As a consequence of our strategy, we prove there are incomparable classes in Keisler's order just in ZFC; this was known to be true under the existence of a supercompact cardinal, as noticed independently by Ulrich [**18**] and the authors [**16**]. Along the way we give a gentle

2010 *Mathematics Subject Classification.* Primary .

The first author was partially supported by DMS-1553653 and by a Minerva Research Foundation membership at the IAS..

The second author was partially supported by European Research Council grant 338821. Both authors thank NSF grants 136974 and 1833363 (Rutgers) and ERC 338821. This is paper 1140 in the second author's list.

introduction to some key methods of [**12**], [**14**], [**15**], the papers which have been the foundation of our current work on simple theories.

CONTENTS

1. Notation and preliminaries

Keisler's order compares complete countable theories via the difficulty of saturating their regular ultrapowers.

DEFINITION 1.1 (Keisler's order, 1967 [**8**]). Let T_1, T_2 be complete countable theories. We say $T_1 \trianglelefteq T_2$ if for every infinite λ, every regular ultrafilter $\mathcal{D}$ on λ, every model $M_1 \models T_1$, every model $M_2 \models T_2$, if $(M_2)^\lambda/\mathcal{D}$ is λ^+-saturated, then $(M_1)^\lambda/\mathcal{D}$ is λ^+-saturated.

We remind the reader of Keisler's result that if $\mathcal{D}$ is a regular ultrafilter on λ and $M \equiv N$ in a countable language, then $M^\lambda/\mathcal{D}$ is λ^+-saturated iff $N^\lambda/\mathcal{D}$ is λ^+-saturated. *Thus*, the choice of M_1 and M_2 in Definition 1.1 is only important up to elementary equivalence. A further introduction to Keisler's order can be found in the recent lecture notes [**1**] sections 2-3 and in sections 1-2 of [**14**].

CONVENTION 1.2. In what follows:

(1) All theories are complete and countable, unless otherwise stated.
(2) When $\mathcal{D}$ is a regular ultrafilter on I and T is a theory, we will say "$\mathcal{D}$ is good for T" to mean that for some (equivalently every) model $M \models T$, the ultrapower $M^I/\mathcal{D}$ is $|I|^+$-saturated.

Our recent work on simple theories started with an idea in [**12**] to increase the range of ultrafilter construction. The idea is to build regular ultrafilters in two stages: by building a regular filter $\mathcal{D}_0$ on I so that the quotient Boolean algebra $\mathcal{P}(I)/\mathcal{D}_0$ is isomorphic to some specific Boolean algebra $\mathfrak{B}$, and then building an ultrafilter $\mathcal{D}_*$ on $\mathfrak{B}$ (which, a priori, need not be regular), finally combining $\mathcal{D}_0$ and $\mathcal{D}_*$ in the natural way to obtain an ultrafilter $\mathcal{D}$ on I.

A key lemma, "separation of variables," says that in this setup there is a natural translation between realizing types in the ultrapower $M^I/\mathcal{D}$ and showing that certain related patterns, called "possibility patterns," have multiplicative refinements in $\mathcal{D}_*$ (see below). As a result, in many subsequent saturation-of-ultrapowers arguments it is most convenient to work in the Boolean algebra $\mathfrak{B}$, a completion of a free boolean algebra.

DEFINITION 1.3. $\mathfrak{B}^0_{2^\lambda,\mu,\theta}$ is the free Boolean algebra generated by 2^λ independent partitions each of size μ, where intersections of $< \theta$ elements of distinct partitions are nonempty. For a boolean algebra $\mathfrak{B}$, let $\mathfrak{B}^1$ denote its completion, and let $\mathfrak{B}^+$ denote its nonzero elements.

FACT 1.4. Assuming $\mu \leq \lambda = \lambda^{<\theta}$, we have that $\mathfrak{B}^0_{2^\lambda,\mu,\theta}$ and thus its completion exists. In this paper, to find $D_0, \mathcal{D}_*, \mathbf{j}$ as in 1.7 below, we use the completion.

PROOF. See Engelking-Karlowicz [**2**], Fichtenholz and Kantorovich[**4**], Hausdorff [**6**], or Shelah [**17**] Appendix, Theorem 1.5. □

The following discussion may help the picture. Consider $\mathfrak{B}^0_{2^\lambda,\mu,\theta}$ as generated by 2^λ independent antichains each of size μ, which we may enumerate as:

$$\langle \mathbf{x}_{\alpha,\epsilon} : \alpha < 2^\lambda, \epsilon < \mu \rangle.$$

For each fixed α, $\langle \mathbf{x}_{\alpha,\epsilon} : \epsilon < \mu \rangle$ is an antichain, so

$$\mathbf{a}_{\alpha,\epsilon} \cap \mathbf{a}_{\alpha,\epsilon'} = 0 \text{ for } \epsilon < \epsilon' < \mu.$$

Otherwise, the intersection of any $< \theta$ elements no two of which are on the same antichain is nonzero. A compact way to denote such an intersection, say of

$$\bigcap_{\ell<\beta} \mathbf{x}_{\alpha_{i_\ell}} \tag{1.A}$$

where $\beta < \theta$ and the sequence $\langle \alpha_{i_\ell} : \ell < \beta \rangle$ is without repetition, is as follows. Consider the function $f = \{(\alpha_{i_\ell}, \epsilon_{i_\ell}) : \ell < \beta\}$ whose domain is an element of $[2^\lambda]^{<\theta}$ and which assigns to each element of its domain (the index for an antichain) a value less than μ (the specific element of the antichain). Then let

$$\text{“}\mathbf{x}_f\text{”} = \bigcap_{\ell<\beta} \mathbf{x}_{\alpha_{i_\ell}} > 0. \tag{1.B}$$

Now consider the completion $\mathfrak{B}^1_{2^\lambda,\mu,\theta}$. As the generators are dense in the completion, for every nonzero $\mathbf{c}$, there is some nonzero element $\leq \mathbf{c}$ which is an intersection of the form (1.A).

For the present paper, the reader could forget the quotes around $\mathbf{x}_f$ without much loss, but we should also explain the general notation. In 1.4 above, assuming $\mu \leq \lambda = \lambda^{<\theta}$, what one proves is the existence of a family of functions $\mathcal{G} \subseteq {}^\lambda\mu$ of size 2^λ, let us write it as $\{g_\alpha : \alpha < 2^\lambda\}$, which is *independent* in the following sense. Given any distinct $\langle \alpha_{i_\ell} : \ell < \beta < \theta \rangle$ and any values $\langle \epsilon_{i_\ell} : \ell < \beta \rangle$ such that for each $\ell < \beta$, i_ℓ is in the range of $g_{\alpha_{i_\ell}}$, we have that the set

$$\{t < \lambda : g_{\alpha_{i_\ell}}(t) = \epsilon_{i_\ell} \text{ for all } \ell < \beta\} \neq \emptyset. \tag{1.C}$$

In this setup, each function $g \in \mathcal{G}$ has domain λ and range μ and so naturally gives rise to a partition of λ into μ pieces, which we can represent as an antichain as above. Then the intersection (1.C) is naturally described by a function, call it f, which assigns $g_{\alpha_{i_\ell}}$ to ϵ_{i_ℓ} for each $\ell < \theta$, so $\mathbf{x}_f$ is just the nonempty set (intersection) in (1.C). So indeed (1.C) is parallel to (1.B), the only difference is whether we take the domain of f to be the set of indices for the antichains or, more generally, the set of functions giving them. This explains the definition 1.5 in the general case.

We summarize this discussion with the following notation:

DEFINITION 1.5. Let

$$\mathrm{FI}_{\mu,\theta}(2^\lambda) = \{h : h \text{ is a function}, \mathrm{dom}(h) \subseteq 2^\lambda, \mathrm{range}(h) \subseteq \mu, |\mathrm{dom}(h)| < \theta\}.$$

For $g \in \mathrm{FI}_{\mu,\theta}(2^\lambda)$, let $\mathbf{x}_g$ denote the corresponding nonzero element of $\mathfrak{B}$.

CONVENTION 1.6. We will assume that giving $\mathfrak{B} = \mathfrak{B}^1_{\alpha,\mu,\theta}$ determines α, μ, θ and a set of generators $\langle \mathbf{x}_f : f \in \mathrm{FI}_{\mu,\theta}(\alpha)\rangle$.

Note that for any $f \in \mathrm{FI}_{\mu,\theta}(2^\lambda)$, $\mathbf{x}_f > 0$. As the generators are dense in the completion, we will often apply this in the form "as $\mathbf{c} \in (\mathfrak{B}^1_{2^\lambda,\mu,\theta})^+$, we may choose $g \in \mathrm{FI}_{\mu,\theta}(2^\lambda)$ so that $\mathbf{x}_g \leq \mathbf{c}$." $\mathfrak{B}$ is generated freely by its generators except for the following conditions: first, if $f \subseteq f' \in \mathrm{FI}_{\mu,\theta}(2^\lambda)$, then $\mathbf{x}_f \geq \mathbf{x}_{f'}$, because we are adding more conditions to the intersection. Second, for any $f, f' \in \mathrm{FI}_{\mu,\theta}(2^\lambda)$, $\mathbf{x}_f \cap \mathbf{x}_{f'} > 0$ if and only if f, f' agree on their common domain (this includes the case where their common domain is empty).

Next we recall the setup of separation of variables:

DEFINITION 1.7 (Regular ultrafilters built from tuples, from [**12**] Theorem 6.13). Suppose $\mathcal{D}$ is a regular ultrafilter on I, $|I| = \lambda$. We say that $\mathcal{D}$ is built from $(\mathcal{D}_0, \mathfrak{B}, \mathcal{D}_*)$ when:

(1) $\mathcal{D}_0$ is a regular, $|I|^+$-excellent filter on I
 (for our purposes here, it is sufficient to use regular and good)
(2) $\mathfrak{B}$ is a Boolean algebra
(3) $\mathcal{D}_*$ is an ultrafilter on $\mathfrak{B}$
(4) there exists a surjective homomorphism $\mathbf{j} : \mathcal{P}(I) \to \mathfrak{B}$ such that:
 (a) $\mathcal{D}_0 = \mathbf{j}^{-1}(\{1_{\mathfrak{B}}\})$
 (b) $\mathcal{D} = \{A \subseteq I : \mathbf{j}(A) \in \mathcal{D}_*\}$.

THEOREM 1 ("Separation of variables", [**12**] Theorem 6.13). *Suppose that $\mathcal{D}$ is a regular ultrafilter on I built from $(\mathcal{D}_0, \mathfrak{B}, \mathcal{D}_*)$. Then the following are equivalent:*

(A) *$\mathcal{D}_*$ is $(|I|, \mathfrak{B}, T)$-moral, see* [**12**] *Definition* 6.3.
(B) *$\mathcal{D}$ is good for T.*

The definition of "moral" is a bit long to quote, but can be easily summarized by saying that the way we will use Theorem 1 is the following. Suppose $\mathcal{D}$ is a regular ultrafilter on I, $|I| = \lambda$, $\mathcal{D}$ is built from $(\mathcal{D}_0, \mathfrak{B}, \mathcal{D}_*)$, $M \models T$, and $N = M^I/\mathcal{D}$ is the ultrapower. Suppose $p = \{\varphi_\alpha(x, \bar{a}_\alpha) : \alpha < \lambda\}$ is a type in the ultrapower. Let $\bar{B} = \langle B_u : u \in [\lambda]^{<\aleph_0}\rangle$ be a sequence of elements of $\mathcal{D}$ such that for each finite u,

$$B_u = \{t \in I : M \models \exists x \bigwedge_{\alpha \in u} \varphi_\alpha(x, \bar{a}_\alpha[t])\}.$$

Let $\bar{\mathbf{b}} = \langle \mathbf{b}_u : u \in [\lambda]^{<\aleph_0}\rangle$ be a sequence of elements of $\mathcal{D}_*$ such that $\mathbf{j}(B_u) = \mathbf{b}_u$ for each u. Then to show p is realized, it will suffice to show that $\bar{\mathbf{b}}$ has a multiplicative refinement in $\mathfrak{B}$, i.e. there exists a sequence $\langle \mathbf{b}'_u : u \in [\lambda]^{<\aleph_0}\rangle$ of elements of $\mathcal{D}_*$ such that for each finite u, $\mathbf{b}'_u \leq \mathbf{b}_u$, and such that for each finite u and v, $\mathbf{b}'_u \cap \mathbf{b}'_v = \mathbf{b}'_{u\cup v}$.

The next claim verifies that we can generally extract many compatible elements from a large enough subset in our Boolean algebra.

CLAIM 1.8. Let $\mathfrak{B} = \mathfrak{B}^1_{2^\lambda,\mu,\aleph_0}$. Suppose $\langle g_\alpha : \alpha \in \mathcal{U}\rangle$ is a subset of $\mathrm{FI}_{\mu,\aleph_0}(2^\lambda)$, where $\mathcal{U}$ is of cardinality $> \mu$, and $m < \omega$. Then there is $u \subseteq \mathcal{U}$, $|u| = m$ such that $\bigcup_{\alpha\in u} g_\alpha$ is a function.

PROOF. By induction on $n \leq m$ we will choose β_n, $\mathcal{U}_n$, and $\langle f^n_\alpha : \alpha \in \mathcal{U}_n \rangle$ such that:

(1) $\beta_n \in \mathcal{U}_{n-1}$, where $\mathcal{U}_{-1} = \mathcal{U}$
(2) $\mathcal{U}_n \subseteq \mathcal{U}$ is of cardinality $> \mu$
(3) $f^n_\alpha \in \mathrm{FI}_{\mu,\aleph_0}(2^\lambda)$ and $f^n_\alpha \supseteq g_\alpha$, for all $\alpha \in \mathcal{U}$
(4) $n < n'$ implies $f^n_\alpha \subseteq f^{n'}_\alpha$
(5) $n < n'$ implies $\mathcal{U}_{n'} \subseteq \mathcal{U}_n$
(6) for every $\alpha \in \mathcal{U}_n$, $\mathbf{x}_{f^n_\alpha} \leq \mathbf{x}_{f_{\beta_n}}$
(7) $\beta_n \notin \mathcal{U}_n$

We will repeatedly use the fact that no set of size $> \mu$ is covered by a union of $\leq \mu$ sets of cardinality $\leq \mu$.

At stage -1, $f^{-1}_\alpha = g_\alpha$, $\mathcal{U}_{-1} = \mathcal{U}$, and let "$\mathbf{x}_{f_{\beta_{-1}}}$" mean $1_{\mathfrak{B}}$.

At stage $n \geq 0$, by induction on t we try to build a maximal antichain of $\mathbf{x}_{f_{\alpha_{n-1}}}$'s, say $\langle \mathbf{x}_{f_{\alpha_{i_t}}} : t < t_* \rangle$, where each $f_{\alpha_i} \supseteq f^{n-1}_{\alpha_i}$ for some $\alpha_i \in \mathcal{U}_{n-1}$. This must stop at t_* an ordinal $< \mu^+$ as $\mathfrak{B}$ has the μ^+-c.c. For each remaining $\alpha \in \mathcal{U}_{n-1}$, the element $\mathbf{x}_{f^{n-1}_\alpha}$ must be compatible with one of the elements of our antichain (since the construction could not continue), and since $\mathcal{U}_{n-1}$ is of cardinality $> \mu$, there is a subset $\mathcal{U}_n \subseteq \mathcal{U}_{n-1}$ of cardinality $> \mu$ whose elements are all $\neq \beta_n$ compatible with a single element of the antichain. Let β_n be the subscript of this single element. For all $\alpha \in \mathcal{U}_n$, let $f^n_\alpha = f^{n-1}_\alpha \cup f^{n-1}_{\beta_n}$. (For $\alpha \in \mathcal{U} \setminus \mathcal{U}_n$, let $f^n_\alpha = f^{n-1}_\alpha$.) So we can carry the construction.

When we arrive to m, we have defined $f^m_{\beta_0}, \ldots, f^m_{\beta_m}$ and by construction,

$$0 < \mathbf{x}_{f^m_{\beta_m}} \leq \cdots \leq \mathbf{x}_{f^0_{\beta_0}}.$$

Recalling (3) and (4) of the inductive hypothesis, we conclude that

$$\mathbf{x}_{g_{\beta_0}} \cap \cdots \cap \mathbf{x}_{g_{\beta_m}} > 0$$

i.e. when $u = \{\beta_0, \ldots, \beta_m\}$, $\bigcup_{\beta \in u} g_\beta$ is a function, as desired. (This proof is also easy by the Δ-system lemma.) $\square$

NOTATION 1.9. Let $\mathcal{F} = \{f : f$ a strictly increasing function from $\mathbb{N}$ to $\mathbb{N} \setminus \{0\}\}$. In particular, $f(k) > k$ for all $k \in \mathbb{N}$.

2. Construction

In this section we build our example, really a family of examples taking as a parameter an increasing function $f \in \mathcal{F}$ (see 1.9). We begin with the definition, and continue with a leisurely discussion of the types in models of this theory.

NOTATION 2.1. Let $f \in \mathcal{F}$.

(a) For each $k < \omega$, let $\mathcal{T}_k = \mathcal{T}_{f,k} = \{\eta : \eta \in {}^k\omega$ and $\ell < k \to \eta(\ell) \leq f(\ell)\}$.
(b) Let $\mathcal{T}_{[k]} = \mathcal{T}_{f,[k]} = \bigcup_{j \leq k} \mathcal{T}_j$.
(c) Let $\mathcal{T} = \mathcal{T}_f = \bigcup_{k<\omega} \mathcal{T}_k$.
(d) Let $\lim(\mathcal{T}) = \{\eta \in {}^\omega\omega : \eta \restriction k \in \mathcal{T}_k$ for all $k < \omega\}$.
(e) We may write $\mathcal{T}, \mathcal{T}_k$, etc. omitting f when it is clear from context.

Notation 2.1 gives us trees $\mathcal{T}_{[k]}$ of various finite heights whose splitting is finite but growing: for example, the root has two successors, nodes at level 1 each have three successors, nodes at level 2 each have 5 successors. (It isn't crucial that the function is strictly increasing, but it should not be constant on an infinite set.) $\mathcal{T}$ is the tree whose restriction to height k is $\mathcal{T}_k$ for all finite k.

DEFINITION 2.2. We say $s \subseteq \mathcal{T}_{[k]}$ is *k-maximal* if it is nonempty and either $k = 0$, or $k > 0$ and s satisfies: closed under initial segment, every maximal node belongs to $\mathcal{T}_k$, for no $\rho \in \mathcal{T}_{[k]}$ do we have $\{\rho^\frown\langle\ell\rangle : \ell \leq f(\lg(\rho))\} \subseteq s$, *and* s is maximal subject to these conditions.

Definition 2.2 describes a subtree of $\mathcal{T}_{[k]}$ which is maximal subject to containing no full splitting. This means: choose all but one immediate successors of the root; for every element you have chosen at level 1, choose all but one of its immediate successors; and so on. (Once an element is *not* chosen, all of its successors later in the tree are ruled out.) There is a natural partial order on these subtrees, as the next definition points out.

NOTATION 2.3. Let $f \in \mathcal{F}$.

(f) Let $\mathcal{S}_k = \mathcal{S}_{f,k} = \{s : s \subseteq \mathcal{T}_{[k]} \text{ is } k\text{-maximal}\}$.
(g) Let $\mathcal{S} = \mathcal{S}_f = \bigcup_k \mathcal{S}_k$.
(h) Let $\lim(\mathcal{S}) = \{\nu : \nu = \langle s_k : k < \omega\rangle$ is such that $s_k \in \mathcal{S}_k$ for each $k < \omega$ and $s_k \subseteq s_{k'}$ for all $k < k' < \omega\}$.
(i) We may write $\mathcal{S}, \mathcal{S}_k$, etc. omitting f when it is clear from context.

To emphasize that $\lim(\mathcal{S})$ and $\lim(\mathcal{T})$ are parallel, we might think of elements of $\lim(\mathcal{T})$ as sequences $\langle \eta_k : k < \omega\rangle$ where each $\eta_k \in \mathcal{T}_k$ and $k < k' < \omega$ implies $\eta_k \trianglelefteq \eta_{k'}$. Elements of $\lim(\mathcal{T})$ are branches of $\mathcal{T}$ under the ordering $\trianglelefteq$, and elements of $\lim(\mathcal{S})$ are branches of $\mathcal{S}$ under the ordering $\subseteq$. In particular, since all of the $\mathcal{T}_k$'s and $\mathcal{S}_k$'s are finite and f is increasing, both $\mathcal{T}$ and $\mathcal{S}$ are countable, and both $\lim(\mathcal{T})$ and $\lim(\mathcal{S})$ have cardinality $\leq 2^{\aleph_0}$.

We will define our theory T_f as the model completion of the following universal theory $T_{0,f}$. Note that this theory is *not* in the same language as $\mathcal{T}$ and $\mathcal{S}$. We simply use a family of unary predicates to partition the universe of the model into P and Q and then further partition these to mirror $\mathcal{T}$ and $\mathcal{S}$. [For example, if $\mathcal{T}$ has a root at level 0 followed by three nodes at level 1, P is partitioned into one piece by a predicate P_0, which is in turn partitioned into three pieces by predicates called P_{00}, P_{01}, P_{02}, and so forth.] So the tree structure on each side is "hard coded" into the picture without the complexity of having a definable partial order in the language. The remaining ingredient is $R \subseteq P \times Q$, described after the definition.

DEFINITION 2.4. Let $f \in \mathcal{F}$. We define $T_{0,f}$ as the following universal theory.

(1) $\tau(T_{0,f}) = \{P, Q, R, P_\eta, Q_s : \eta \in \mathcal{T}, s \in \mathcal{S}\}$
where R is a binary predicate and the Q_s's, P_η's, P, Q are unary predicates.
(2) for a τ-model M, $M \models T_{0,f}$ iff:
  (a) P^M, Q^M is a partition of M
  (b) for each $k < \omega$, $\langle P^M_\eta : \eta \in \mathcal{T}_k\rangle$ is a partition of P^M
  (c) for each $k < \omega$, $\langle Q^M_s : s \in \mathcal{S}_k\rangle$ is a partition of Q^M
  (d) $\eta \trianglelefteq \eta' \in \mathcal{T}$ implies $P^M_{\eta'} \subseteq P^M_\eta$ and $s \subseteq s' \in \mathcal{S}$ implies $Q^M_{s'} \subseteq Q^M_s$
  (e) $R^M \subseteq Q^M \times P^M$ satisfies:
    - if $\eta \in \mathcal{T}_k$, and $b_\ell \in P^M_{\eta^\frown\langle\ell\rangle}$ for $\ell \leq f(k)$, then

$$M \models \neg(\exists x)(Q(x) \wedge \bigwedge_{\ell \leq f(k)} R(x, b_\ell)).$$

- if $(b,a) \in R^M$ and $b \in Q_s$, $s \in \mathcal{S}_k$, $a \in P_\eta$, $\eta \in \mathcal{T}_k$ then $\eta \in s$.

(3) Let T_f be the model completion of $T_{0,f}$.

A more detailed informal description:

Q and P partition the domain. If we think of the elements of P as being indexed by the leaves of some infinite, finitely branching tree whose branching at height k is $f(k)$, then for each $\eta \in \mathcal{T}_k$, P_η names the elements in the cone above η. In particular, for each finite k and $\eta \in \mathcal{T}_k$, $P^M_{\eta^\frown\langle\ell\rangle}$ for $\ell \leq f(k)$ partition P^M_η.

R, the edge relation, relates elements of Q to elements of P. The restriction on R is that for each element $b \in Q$ and each $k < \omega$ and $\eta \in \mathcal{T}_k$, b cannot connect to some element in each piece of the partition $\{P^M_{\eta^\frown\langle\ell\rangle} : \ell \leq f(k)\}$. So for each $b \in Q^M$, each $k < \omega$ and each $\eta \in \mathcal{T}_k$, either $R(b,x)$ misses the set P^M_η entirely, or else it misses one of $P^M_{\eta^\frown\langle\ell\rangle}$ for $\ell \leq f(k)$.

If we think of the elements of Q_s as being indexed by the elements of $\mathcal{S}$, in other words, by some subtree of finite height which is not full (and more), then the elements of Q_s are only allowed to connect via R to elements in P_η where $\eta \in s$. Notice that these predicates name already definable sets (and so will help with quantifier elimination): $Q^M_s(b)$ means

$$\{\eta \in \mathcal{T}_k : M \models (\exists x)(R(b,x) \wedge P_\eta(x))\} = s \in \mathcal{S}_k.$$

Of course we can't formally quantify over $\eta \in \mathcal{T}_k$, but for each k, there are only finitely many P_η's to consider, so the set named by Q_s is indeed a definable set. Condition (2)(c) uses that the s are maximal in the sense of 2.2.

Finally, note that in models of T_f, the model completion, whenever $R(b,x)$ is consistent with $P_\eta(x)$, *exactly* one of the formulas $\{P_{\eta^\frown\langle\ell\rangle}(x) : \ell \leq f(k)\}$ will be inconsistent with $R(b,x)$. So the Q_s's with s maximal are indeed the relevant ones.

DEFINITION 2.5. For each $k < \omega$, let $T_{0,f,k}$ be the universal theory which is the restriction of $T_{0,f}$ to $\tau_k = \{P, Q, R, P_\eta, Q_s : \eta \in \mathcal{T}_k, s \in \mathcal{S}_k\}$.

OBSERVATION 2.6. For each $k < \omega$, $T_{0,f,k}$ is a universal theory in a finite relational language. It has the joint embedding property and amalgamation: given models M, N, the model whose domain is the disjoint union of $\mathrm{dom}(M) \cap \mathrm{dom}(N)$, $\mathrm{dom}(M) \setminus \mathrm{dom}(N)$, and $\mathrm{dom}(N) \setminus \mathrm{dom}(M)$ and the interpretation of a predicate is the union of the interpretations is also a model of the theory. As a result, a model completion $T_{f,k}$ exists and is well defined, and complete.

CLAIM 2.7. $T_{0,f}$ has a model completion T_f. In fact $T_{0,f}$ is relational, has amalgamation and the joint embedding property. Moreover T_f is simple, has elimination of quantifiers has no algebraicity, and has trivial forking.

PROOF SKETCH. There are two parts.

Part I. First, we briefly describe the two key cases for quantifier elimination, ignoring the atomic formulas $x = a$.

(a) given $P(y_0), \ldots, P(y_n)$, does there exist x s.t. $Q(x) \wedge \bigwedge_{i \leq n} R(x, y_i)$?

Informally, the answer is no if and only if the y_i's fall in all pieces of some successor partition, that is, for some $k < \omega$ and $\eta \in \mathcal{T}_k$, for each $\ell \leq f(k+1)$ there is at least one $i \leq n$ such that $P^M_{\eta^\frown\langle\ell\rangle}(y_i)$. Since f is increasing (1.9), we can choose the minimal k_* such that $f(k_*) > n$ and it is then sufficient to ensure that this doesn't happen for any $\eta \in \mathcal{T}_{[k_*]}$. Since the number of pieces in each successor partition is finite and n is finite, this can be expressed by a disjunction of all legal possibilities.

If we now modify (a) by adding finitely may conditions on the y_i's, of the form $P_\eta(y_i)$ or $\neg P_\eta(y_i)$ for some $\eta \in \mathcal{T}$, and by adding finitely many conditions on x of the form $Q_s(x)$ or $\neg Q_s(x)$ for some $s \in \mathcal{S}$, these conditions simply affect which of all the possible "legal possibilities" remain legal.

(b) given $Q(y_0), \dots, Q(y_n)$, does there exist x s.t. $P(x) \wedge \bigwedge_{i \leq n} R(y_i, x)$?

Let k_* be minimal so that $f(k_*) > n$. Informally, the answer to (b) is yes if and only if for some $\ell \leq f(k_* + 1)$, and some $\eta \in \mathcal{T}_{k_*}$, $\bigwedge_{i \leq n}(R(y_i, x) \wedge P_{\eta^\frown\langle\ell\rangle}(x))$ is consistent.[1] (Why? We may verify by induction on $k \geq k_*$ that there is $\eta_k \in \mathcal{T}_k$ such that $\bigwedge_{i \leq n}(R(y_i, x) \wedge P_{\eta_k}(x))$ is consistent. Let $\eta_{k_*} = \eta$. At $k > k_*$, for each $i \leq n$, $R(y_i, x)$ will be inconsistent with precisely one of the formulas $\{P^M_{\eta_k{}^\frown\langle\ell\rangle} : \ell \leq f(k+1)\}$, so n is not large enough to rule out one which works for all. Conversely, if the answer to (b) is no, we can bound the height of an inconsistency by noting that each y_i which is consistent with P_ν can only miss one piece of the successor partition $\{P^M_{\nu^\frown\langle\ell\rangle} : \ell \leq f(\lg(\nu))\}$.)

Since the number of pieces in each successor partition is finite and n is finite, (b) can be expressed by a disjunction over all legal possibilities, using the possible types for the y_i as expressed by the Q_s's. As before, if we modify (b) by adding conditions on the y_i's expressing which pieces of which successor partitions they may miss (expressible by the Q_s's) and by adding conditions on x expressing in which pieces of which partitions it may be (expressible by the P_η's), these conditions simply affect which of all the possible "legal possibilities" remain legal.

In both (a) and (b) above, recall 1.9 tells us that $k_* \leq n$.

Part II. Second, let us extend this analysis to show that

> ($\star$) if M is a model of $T_{f,k}$, N is a model of $T_{f,k+1}$, and ψ is a sentence of τ_k of length $\leq k$, then $M \models \psi \Leftrightarrow N \models \psi$.

This will tell us that the sequence $\langle T_{f,k} : k < \omega \rangle$ from 2.6 converges to a theory we may call T_f, and we can conclude that for every formula $\varphi(\bar{x})$ of τ, for some quantifier free $\psi(\bar{x})$, for every $k < \omega$ large enough, we have

$$(\forall \bar{x})(\ \psi(\bar{x}) \equiv \varphi(\bar{x})\ ) \in T_{fk}.$$

Fix $k < \omega$. Let $f = \{(a_\ell, b_\ell) : \ell < k\}$ be a partial one to one map from M into N with $|\operatorname{dom}(f)| \leq k$ such that for every atomic, or equivalently, quantifer-free formula $\psi(x_0, \dots, x_{k-1})$ of τ_k, we have that $M \models \psi[a_0, \dots, a_{k-1}]$ if and only if $N \models \psi[b_0, \dots, b_{k-1}]$. The point is that N and M agree on the information up to level k, but N has more information as it is a τ_{k+1}-model. Let us show that for every $a \in M$ there is $b \in N$ such that $f \cup \{(a, b)\}$ is a partial one to one map from M into N such that for every quantifier-free formula $\psi(x_0, \dots, x_k)$ of τ_{k+1}, we have that $M \models \psi[a_0, \dots, a_{k-1}, a]$ if and only if $N \models \psi[b_0, \dots, b_{k-1}, b]$. (The proof that for every $b \in N$, there is $a \in M$ such that the same holds, is similar.)

As equalities and non-connections are easy to handle, the two main cases to consider are positive instances of R, first when $Q(a)$ and second when $P(a)$. We may assume $a \notin \{a_i : i < k\}$ as otherwise it is trivial.

[1]Note this is equivalent to saying: for every $k \leq k_*$, for some $\ell \leq f(k+1)$, and some $\eta \in \mathcal{T}_k$, we have that $\bigwedge_{i \leq n}(R(y_i, x) \wedge P_{\eta^\frown\langle\ell\rangle}(x))$ is consistent.

Suppose that $Q(a)$ and $\{P(a_i) : i < i_* \leq k\}$ (we can safely ignore the others in this case). Since M is a model, $Q_s(a)$ for some $s \in \mathcal{S}_k$, and also for each $i < i_*$ there is $\eta_i \in \mathcal{T}_k$ such that $P_{\eta_i}(a_i)$. Suppose $R(a, a_i)$ for each $i < i_*$. Then in N, $P_{\eta_i}(b_i)$ for each $i < i_*$ by hypothesis, where we know that $\{\eta_i : i < k\} \subseteq s$ so do not include any full splitting. Now in N, which is a τ_{k+1}-structure, for each $i < i_*$ there is $\ell_i < f(k+1)$ such that

$$P_{\eta_i{}^\frown\langle \ell_i \rangle}(b_i).$$

Then $|\{\ell_i : i < i_*\}| \leq k$, so recalling (a) earlier in the proof, there must be s' such that $s \subseteq s' \in \mathcal{S}_k$ and $\{\eta_i{}^\frown\langle \ell_i \rangle : i < k\} \subseteq \mathcal{S}'$. Then we may choose $b \in Q^N_{s'}$ which relates to the b_i's as desired.

Suppose that $P(a)$ and suppose $\{Q(a_i) : i < i_* \leq k\}$ (we can safely ignore the others in this case). Since M is a model, $P_\eta(a)$ for some $\eta \in \mathcal{T}_k$ and likewise for each $i < i_*$ $Q_{s_i}(a_i)$ for some $s_i \in \mathcal{S}_k$. Suppose $R(a_i, a)$ for each $i < i_*$. Then in N, being a τ_{k+1}-structure, for each $i < i_*$ there is $s'_i \in \mathcal{S}_{k+1}$ such that $s_i \subseteq s'_i$ and

$$\{Q_{s'_i}(b_i) : i < i_*\}.$$

The existence of a tells us that $\eta \in s'_i$ for each $i < i_*$, so as $i_* \leq k$, we may find $\ell < f(k+1)$ such that $\eta^\frown\langle \ell \rangle \in s'_i$ for all $i < i_*$ Recalling (b) earlier in the proof, this tells us that

$$\bigwedge_{i<i_*} Q_{s'_i}(y_i) \wedge R(y_i, x) \wedge P_{\eta^\frown\langle \ell \rangle}(x)$$

is consistent, and so we may choose b in $P^N_{\eta^\frown\langle \ell \rangle}$ to complete the proof. □

Before proceeding to an analysis of saturation and non-saturation in ultrapowers of models of T_f, let us briefly describe the types we will be dealing with. Recall that $T_{\mathbf{rg}}$ is the theory of the random graph.

Claim 2.8. *Let $M \models T_f$. Let I be any infinite set, $|I| = \lambda$, and $\mathcal{D}$ a regular ultrafilter on I which is good for $T_{\mathbf{rg}}$. Then to show $N = M^I/\mathcal{D}$ is λ^+-saturated, it suffices to show that N realizes all partial types of the following form:*

(a) $\{Q(x)\} \cup \{R(x, a) : a \in A\}$ *for* $A \subseteq P^N$, $|A| \leq \lambda$.
(b) $\{P(x)\} \cup \{R(b, x) : b \in B\}$ *for* $B \subseteq Q^N$, $|B| \leq \lambda$.

Proof. Fix a model N. To analyze saturation of N, it suffices to consider 1-types, so there are two cases: the type contains $P(x)$, or it contains $Q(x)$. Since the ultrafilter is good for the random graph, necessarily $\mu(\mathcal{D}) \geq \lambda^+$ (i.e. any pseudofinite set has size at least λ^+). Recall that for models with countable vocabularies, which is always our case here, saturation of ultrapowers reduces to saturation of φ-types ([**10**], Theorem 12). [So we may hope to not need to consider types with infinitely many distinct Q_s's or P_η's.]

Let $p \in \mathbf{S}(C, N)$, $C \subseteq N$, $|C| \leq \lambda$, p not algebraic.

Without loss of generality,

(i) $N \restriction C \preceq N$.
(ii) if $\eta \in \mathcal{T}$, $Q(x) \in p(x)$ and $(\exists z)(P_\eta(z) \wedge R(x, z)) \in p$, then for some $a_\eta \in P^N_\eta$ we have $R(x, a_\eta) \in p$.
(iii) if $s \in \mathcal{S}$, $P(x) \in p(x)$ and $(\exists z)(Q_s(z) \wedge R(z, x)) \in p$, then for some $b_s \in Q^N_s$ we have $R(b_s, x) \in p$.

[We can assume this because for all $\eta \in \lim(\mathcal{T})$ and all $s \in \lim(\mathcal{S})$, $|\bigcap_n P^N_{\eta\restriction n}|$, $|\bigcap_n Q^N_{s\restriction n}|$ are $> \lambda$ because $\mathcal{D}$ is regular and $N = M^I/\mathcal{D}$.]

Now let:

(a) $A_1 = \{a \in C : P(x) \wedge R(a,x) \in p \text{ or } Q(x) \wedge R(x,a) \in p\}$.
(b) $A_0 = \{a \in C : P(x) \wedge \neg R(a,x) \in p \text{ or } Q(x) \wedge \neg R(x,a) \in p\}$.
(c) if $P(x) \in p(x)$ then let:
- $\eta = \eta_p \in \lim(\mathcal{T})$ be such that $p_2(x) := \{P_{\eta\restriction\ell}(x) : \ell < \omega\} \subseteq p(x)$,
- $p_1(x) = \{R(a,x) : a \in A_1\}$,
- $p_0(x) = \{\neg R(a,x) : a \in A_0\}$, and
- $p_3(x) = \{x \neq c : c \in C \text{ and } c \text{ realizes } \{P_{\eta_p\restriction n}(x) : n < \omega\}\}$.

(d) if $Q(x) \in p(x)$ then let:
- $\nu = \nu_p \in \lim(\mathcal{S})$ be such that $p_2(x) := \{Q_{\nu(\ell)}(x) : \ell < \omega\} \subseteq p(x)$,
- $p_1(x) = \{R(x,a) : a \in A_1\}$, and
- $p_0(x) = \{\neg R(x,a) : a \in A_0\}$.
- $p_3(x) = \{x \neq c : c \in C \text{ and } c \text{ realizes } \{Q_{\nu_p\restriction n}(x) : n < \omega\}\}$.

The p is equivalent to $p_1(x) \cup p_0(x) \cup p_2(x) \cup p_3(x)$, by elimination of quantifiers and our assumptions (i),(ii),(iii).

Now if $\eta \in \mathcal{T}$, $P_\eta(x) \in p$ then $p_1(x) \vdash P_\eta(x)$. [Why? Let $k = \lg(\eta)$ and let $\mathcal{S}^* = \{s \in \mathcal{S}_k : \eta \in s\} \subseteq \mathcal{S}_k$. So for each $s \in \mathcal{S}^*$, there is $b_s \in Q^N_s$ such that $R(b_s, x) \in p$ by (ii), so it is easy to see that $\{R(b_s,x) : s \in \mathcal{S}^*\} \vdash P_\eta(x)$.]

Similarly, if $Q_s(x) \in p(x)$ for $s \in \mathcal{S}$ then $p_1(x) \vdash Q_s(x)$. So in both cases,

$$p_1(x) \vdash p_2(x).$$

It remains to handle $p_0 \cup p_3$. Suppose, then, that our ultrapower realizes $p_1(x)$. The sets $A_1, A_0 \subseteq N$ defined in (a), (b) are disjoint and both of size $\leq \lambda$. As $\mathcal{D}$ is regular and $T_{\mathbf{rg}}$-good, there are disjoint pseudo-finite internal sets X_0, X_1, X_2 such that $A_\ell \subseteq X_\ell$ for $\ell = 0,1$ and $X_0 \cap X_1 = \emptyset$ and X_2 disjoint to $X_0 \cup X_1$ (even, $\subseteq P$ if $X_0 \cup X_1 \subseteq Q$, etc). Suppose $c \in N$ realizes $p_1(x)$. There are two cases depending on whether $Q(x) \in p$ or $P(x) \in p$. If $Q(x) \in p$, there is $c' \in N$ such that:

- $(\forall y \in X_1)(R(c',y) \Leftrightarrow R(c,y))$
- $(\forall y \in X_0)(\neg R(c',y))$.

So c' realizes $p_1(x)$, $p_2(x)$, $p_0(x)$ and even $p_3(x)$, so we are done. If $P(x) \in p$, replace $R(c',y)$ by $R(y,c')$ and $R(c,y)$ by $R(y,c)$ in the above quotation. This completes the proof. □

3. A non-saturation result for T_f

NOTATION 3.1. When $\langle a_\alpha : \alpha < \kappa\rangle$, and $w \subseteq \kappa$, write $\bar{a}_w$ to mean $\langle a_\alpha : \alpha \in w\rangle$.

Recall also our notation for $\mathcal{T}$, $\mathcal{T}_{[k]}$, etc. from 2.1.

LEMMA 3.2. *Let $f \in \mathcal{F}$. Let λ be any infinite cardinal and $\mu < \kappa$, $\mu \leq \lambda$. Then no ultrafilter $\mathcal{D}$ of $\mathfrak{B} = \mathfrak{B}^1_{2^\lambda,\mu,\aleph_0}$ is (κ^+, T_f)-moral.*

PROOF. We use the setup of separation of variables, so we have in mind the background set $I = \lambda$, a homomorphism $\mathbf{j} : \mathcal{P}(\lambda) \to \mathfrak{B}$, an ultrafilter $\mathcal{D}_*$ on $\mathfrak{B}$, and also a model $M \models T_f$ and an ultrapower $M^I/\mathcal{D}$. We will build a possibility pattern in the variables x and $\langle x_\alpha : \alpha < \kappa \rangle$ and show it doesn't have a multiplicative refinement.[2]

Recall that $\mathfrak{B}^+ = \mathfrak{B} \setminus \{0\}$.

(3.A) For each $\rho \in \mathcal{T}_k$ and $\alpha < \kappa$, define $\mathbf{a}[P_\rho(x_\alpha)] \in \mathfrak{B}^+$ by induction on $k < \omega$:

(a) if $k = 0$, i.e. $\rho \restriction k = \langle\rangle$, $\mathbf{a}[P_{\langle\rangle}(x_\alpha)] = 1_{\mathfrak{B}}$.
(b) for k and $i \leq f(k)$, let $g_{\alpha,k,i}$ be the function with domain $\{\omega\alpha + k\}$ such that $g_{\alpha,k,i}(\omega\alpha + k) = i$. Then for $\rho \in \mathcal{T}_k$, $\ell \leq f(k)$ define

$$\mathbf{a}[P_{\rho^\frown\langle\ell\rangle}(x_\alpha)] = \mathbf{a}[P_\rho(x_\alpha)] \cap \mathbf{x}_{g_{\alpha,k,\ell}} \text{ if } \ell < f(k)$$

$$\text{and } = \mathbf{a}[P_\rho(x_\alpha)] \setminus \bigcup_{i<f(k)} \mathbf{x}_{g_{\alpha,k,i}} \text{ if } \ell = f(k).$$

(c) for $\alpha \neq \beta < \kappa$,

$$\mathbf{a}[x_\alpha \neq x_\beta] = 1_{\mathfrak{B}}.$$

Without loss of generality we assume

(3.B) $$\mathbf{x}_{g_{\alpha,k,0}} \in \mathcal{D}_* \text{ for } \alpha < \kappa.$$

By (3.A)(c),

(3.C) $$\mathbf{a}[x_\alpha \neq x_\beta] \in \mathcal{D}_*$$

and by quantifier elimination, $\mathbf{a}[\varphi(\bar{x}_v)]$ is determined for any finite $v \subseteq \kappa$ and any $\varphi(\bar{x}_v)$ in the language of T. It follows that for each $\alpha < \kappa$ and each $k < \omega$, $\mathbf{a}[P_{\langle 0_k\rangle}(x_\alpha)] \in \mathcal{D}_*$, where $\langle 0_k \rangle$ is $\langle 0, \ldots, 0\rangle$ (k times). By (3.A), (3.B), (3.C), the sequence

(3.D) $$\bar{\mathbf{b}} = \langle \mathbf{b}_u : u \in [\kappa]^{<\aleph_0}\rangle \text{ where } \mathbf{b}_u = \mathbf{a}[\exists x \bigwedge_{\alpha \in u} R(x, x_\alpha)]$$

is a possibility pattern $\mathbf{b}_u \in \mathcal{D}_*$. [That is, in the related ultrapower we are considering the type $\{R(x, a_\alpha) : \alpha < \kappa\}$ where the a_α's are pairwise distinct realizing the type $\{P_{\rho_* \restriction k}(x) : k < \omega\}$ where $\rho_* = (0, 0, \ldots)$.[3]]

Towards contradiction assume

(3.E) $$\bar{\mathbf{b}}' = \langle \mathbf{b}'_u : u \in [\kappa]^{<\aleph_0}\rangle$$

is a multiplicative refinement of $\bar{\mathbf{b}}$.

For $\alpha < \kappa$ choose $g_\alpha \in \mathrm{FI}_{\aleph_0}(2^\lambda)$ such that $\mathbf{x}_{g_\alpha} \leq \mathbf{b}'_{\{\alpha\}}$. We will repeatedly use the fact that no set of size $> \mu$ (e.g. no set of size κ) is covered by a countable union of sets of cardinality $\leq \mu$. Given an ordinal β, let us say "the remainder of β mod ω is k" to mean that $\beta = \omega\gamma + k$ for some ordinal γ and integer k.

Since each g_α has finite domain, there is a smallest integer k_α such that for every $\beta \in \mathrm{dom}(g_\alpha)$, the remainder of β mod ω is $\leq k_\alpha$. As $\mathcal{T}_*$ has cardinality $> \mu$, there is some finite k_* such that $\mathcal{U}_1 = \{\alpha < \kappa : k_\alpha = k_*\}$ is of cardinality $> \mu$.

---

[2]The proof will take place entirely in $\mathfrak{B}$, but it is simply asserting that we could build a type $\{R(x, a_\alpha) : \alpha < \kappa\}$ in the ultrapower $M^I/\mathcal{D}$, i.e. we could choose our parameters a_α, in such a way that each $\mathbf{a}[\varphi(\bar{x}_w)]$ in the proof below is really the image of $\{t \in I : M \models \varphi[\bar{a}_w[t]]\}$ under $\mathbf{j}$, and likewise for each finite $u \subseteq \kappa$, $\mathbf{b}_u$ is the image of $\{t \in I : M \models \exists x \bigwedge_{\alpha \in u} R(x, a_\alpha[t])\}$ under $\mathbf{j}$.

[3]The intent is that the elements a_α all satisfy the "constant zero" branch in the ultrapower, so $p(x)$ is clearly a type. The potential problem in realizing the type is that in each index model, by our construction (3.A), the elements $a_\alpha[t]$ may project to different branches "$\mathcal{D}$-rarely", possibly violating the condition on splitting.

For each $\alpha \in \mathcal{U}_1$, the elements

$$\{\mathbf{a}[P_\nu(x_\alpha)] : \nu \in \mathcal{T}_{k_*}\}$$

form a maximal antichain of $\mathbf{b}_\alpha$. For each $\alpha \in \mathcal{U}_1$, choose $\nu_\alpha \in \mathcal{T}_{k_*}$ such that

$$\mathbf{x}_{g_\alpha} \cap \mathbf{a}[P_{\nu_\alpha}(x_\alpha)] > 0.$$

By the construction in (3.A) above, we can translate as follows: letting

$$g^*_\alpha = g_\alpha \cup \bigcup_{k \leq k_*} \{(\omega\alpha + k, \nu_\alpha(k))\}$$

we have that

$$\text{(3.F)} \qquad \mathbf{x}_{g^*_\alpha} = \mathbf{x}_{g_\alpha} \cap \mathbf{a}[P_{\nu_\alpha}(x_\alpha)] > 0.$$

Moreover, for each $\alpha \in \mathcal{U}_1$, it is still the case that for every $\beta \in \operatorname{dom}(g^*_\alpha)$, the remainder of β mod ω is still $\leq k_*$. Since $\mathcal{T}_{k_*}$ is finite, for some $\nu_* \in \mathcal{T}_{k_*}$,

$$\mathcal{U}_2 = \{\alpha \in \mathcal{U}_1 : \nu_\alpha = \nu_*\}$$

is of cardinality $> \mu$.

Choose some very large finite m (at least $f(k_* + 1)$ suffices, see below). Apply Claim 1.8 to $\langle g^*_\alpha : \alpha \in \mathcal{U}_2\rangle$ and m, so the Claim says: *Suppose* $\langle g^*_\alpha : \alpha \in \mathcal{U}_2\rangle$ *is a subset of* $\mathrm{FI}_{\mu,\aleph_0}(2^\lambda)$, *where* $\mathcal{U}_2$ *is of cardinality* $> \mu$, *and* $m < \omega$. *Then there is* $u \subseteq \mathcal{U}_2$, $|u| = m$ *such that* $\bigcup_{\alpha \in u} g^*_\alpha$ *is a function.*

Let $u \subseteq \mathcal{U}_2$ be as returned by Claim 1.8. Then $g^* = \bigcup_{\alpha \in u} g^*_\alpha$ is a function, and it is still the case that for all $\beta \in \operatorname{dom}(g^*)$, the remainder of β mod ω is $\leq k_*$. Since we assumed (for a contradiction) that $\bar{\mathbf{b}}'$ is multiplicative and refines $\bar{\mathbf{b}}$, and each $\mathbf{x}_{g_\alpha} \leq \mathbf{b}'_\alpha$, we have that for every $v \subseteq u$,

$$\text{(3.G)} \qquad 0 < \mathbf{x}_{g^*} \leq \bigcap_{\alpha \in v} \mathbf{x}_{g_\alpha} \leq \bigcap_{\alpha \in v} \mathbf{b}'_\alpha = \mathbf{b}'_v \leq \mathbf{b}_v.$$

Thus, for every $\ell \leq f(k_*)$ and every $\alpha \in u$,

$$\operatorname{dom}(g^*) \cap \{\omega\alpha + k_* + 1\} = \emptyset$$

and clearly for $\alpha \neq \alpha' \in u$,

$$\{\omega\alpha + k_* + 1\} \cap \{\omega\alpha' + k_* + 1\} = \emptyset.$$

So for any $f(k_*+1)$ distinct α's in u, say we enumerate them as $\langle \alpha_{i_\ell} : \ell \leq f(k_*+1)\rangle$,

$$\mathbf{x}_{g_*} \cap \bigcap_{\ell \leq f(k_*+1)} \mathbf{a}[P_{\nu_* {}^\frown \langle \ell \rangle}(x_{\alpha_{i_\ell}})] > 0$$

in other words, recalling (3.F),

$$\bigcap_{\ell \leq f(k_*)} \mathbf{x}_{g_{\alpha_\ell}} \cap \bigcap_{\ell \leq f(k_*)} \mathbf{a}[P_{\nu^*}(x_{\alpha_{i_\ell}})] \cap \bigcap_{\ell \leq f(k_*+1)} \mathbf{a}[P_{\nu_* {}^\frown \langle \ell \rangle}(x_{\alpha_{i_\ell}})] > 0$$

Rewriting $v = \{\alpha_{i_\ell} : \ell \leq f(k_*)\} \subseteq u$, this contradicts the fact from (3.G) that $\bigcap_{\alpha \in v} \mathbf{x}_{g^*_\alpha} \subseteq \bigcap_{\alpha \in v} \mathbf{x}_{g_\alpha} \leq \mathbf{b}_v$. So there can be no such $\bar{\mathbf{b}}'$, which completes the proof. $\square$

4. T_f is explicitly simple

This section proves Theorem 4.2, but its aim is equally or even primarily pedagogical, to exposit a way of measuring simple theories. In [**14**] we defined "explicit simplicity," a way of stratifying the complexity of simple theories using cardinals $(\lambda, \mu, \theta, \sigma)$ satisfying Definition 4.1 below. This is motivated in the introduction to [**14**], §3. The definition of "$(\lambda, \mu, \theta, \sigma)$-explicitly simple" implies that T is simple, and it follows from the definitions that this becomes weaker as μ increases. [**14**] Theorem 4.10 had shown that this definition holds for any simple theory T when $|T| < \sigma$ and we use the largest nontrivial number of "colors," $\mu^+ = \lambda$.

This new characterization of simplicity suggested a program of stratifying simple theories according to the necessary value of μ. For the random graph one color is enough, but it turned out in [**15**] that the case, of, say, the tetrahedron-free three-hypergraph, needs either $\mu^+ = \lambda$ or $\mu^{++} = \lambda$. Moreover, the idea of the infinite descending chain of [**15**] is essentially to look for theories whose μ must satisfy $\mu < \lambda \leq \mu^{+n}$ for larger and larger finite n,[4] inspired by ideas on free sets in set mappings and the Kuratowski-Sierpinski characterization of the $\aleph_n$'s (for more on set mappings, see [**3**], [**9**]).

We left as an open question, Question 10.1 of [**14**], whether it was possible to build a simple theory where μ, the range of the coloring function, is truly uncountable but does not depend on λ.

In this section we show that T_f is *not* such an example. In light of §3, Theorem 4.2 tells us that when $\sigma > |T|$, uncountable μ is not necessary to be in a class strictly above the random graph. This is a delicate point: it highlights that the definition of explicit simplicity *requires* $\sigma > |T|$, with the closure of a finite set (in the relevant algebra) giving rise to an elementary submodel. So $\sigma > |T|$ is an assumption in Theorem 4.2 even though $\theta > |T|$ would seem more natural for our case. For, as we will see, $\sigma > |T|$ and knowing a type of T_f over an elementary submodel is already enough to control consistency of its automorphic images. Indeed, if σ were finite, the story would be different[5]: we will see at the end of the proof below a suggested strategy for saturation of ultrapowers in §5.

We now review the setup.

DEFINITION 4.1 ([**14**] Definition 1.1). Call $(\lambda, \mu, \theta, \sigma)$ *suitable* when:

(1) $\sigma \leq \theta \leq \mu < \lambda$
(2) θ is regular, $\mu = \mu^{<\theta}$ and $\lambda = \lambda^{<\theta}$
(3) $(\forall \alpha < \theta)(2^{|\alpha|} < \mu)$.

For Theorem 4.2, the needed definitions will be given in the course of the proof.

THEOREM 4.2. *The theory T_f is $(\lambda, \mu, \theta, \sigma)$-explicitly simple whenever $(\lambda, \mu, \theta, \sigma)$ is suitable and $\sigma > |T_f|$.*

PROOF. We will simply follow [**14**], Section 3, but all relevant definitions have been quoted for ease of reading. We are aiming for Definition 4.3; all the terms mentioned in the definition will be defined subsequently.

---

[4]This is motivation, not a mathematical statement: [**15**] isn't carried out in the language of explicit simplicity, since there we had a concrete family of theories with trivial forking and so could simplify the definitions somewhat, e.g. dropping the requirement that closures be submodels.

[5]It is natural to give a general definition of explicit simplicity for $\sigma = \aleph_0$ which allows the closure of finite sets to be finite, and applies to theories with trivial forking, see after the proof.

DEFINITION 4.3 (Explicitly simple, [**14**] Definition 3.2). Assume $(\lambda, \mu, \theta, \sigma)$ are suitable. We say T is $(\lambda, \mu, \theta, \sigma)$-*explicitly simple* if T is simple and for every $N \models T$, $||N|| = \lambda$, $p \in \mathbf{S}(N)$ nonalgebraic,

(a) there exists a $(\lambda, \theta, \sigma)$-presentation $\mathfrak{m}$ of p.
(b) for every $(\lambda, \theta, \sigma)$-presentation $\mathfrak{m}$ of p, there is a presentation $\mathfrak{n}$ of p refining $\mathfrak{m}$ and a function $G : \mathcal{R}_{\mathfrak{n}} \to \mu$ such that G is an intrinsic coloring of $\mathcal{R}_{\mathfrak{n}}$.

T_f is indeed simple. So suppose we are given $N \models T$, $||N|| = \lambda$, $p \in \mathbf{S}(N)$ nonalgebraic. We will have two cases: the case where $P(x) \in p$, and the case where $Q(x) \in p$. The only difference will come at the end.

DEFINITION 4.4 (Presentation, [**14**] Definition 3.3). Suppose we are given $N \models T$, $||N|| = \lambda$, and $p \in \mathbf{S}(N)$. A $(\lambda, \theta, \sigma)$-*presentation* for p is the data of an enumeration and an algebra,

$$\mathfrak{m} = (\langle \varphi_\alpha(x, a^*_\alpha) : \alpha < \lambda \rangle, \mathcal{M})$$

where these objects satisfy:

(1) $p = \langle \varphi_\alpha(x; a^*_\alpha) : \alpha < \lambda \rangle$ is an enumeration of p, which induces an enumeration $\langle a^*_\alpha : \alpha < \lambda \rangle$ of $\mathrm{dom}(N)$, possibly with repetitions, and with the a^*_α possibly imaginary.
(2) $\mathcal{M}$ is an algebra on λ with $< \theta$ functions.
(3) For any finite $u \subseteq \lambda$, $|\,\mathrm{cl}_{\mathcal{M}}(u)| < \sigma$. Thus, for any $u \subseteq \lambda$, if $|u| < \sigma$ then $|\,\mathrm{cl}_{\mathcal{M}}(u)| < \sigma$, and if $|u| < \theta$ then $|\,\mathrm{cl}_{\mathcal{M}}(u)| < \theta$.
(4) $\mathrm{cl}_{\mathcal{M}}(\emptyset)$ is an infinite cardinal $\leq |T|$, so an initial segment of λ. $M_* := N \restriction \{a^*_\alpha : \alpha < \mathrm{cl}_{\mathcal{M}}(\emptyset)\}$ is a distinguished elementary submodel of N, and we require that p does not fork over M_*.
(5) Moreover, for each $u \in [\lambda]^{<\sigma}$, $N_u := N \restriction \{a^*_\alpha : \alpha \in \mathrm{cl}_{\mathcal{M}}(u)\}$ is an elementary submodel of N, and $\{\varphi_\alpha(x, a^*_\alpha) : \alpha \in \mathrm{cl}_{\mathcal{M}}(u)\}$ is a complete type over this submodel which dnf over M_*. (In particular, $\{\varphi_\alpha(x, a^*_\alpha) : \alpha \in \mathrm{cl}_{\mathcal{M}}(\emptyset)\}$ is a complete type over M_*.)
(6) If $\alpha \in \mathrm{cl}_{\mathcal{M}}(u)$, $\beta \leq \alpha$, writing $A_\beta = \{a^*_\gamma : \gamma < \beta\}$, we have that $\mathrm{tp}(a^*_\alpha, A_\beta \cup M_*)$ does not fork over $\{a^*_\gamma : \gamma \in \mathrm{cl}_{\mathcal{M}}(u) \cap \beta\} \cup M_*$.

In a context where $(\lambda, \theta, \sigma)$ are given, "presentation" means "$(\lambda, \theta, \sigma)$-presentation."

To show p has a presentation, first fix a countable elementary submodel $M_* \preceq N$. Choose an enumeration $\langle \varphi_\alpha(x, a^*_\alpha) : \alpha < \lambda \rangle$ of p so that:

- $\{a^*_\alpha : \alpha < \omega\} = \mathrm{dom}(M_*)$,
- $\{\varphi(x, a^*_\alpha) : \alpha < \omega\} = p \restriction M_*$,
- and $\{a^*_\alpha : \alpha < \lambda\} = \mathrm{dom}(N)$,

noting the sequence $\langle a^*_\alpha : \alpha < \lambda \rangle$ may have repetitions. This is easily done as $x \neq a$ belongs to p for every $a \in N$. For the algebra $\mathcal{M}$, we add three kinds of functions.

- First choose countably many unary functions $\{g_n : n < \omega\}$, where g_n is the constant function n, to ensure that the "closure of the empty set"[6] is ω.

[6]we never consider $\emptyset$ as a base set, so this effectively is the set which is contained in every closure of every nontrivial set; or if you prefer, consider an algebra to be a structure on λ with functions and no relations *and a constant* (interpreted as any element of ω).

- Second, choose countably many functions which are analogues of Skolem functions for T_f. That is, for each formula $\varphi(x, \bar{y})$ of $\mathcal{L}(\tau(T_f))$, let $f_{\varphi(\bar{y})}$ be a new function symbol, and interpret these countably many new function symbols as Skolem functions for T_f in N, in each case choosing the witness a^*_α of smallest index $\alpha < \lambda$. Then for each $\varphi(x, \bar{y})$ add to the algebra a new $\ell(\bar{y})$-place function symbol $g_{f_{\varphi(x,\bar{y})}}$ which mirrors the action of the Skolem function on the indices λ: $g_{f_\varphi}(\bar{u}) = v$ only if $f_\varphi(\bar{a}^*_u) = a^*_v$.
- Third, we want to ensure that the type restricted to closed sets is complete. For each $\psi(x, y) \in \{R(x, y), x = y, Q(x) \wedge y = y, P(y)\} \cup \{Q_s(x) : s \in \mathcal{S}\} \cup \{P_\eta(y) : \eta \in \mathcal{T}\}$, let $h_\psi(y)$ be defined so that $h_\psi(\alpha) = \beta$ if $\beta < \lambda$ is the least ordinal such that $\varphi_\beta(x, a^*_\beta)$ is equivalent mod T to either $\psi(x, a^*_\alpha)$ or its negation.[7]

Since there is no nontrivial forking, this suffices, and $(\langle \varphi_\alpha(x, a^*_\alpha) : \alpha < \lambda \rangle, \mathcal{M})$ is a presentation.

DEFINITION 4.5 (Refinements of presentations, [**14**] Definition 3.6). Suppose we are given $N \models T$, $||N|| = \lambda$, and $p \in \mathbf{S}(N)$. Let $\mathfrak{m} = (\bar{\varphi}_\mathfrak{m}, \mathcal{M}_\mathfrak{m})$, $\mathfrak{n} = (\bar{\varphi}_\mathfrak{n}, \mathcal{M}_\mathfrak{n})$ be presentations of p. We say that $\mathfrak{n}$ *refines* $\mathfrak{m}$ when:

(a) $\bar{\varphi}_\mathfrak{m} = \bar{\varphi}_\mathfrak{n}$.
(b) $\mathrm{cl}_{\mathcal{M}_\mathfrak{m}}(\emptyset) = \mathrm{cl}_{\mathcal{M}_\mathfrak{n}}(\emptyset)$.
(c) $\mathcal{M}_\mathfrak{m} \subseteq \mathcal{M}_\mathfrak{n}$, i.e. the algebra of $\mathfrak{n}$ extends that of $\mathfrak{m}$.

In a refinement, the enumeration stays the same, the distinguished elementary submodel stays the same, but we may add a few more functions to the algebra if we wish. In our case it isn't necessary; we'll just show directly that every presentation has a coloring. For the rest of the proof, then, assume we have been given some fixed presentation $\mathfrak{m}$.

DEFINITION 4.6. *(The set of quadruples $\mathcal{R}_\mathfrak{m}$, [**14**] Definition 3.9)* Let $\mathfrak{m}$ be a presentation of a given type $p = p_\mathfrak{m}$. Then $\mathcal{R} = \mathcal{R}_\mathfrak{m}$ is the set of $\mathfrak{r} = (u, w, q, r)$ such that:

(1) $u \in [\lambda]^{<\sigma}$, $w \in [\lambda]^{<\theta}$ and $w = \mathrm{cl}_\mathcal{M}(w)$.
(2) $u \subseteq \mathrm{cl}_\mathcal{M}(u) \subseteq w$.
(3) $q = q(\overline{x}_w)$ is a complete type in the variables $\overline{x}_w$ such that:
  (a) for any finite $v \subseteq \mathrm{cl}_\mathcal{M}(\emptyset)$, if $M_* \models \psi(\bar{a}^*_v)$ then $\psi(\overline{x}_v) \in q$.
  (b) for any finite $\{\alpha_0, \dots, \alpha_n\} \subseteq u$, $\exists x \bigwedge_{i \leq n} \varphi_\alpha(x, a^*_\alpha) \in q$.
(4) $r = r(x, \overline{x}_w)$ is a complete type in the variables $x, \overline{x}_w$, extending

$$q(\overline{x}_w) \cup \{\varphi_\alpha(x, x_\alpha) : \alpha \in u\}.$$

(5) <u>if</u> $\overline{b}^*_w$ realizes $q(\overline{x}_w)$ in $\mathfrak{C}_T$ and $\alpha < \mathrm{cl}_\mathcal{M}(\emptyset) \implies b^*_\alpha = a^*_\alpha$, <u>then</u>
  (a) $r(x, \overline{b}^*_w)$ is a type which does not fork over M_* and extends $p \restriction M_*$.
  (b) if $w' \subseteq w$ is $\mathcal{M}_\lambda$-closed, $\mathfrak{C}_T \restriction \{b^*_\alpha : \alpha \in w'\} \preceq \mathfrak{C}_T$ and $r(x, \overline{b}^*_w) \restriction \overline{b}^*_{w'}$ is a complete type over this elementary submodel.

[7] It's sufficient if the restriction of p to a closed set generates a complete type. The reason to ask that $\{\varphi(x, a^*_\alpha) : \alpha < \omega\} = p \restriction M_*$ above with equality instead of $\vdash$ was just to ensure the closure of the empty set didn't grow.

(c) if $w' \subseteq w$ is $\mathcal{M}_\lambda$-closed and $\alpha \in w'$ then $tp(b^*_\alpha, \{b^*_\beta : \beta \in w \cap \alpha\})$ dnf over $\{b^*_\beta : \beta \in w' \cap \alpha\}$.

Note that u need not be closed. So in our case, r will describe a type in the variables $x, \bar{x}_w$ which agrees with $p \restriction M_*$ on $\{x_\alpha : \alpha < \omega\}$; it will then contain new conditions stating that x connects to additional elements x_α and stating in which "leaves" of the tree these x_α's fall. [For example, if $\{\varphi_\alpha(x, x_\alpha) : \alpha \in u\} = \{R(x, x_\alpha) : \alpha \in u\}$, a priori x_α need not be in the same leaf as a^*_α for $\alpha \geq \omega$.]

Definition 4.7 (A non-triviality condition, [**14**] Definition 3.10). Suppose we are given $\bar{\mathfrak{r}} = \langle \mathfrak{r}_t = (u_t, w_t, q_t, r_t) : t < t_* < \sigma \rangle$ from $\mathcal{R}_\mathfrak{m}$. Say that $\overline{b}^* = \langle b^*_\alpha : \alpha \in \bigcup_t w_t \rangle$, with each $b^*_\alpha \in \mathfrak{C}$ (possibly imaginary), is a *good instantiation* of $\bar{\mathfrak{r}}$ when the following conditions hold.

(1) $\alpha \in \mathrm{cl}_\mathcal{M}(\emptyset) \implies b^*_\alpha = a^*_\alpha$.
(2) for each $t < t_*$, $\overline{b}^* \restriction_{w_t}$ realizes $q_t(\overline{x}_{w_t})$.
(3) for each $t < t' < t_*$, if $v \subseteq w_t \cap w_{t'}$ is finite, then:
    (a) for each formula $\psi(\overline{x}_v)$, $\psi(\overline{b}^*_v) \in q_t \Leftrightarrow \psi(\overline{b}^*_v) \in q_{t'}$.
    (b) for each formula $\psi(x, \overline{x}_v)$, $\psi(x, \overline{b}^*_v) \in r_t \Leftrightarrow \psi(x, \overline{b}^*_v) \in r_{t'}$.
(4) if $\beta \in w_t$ for some $t < t_*$ then

$$\mathrm{tp}(b^*_\beta, \{b^*_\gamma : \gamma \in \bigcup_{s \leq t} w_s \text{ and } \gamma < \beta\}) \text{ dnf over } \{b^*_\gamma : \gamma \in w_t \cap \beta\}.$$

(5) for each $t < t_*$, if $w' \subseteq w$ and $\mathrm{cl}_\mathcal{M}(w') = w'$ then $\mathfrak{C}_T \restriction \{b^*_\alpha : \alpha \in w'\} \preceq \mathfrak{C}_T$ and $r_t(x, \overline{b}^*_{w'})$ is a complete type over this elementary submodel which does not fork over M_* (noting that the domain of M_* is $\{b^*_\alpha : \alpha \in \mathrm{cl}_\mathcal{M}(\emptyset)\}$ by the first item).

Definition 4.7 is simply to make the definition of coloring meaningful by ruling out trivial inconsistency, as will be clear from the next definition.

Definition 4.8 (Coloring, [**14**] Definition 3.11). Let $\mathfrak{m}$ be a $(\lambda, \theta, \sigma)$-presentation and $\mathcal{R} = \mathcal{R}_\mathfrak{m}$ be from 4.6. Call $G : \mathcal{R}_\mathfrak{m} \to \mu$ *an intrinsic coloring of* $\mathcal{R}_\mathfrak{m}$ if: whenever

$$\bar{\mathfrak{r}} = \langle \mathfrak{r}_t = (u_t, w_t, q_t, r_t) : t < t_* < \sigma \rangle$$

is a sequence of elements of $\mathcal{R}_\mathfrak{m}$ and $\overline{b}^* = \langle b^*_\alpha : \alpha \in \bigcup_{t<t_*} w_t \rangle$ is a good instantiation of $\bar{\mathfrak{r}}$,

if $G \restriction \{\mathfrak{r}_t : t < t_*\}$ is constant,
then the set of formulas

$$\{\varphi_\alpha(x, b^*_\alpha) \colon \alpha \in u_t,\ \varphi_\alpha \in r_t,\ t < t_*\}$$

is a consistent partial type which does not fork over M_*.

END OF QUOTATIONS

It remains to find a coloring given our fixed $\mathfrak{m}$, and therefore $\mathcal{R}$.

Case 1 ($Q(x) \in p$). The surprise in this case is that since $p \restriction M_*$ is determined, we know whether $Q_s(x)$ for all $s \in \mathcal{S}_k$ and all $k < \omega$. This means the set Λ of possible leaves η such that $p \restriction M_* \cup \{R(x, a)\} \cup \{P_{\eta \restriction k}(a) : k < \omega\}$ is fixed by $p \restriction M_*$ and inherited by any r from some $\mathfrak{r} \in \mathcal{R}$. So *whenever*

$$\bar{\mathfrak{r}} = \langle \mathfrak{r}_t = (u_t, w_t, q_t, r_t) : t < t_* < \sigma \rangle$$

is a sequence of elements of $\mathcal{R}_{\mathfrak{m}}$ and $\overline{b}^* = \langle b^*_\alpha : \alpha \in \bigcup_{t<t_*} w_t \rangle$ is a good instantiation of $\bar{\mathfrak{r}}$, it must be the case that for each $t < t_*$ and each $\alpha \in w_t$, r_t determines that the leaf of b^*_α must be η for some $\eta \in \Lambda$. It follows that

$$\{\varphi_\alpha(x, b^*_\alpha) \colon \alpha \in u_t,\ \varphi_\alpha \in r_t,\ t < t_*\}$$

is a consistent partial type which does not fork over M_*.

CASE 2 ($P(x) \in p$). Since p is a type, there will already be $\eta_* \in \lim(\mathcal{T})$ such that $p \vdash P_{\eta \restriction k}(x)$ for all $k < \omega$. This information will be part of $p \restriction M_*$. So only a single color is needed.

That is, suppose we are given a sequence $\bar{\mathfrak{r}}$ of elements of $\mathcal{R}$, on which G is constant and equal to β, and $\overline{b}^* = \langle b^*_\alpha : \alpha \in \bigcup_{t<t_*} w_t \rangle$ which is a good instantiation of $\bar{\mathfrak{r}}$. As we have ruled out trivial inconsistency, by our observation,

$$\{\varphi_\alpha(x, b^*_\alpha) \colon \alpha \in u_t,\ \varphi_\alpha \in r_t,\ t < t_*)\} \cup \{P_{\eta_* \restriction k}(x) : k < \omega\}$$

is a consistent partial type which does not fork over M_*, and this suffices. □

DISCUSSION 4.9. It is still interesting to ask what would happen if we had available only finitely much information from $p \restriction M_*$. Would some coloring work, which does not rely on having already determined the predicates Q_S or P_ν? In the remainder of this section we consider this, which will give the key idea for dealing with ultrapowers in the next section. We handle just the case of $Q(x) \in p$ as an illustration, since both cases are worked out in detail in the next section.

For each $a \in P^{\mathfrak{C}_{T_f}}$, let leaf$(a)$ denote the "leaf of a," i.e. the unique $\eta \in \lim(\mathcal{T})$ such that $\models P_{\eta \restriction k}(a)$ for all $k < \omega$.

DEFINITION 4.10. Call $B \subsetneq \lim(\mathcal{T})$ a *blocking set* when: for every $A \subseteq \mathfrak{C}_{T_f}$,

if $\{\text{leaf}(a) : a \in A\} \cap B = \emptyset$ then $\{R(x, a) : a \in A\}$ is a partial type.

[We can also give a direct definition, using 2.4: B is a blocking set if $\emptyset \subsetneq B \subsetneq \lim(\mathcal{T})$, and for all $\eta \in B$ and $k < \omega$ and $\ell \leq f(k)$, there is $\eta' \in B$ such that $\eta \restriction k \frown \langle \ell \rangle \trianglelefteq \eta'$.] The number of blocking sets is no more than $2^{2^{\aleph_0}}$.[8] Let $\bar{B} = \langle B_\alpha : \alpha < 2^{2^{\aleph_0}} \rangle$ enumerate them, possibly with repetition. Now for each $\mathfrak{r} = (u, w, q, r) \in \mathcal{R}$, and each $\alpha \in w$, as q is a type, each x_α is either determined to belong to P or to Q. If x_α belongs to P, then (again since q is a type) there is $\eta_\alpha \in \lim(\mathcal{T})$ such that $q \vdash P_{\eta_\alpha \restriction k}(x_\alpha)$ for $k < \omega$. Moreover, as r is a type, there is at least one blocking set B such that

$$\{\text{leaf}_{\mathfrak{r}}(x_\alpha) : \alpha \in w, R(x, x_\alpha) \in q\} \cap B = \emptyset.$$

Let $\beta_{\mathfrak{r}}$ be an index for this B in the enumeration $\bar{B}$, say for definiteness a minimal index. Choose the coloring function G so that $G(\mathfrak{r}) = \beta_{\mathfrak{r}}$ for all $\mathfrak{r} \in \mathcal{R}$.

Let's verify that this works. Suppose we are given a sequence $\bar{\mathfrak{r}}$ of elements of $\mathcal{R}$, on which G is constant and equal to β, and $\overline{b}^* = \langle b^*_\alpha : \alpha \in \bigcup_{t<t_*} w_t \rangle$ which is a good instantiation of $\bar{\mathfrak{r}}$. Since we have ruled out trivial inconsistency with 4.7, inconsistency cannot come from direct disagreement in the sense that, say, $R(x, b^*_\alpha)$ appears in one instance and $\neg(R(x, b^*_\alpha))$ appears in another. It will suffice to show

[8] But see Comment 4.11.

that the type restricted to positive instances of R is consistent. By definition of G, $\{\mathrm{leaf}(b^*_\alpha) : \alpha \in w_t, t < t_*, R(x, x_\alpha) \in q_t\} \cap B_\beta = \emptyset$, hence the set of formulas

$$\{\varphi_\alpha(x, b^*_\alpha) : \alpha \in u_t,\ \varphi_\alpha \in r_t,\ t < t_*, \varphi_\alpha(x, y) = R(x, y)\}$$

is a consistent partial type which does not fork over M_*, and this suffices.

COMMENT 4.11. In fact, as the theory has trivial forking, we may use $\sigma = \aleph_0$, $\theta = \aleph_1$ and various natural changes to the definition to accommodate this, such as having the closure of a set be itself; see Observation 3.5 of [**14**] and the paragraph before it. With these modifications, we can use co-finite blocking sets only, hence we can replace $2^{2^{\aleph_0}}$ above by $2^{\aleph_0}$. We plan to address this in future work (but see also the proof of Theorem 5.5).

5. A saturation result for T_f

OBSERVATION 5.1 (see e.g. Jech Theorem I.5.20). To satisfy Definition 4.1, we may take e.g. $\sigma = \theta = \aleph_1$, $\mu = 2^{\aleph_0}$ and $\lambda = \mu^{+\ell}$ for any finite $\ell > 0$.

Perfect ultrafilters were defined and shown to exist in [**14**] §9, for the case of suitable $(\lambda, \mu, \aleph_0, \aleph_0)$. [These were called (λ, μ)-perfect, with θ, σ omitted when countable.] Here we make the essentially cosmetic changes to extend this definition to allow for possibly uncountable θ, starting with the definition.

DEFINITION 5.2 (Support of a sequence, [**14**] Definition 5.6.1). Let $\bar{\mathbf{b}} = \langle \mathbf{b}_u : u \in [\lambda]^{<\aleph_0}\rangle$ be a sequence of elements of $\mathfrak{B} = \mathfrak{B}^1_{2^\lambda,\mu,\theta}$. We say X is a support of $\bar{\mathbf{b}}$ in $\mathfrak{B}$ when $X \subseteq \{\mathbf{x}_f : f \in \mathrm{FI}_{\mu,\theta}(\alpha)\}$ and for each $u \in [\lambda]^{<\aleph_0}$ there is a maximal antichain of $\mathfrak{B}$ consisting of elements of X all of which are either $\leq \mathbf{b}_u$ or $\leq 1 - \mathbf{b}_u$. Though there is no canonical choice of support we will write $\mathrm{supp}(\bar{\mathbf{b}})$ to mean *some* support.

Defintion 5.3 extends [**14**], Definition 9.1 to possibly uncountable θ.

DEFINITION 5.3 (Perfect ultrafilters, for suitable $(\lambda, \mu, \theta, \aleph_0)$). Let $(\lambda, \mu, \theta, \aleph_0)$ be suitable. We say that an ultrafilter $\mathcal{D}_*$ on $\mathfrak{B} = \mathfrak{B}^1_{2^\lambda,\mu,\theta}$ is $(\lambda, \mu, \theta, \aleph_0)$-*perfect* when (A) implies (B):

(A) $\langle \mathbf{b}_u : u \in [\lambda]^{<\aleph_0}\rangle$ is a monotonic sequence of elements of $\mathcal{D}_*$
and $\mathrm{supp}(\bar{\mathbf{b}})$ is a support for $\bar{\mathbf{b}}$ of cardinality $\leq \lambda$, see 5.2, such that for every $\alpha < 2^\lambda$ with $\bigcup\{\mathrm{dom}(f) : \mathbf{x}_f \in \mathrm{supp}(\bar{\mathbf{b}})\} \subseteq \alpha$,
there exists a multiplicative sequence

$$\langle \mathbf{b}'_u : u \in [\lambda]^{<\aleph_0}\rangle$$

of elements of $\mathfrak{B}^+$ such that
   (a) $\mathbf{b}'_u \leq \mathbf{b}_u$ for all $u \in [\lambda]^{<\aleph_0}$,
   (b) for every $\mathbf{c} \in \mathfrak{B}^+_{\alpha,\mu,\theta} \cap \mathcal{D}_*$, no intersection of finitely many members of $\{\mathbf{b}'_{\{i\}} \cup (1 - \mathbf{b}_{\{i\}}) : i < \lambda\}$ is disjoint to $\mathbf{c}$.

(B) there is a multiplicative sequence $\bar{\mathbf{b}}' = \langle \mathbf{b}'_u : u \in [\lambda]^{<\aleph_0}\rangle$ of elements of $\mathcal{D}_*$ which refines $\bar{\mathbf{b}}$.

Definition 5.4 extends [**15**], Definition 3.11 to possibly uncountable θ.

DEFINITION 5.4. Suppose $(\lambda, \mu, \theta, \aleph_0)$ are suitable. If $\mathcal{D}$ is built from $(\mathcal{D}_0, \mathfrak{B}, \mathcal{D}_*)$ where $\mathcal{D}_0$ is a regular filter on I, $|I| = \lambda$, $\mathfrak{B} = \mathfrak{B}^1_{2^\lambda,\mu,\theta}$ and $\mathcal{D}_*$ is $(\lambda, \mu, \theta, \aleph_0)$-perfect, say $\mathcal{D}$ is $(\lambda, \mu, \theta, \aleph_0)$-*perfected.*

In the Appendix below, we update [**14**] Theorem 9.4 to allow for the possibility of uncountable θ:

THEOREM (Existence, Theorem 9.6 below). *Let $(\lambda, \mu, \theta, \aleph_0)$ be suitable. Let $\mathfrak{B} = \mathfrak{B}^1_{2^\lambda, \mu, \theta}$. Then there exists a $(\lambda, \mu, \theta, \aleph_0)$-perfect ultrafilter on $\mathfrak{B}$.*

The main result of this section is that perfect ultrafilters saturate T_f for an uncountable but constant value of μ.

THEOREM 5.5. *Let $(\lambda, \mu, \theta, \aleph_0)$ be suitable, and in addition suppose $\mu \geq 2^{\aleph_0}$ and $\theta \geq \aleph_1$ (e.g. let $\theta = \aleph_1$, $\mu = 2^{\aleph_0}$, and $\lambda = \mu^{+n}$ for any finite n). Let $\mathcal{D}$ be a $(\lambda, \mu, \theta, \aleph_0)$-perfected ultrafilter on I, $|I| = \lambda$. Then $\mathcal{D}$ is good for T_f, i.e. for any $M \models T_f$, the ultrapower $M^I/\mathcal{D}$ is λ^+-saturated.*

PROOF. We begin with the usual setup. We fix $\mathcal{D}_0$, $\mathfrak{B} = \mathfrak{B}^1_{2^\lambda, \mu, \theta}$, $\mathbf{j}$ and a $(\lambda, \mu, \theta, \aleph_0)$-perfect ultrafilter $\mathcal{D}_*$ on $\mathfrak{B}$ such that $\mathcal{D}$ is built from $(\mathcal{D}_0, \mathfrak{B}, \mathcal{D}_*)$ via $\mathbf{j}$. We choose $M \models T_f$ as the index model, without loss of generality λ^+-saturated (by regularity of $\mathcal{D}$ the choice of M will not matter). We fix some lifting from $M^I/\mathcal{D}$ to M^I, so that for each $a \in M^I/\mathcal{D}$ and each index $t \in I$ the projection $a[t]$ is well defined. If $\bar{c} = \langle c_i : i < m \rangle \in {}^m(M^I/\mathcal{D})$ then we use $\bar{c}[t]$ to denote $\langle c_i[t] : i < m \rangle$.

Following Claim 2.8, it suffices to consider partial types of the following form. (Moreover, since R is not symmetric, $Q(x)$ and $P(x)$ are implied by the rest of the partial type in each case, so we may omit them.)

(1) $\{Q(x)\} \cup \{R(x, a) : a \in A\}$ for $A \subseteq P^N$, $|A| \leq \lambda$.
(2) $\{P(x)\} \cup \{R(b, x) : b \in B\}$ for $B \subseteq Q^N$, $|B| \leq \lambda$.

Fix a partial type $p = p(x)$ which is either of type (1) or type (2). Depending on which there will be some minor choices to make in the proof below. Recall two useful facts from the proof of Claim 2.7 above: in models of T_f, for each finite n,

(Fact A) For elements $a_0, \dots, a_n \in N$, $N \models (\exists x) \bigwedge_{i \leq n} R(x, a_i)$ if and only if there exist $\eta_0, \dots, \eta_n \in \mathcal{T}_{k_*}$ such that $N \models (\exists x) \bigwedge_{i \leq n} (R(x, a_i) \wedge P_{\eta_i}(a_i))$, where k_* is minimal such that $f(k_*) > n$.

(Fact B) For elements $b_0, \dots, b_n \in N$, $N \models (\exists x) \bigwedge_{i \leq n} R(b_i, x)$ if and only if there exists $\eta \in \mathcal{T}_{k_*}$ such that $N \models (\exists x) \bigwedge_{i \leq n} (R(b_i, x) \wedge P_\eta(x))$, where k_* is minimal such that $f(k_*) > n$.

We'll follow the strategy of [**15**], Theorem 4.1.

We begin with the case where p is of type (1).

Without loss of generality (possibly $|A| < \lambda$, but $||N|| \geq 2^\lambda$ so this is no problem),

(5.A) let $\langle a_i : i < \lambda \rangle$ list the elements of A without repetition.

This induces an enumeration of p as

(5.B) $$\langle R(x, a_i) : i < \lambda \rangle.$$

As usual, for each finite $u \subseteq \lambda$, let

(5.C) $$B_u := \{t \in I : M \models \exists x \bigwedge_{i \in u} R(x, a_i[t])\} \text{ and } \mathbf{b}_u = \mathbf{j}(B_u).$$

and let

(5.D) $$\bar{\mathbf{b}} = \langle \mathbf{b}_u : u \in [\lambda]^{<\aleph_0} \rangle.$$

First we build an appropriate support for $\bar{\mathbf{b}}$. This will require handling equality and leaves. For equality, for each $i, j \in \lambda$ let

(5.E) $$A_{a_i = a_j} := \{t \in I : a_i[t] = a_j[t]\} \text{ and let } \mathbf{a}_{a_i=a_j} := \mathbf{j}(A_{a_i=a_j}).$$

For leaves, for each $i \in \lambda$, and each $\eta \in \mathcal{T}$, let

(5.F) $$\mathbf{a}_{P_\eta(a_i)} = \mathbf{j}(\ \{t \in I : M \models P_\eta(a_i[t])\}\ ).$$

Remembering that $\theta > \aleph_0$, for each $\eta \in \lim(\mathcal{T})$, define

(5.G) $$\mathbf{a}_{\text{``leaf}(a_i)=\eta\text{''}} := \bigcap_{k<\omega} \mathbf{a}_{P_{\eta \restriction k}(a_i)}.$$

Then (5.G) will be nonzero for some, but not necessarily all, η, however, for each i,

(5.H) $$\langle \mathbf{a}_{\text{``leaf}(a_i)=\eta\text{''}} : \eta \in \lim(\mathcal{T}) \rangle$$

is a maximal antichain of $\mathfrak{B}$.

For each $i < \lambda$ let $\mathcal{F}_{\{i\}}$ be the set of all $f \in \mathrm{FI}_{\mu,\theta}(2^\lambda)$ such that for some $j \leq i$, the three condtions (5.I), (5.J), (5.K) hold:

(5.I) $$\mathbf{x}_f \leq \mathbf{a}_{a_i=a_j}.$$

(5.J) $$\text{for all } k < j,\ \ \mathbf{x}_f \cap \mathbf{a}_{a_i=a_k} = 0.$$

(5.K) $$\text{for some } \eta,\ \ \mathbf{x}_f \leq \mathbf{a}_{\text{``leaf}(a_i)=\eta\text{''}}.$$

For each finite $u \subseteq \lambda$, define $\mathcal{F}_u$ to be $\bigcap\{\mathcal{F}_{\{i\}} : i \in u\}$. Each $\mathcal{F}_u$ is upward closed, i.e. $f \in \mathcal{F}_u$ and $g \in \mathrm{FI}_{\mu,\theta}(2^\lambda)$ and $g \supseteq f$ implies $g \in \mathcal{F}_u$ and is dense in $\mathcal{F}$.

For each $u \in [\lambda]^{<\aleph_0}$, by induction on $\zeta < \lambda$, choose a maximal antichain $\langle \mathbf{x}_{f_\epsilon} : \epsilon < \epsilon_* \rangle$ of elements of $\mathfrak{B}$ such that (i) each $\mathbf{x}_{f_\epsilon}$ is either $\leq \mathbf{b}_u$ or $\leq 1 - \mathbf{b}_u$, and (ii) each $f_\epsilon \in \mathcal{F}_u$ and $0 \in \mathrm{dom}(f_\epsilon)$. Necessarily the construction will stop at an ordinal $< \mu^+$, but $\geq \mu$ because $0 \in \mathrm{dom}(f_\epsilon)$. Re-index this antichain as

(5.L) $$\langle \mathbf{x}_{f_{u,\zeta}} : \zeta < \mu \rangle$$

where w.l.o.g. if $u \neq v$ and $\zeta, \xi < \mu$ then $f_{u,\zeta} \neq f_{v,\xi}$.
Then

(5.M) $$\{\ \mathbf{x}_{f_{u,\zeta}} : \zeta < \mu, u \in [\lambda]^{<\aleph_0}\ \}$$

is a support of $\bar{\mathbf{b}}$ in the sense of Definition 5.2. When $u = \{i\}$, we will often write

$$f_{i,\zeta} \text{ to mean } f_{\{i\},\zeta}.$$

Second, we build a multiplicative refinement for $\bar{\mathbf{b}}$.

(5.N) $$\text{Fix } \alpha < 2^\lambda \text{ such that } \bigcup\{\mathrm{dom}(f_{u,\zeta}) : \zeta < \mu, u \in [\lambda]^{<\aleph_0}\} \subseteq \alpha\ .$$

As before, let leaf(a) denote the unique $\eta \in \lim(\mathcal{T})$ such that $\models P_{\eta\restriction k}(a)$ for all $k < \omega$, and call $X \subsetneq \lim(\mathcal{T})$ a *blocking set* when: for every $A \subseteq P^{\mathfrak{C}_{T_f}}$,

$$\text{if } \{\mathrm{leaf}(a) : a \in A\} \cap X = \emptyset \text{ then } \{R(x,a) : a \in A\} \text{ is a partial type.}$$

As $\mu \geq 2^{\aleph_0}$, let

(5.O) $$\langle X_\epsilon : \epsilon < \mu \rangle$$

be an enumeration, possibly with repetitions, of all co-finite blocking sets.[9] Let H be the function from $\{f_{i,\zeta} : i < \lambda, \zeta < \mu\} \times \mu$ to $\{0, 1\}$ given by

$$H(f_{i,\zeta}, \epsilon) = 1 \text{ iff } \eta \notin X_\epsilon \tag{5.P}$$

where η is the unique element of $\lim(\mathcal{T})$ such that $\mathbf{x}_{f_{i,\zeta}} \leq \mathbf{a}_{\text{“leaf}(a_i)=\eta\text{”}}$. (Very informally, H returns 1 if a type avoiding the blocking set B_ϵ may contain a_i as it appears on $\mathbf{x}_{f_{i,\zeta}}$.)

We'll need a new antichain to help us divide the work:

$$\text{let } \bar{\mathbf{c}} = \langle \mathbf{c}_\epsilon : \epsilon < \mu \rangle \text{ be given by } \mathbf{c}_\epsilon = \mathbf{x}_{\{(\alpha,\epsilon)\}}. \tag{5.Q}$$

Any element of this antichain will have nonzero intersection with any of the elements from $\mathfrak{B}^+_{\alpha,\mu,\theta}$, our protagonists so far.

Finally, for each $i < \lambda$, define

$$\mathbf{b}'_{\{i\}} = \Big(\bigcup \{\mathbf{x}_{f_{i,\zeta}} \cap \mathbf{c}_\epsilon : \zeta < \mu \text{ and } H(f_{i,\zeta}, \epsilon) = 1\}\Big) \cap \mathbf{b}_{\{i\}}. \tag{5.R}$$

Define

$$\mathbf{b}'_u = \bigcap_{i \in u} \mathbf{b}'_{\{i\}}, \text{ and let } \bar{\mathbf{b}}' = \langle \mathbf{b}'_u : u \in [\lambda]^{<\aleph_0} \rangle. \tag{5.S}$$

By definition, $\bar{\mathbf{b}}'$ is multiplicative. Our final task is to make sure the hypotheses of Definition 5.3 are satisfied, i.e. that for our multiplicative sequence $\bar{\mathbf{b}}'$,

(a) $\mathbf{b}'_u \leq \mathbf{b}_u$ for all $u \in [\lambda]^{<\aleph_0}$,
(b) for every $\mathbf{c} \in \mathfrak{B}^+_{\alpha,\mu,\theta} \cap \mathcal{D}_*$, no intersection of finitely many members of $\{\mathbf{b}'_{\{i\}} \cup (1 - \mathbf{b}_{\{i\}}) : i < \lambda\}$ is disjoint to $\mathbf{c}$.

For (a), suppose for a contradiction that for some $u \in [\lambda]^{<\aleph_0}$ there were a nonzero

$$\mathbf{c} \leq \mathbf{b}'_u \setminus \mathbf{b}_u.$$

Without loss of generality,

$$\mathbf{c} \leq \mathbf{c}_\epsilon \text{ for some } \epsilon < \mu, \tag{5.T}$$

and also, since u is finite,

$$\mathbf{c} \text{ is either below or disjoint to all elements in } \{\mathbf{x}_{f_{i,\zeta}} : i \in u, \zeta < \mu\}. \tag{5.U}$$

So for each $i \in u$ there is $\zeta_i < \mu$ with $\mathbf{c} \leq \mathbf{x}_{f_{i,\zeta_i}}$. By (5.T) and the definition (5.R),

$$\mathbf{c} \leq \bigcap_{i \in u} \mathbf{x}_{f_{i,\zeta_i}} \text{ and } \bigwedge_{i \in u} H(f_{i,\zeta_i}, \epsilon) = 1. \tag{5.V}$$

Now by our hypothesis, $\mathbf{c} \leq \mathbf{b}_{\{i\}}$, meaning

$$\mathbf{c} \leq \mathbf{a}[\exists x R(x, a_i)] \text{ for } i \in u\ . \tag{5.W}$$

Let k_* be minimal so that $f(k_*) > |u|$. Then as $H(f_{i,\zeta_i}, \epsilon) = 1$ for $i \in u$,

$$\mathbf{c} \leq \bigcap \{\mathbf{a}[\neg P_{\rho \restriction k}(a_i)] : i \in u, \rho \in X_\epsilon, k \leq k_*\}. \tag{5.X}$$

---

[9] A priori we could use all blocking sets and $\mu = 2^{2^{\aleph_0}}$, but the nice point is that in our present setup the co-finite blocking sets suffice. Note that here $p(x)$ being a set of *positive* instances of $R(x, y)$ helps. With negation, we'd need $f_{u,\zeta}$'s deciding all cases of $R(x, a_i)$, $\neg R(x, a_i)$, $a_i = a_j$.

Informally, on $\mathbf{c}$ none of the parameters a_i fall into the predicates forbidden by the blocking set, at least up to level k_* (this suffices for our contradiction, recalling Fact A from the beginning of the proof). And $\mathbf{c} \cap \mathbf{b}_u = \emptyset$ means

$$\mathbf{c} \leq \mathbf{a}[\neg \exists x \bigwedge_{i \in u} R(x, a_i)]. \tag{5.Y}$$

Since $\bar{\mathbf{b}}$ is a possibility pattern and $\mathbf{c} > 0$, this means we should be able to find values for a_i in $\mathfrak{C}_{T_f}$ which would make this combination of formulas true, but this is impossible because X_ϵ is a blocking set (so avoiding it gives a type). This contradiction shows that (a) holds, so $\bar{\mathbf{b}}'$ is a multiplicative refinement of $\bar{\mathbf{b}}$.

For (b), it will suffice to show that for any $\mathbf{a} \in \mathcal{D}_*$ such that $\operatorname{supp}(\mathbf{a}) \subseteq \alpha$, and any finite $u \subseteq \lambda$,

$$\mathbf{a} \cap \bigcap \{\mathbf{b}'_{\{i\}} \cup (1 - \mathbf{b}_{\{i\}}) : i \in u\} > 0.$$

Without loss of generality, we can write $u = v \cup w$ where for each $i \in v$, $\mathbf{a} \leq 1 - \mathbf{b}_{\{i\}}$ and for each $i \in w$, $\mathbf{a} \leq \mathbf{b}_{\{i\}}$. If w is empty we are done, so suppose not, and it will suffice to show that

$$\mathbf{a} \cap \bigcap \{\mathbf{b}'_{\{i\}} : i \in w\} > 0.$$

As $\mathbf{b}_w \in \mathcal{D}_*$, without loss of generality $\mathbf{a} \leq \mathbf{b}_w$, and we may choose $g \in \mathrm{FI}_{\mu,\theta}(\lambda)$ such that $\mathbf{x}_f \leq \mathbf{a}$. Moreover, for each $i \in w$, we may assume that there is some $\zeta_i < \mu$ such that $\mathbf{x}_g \leq \mathbf{x}_{f_{i,\zeta_i}}$. So we have that

$$0 < \mathbf{x}_g \leq \bigcap_{i \in w} \mathbf{x}_{f_{i,\zeta_i}} \leq \mathbf{b}_w.$$

Recall that by our choice of partitions, for each f_{i,ζ_i}, there is a unique $\eta = \eta_i \in \lim(\mathcal{T})$ such that $\mathbf{x}_{f_{i,\zeta}} \leq \mathbf{a}_{\text{“}\mathrm{leaf}(a_i)=\eta\text{”}}$. Because this intersection is $\leq \mathbf{b}_w$, there is some blocking set X_ϵ such that $X_\epsilon \cap \{\eta_i : i \in w\} = \emptyset$. Then

$$0 < \mathbf{c}_\epsilon \cap \bigcap_{i \in w} \mathbf{x}_{f_{i,\zeta_i}} \cap \mathbf{x}_g \leq \bigcap \{\mathbf{b}'_{\{i\}} : i \in w\}$$

which completes the proof of (b). This completes the proof of Case 1.

For case (2), the strategy is similar, with a few changes to reflect the dual type. For clarity we give the entire argument, renaming the parameter set as B.

Without loss of generality,

(6.A) let $\langle b_i : i < \lambda \rangle$ list the elements of B without repetition.

This induces an enumeration of p as

$$\langle R(b_i, x) : i < \lambda \rangle. \tag{6.B}$$

For each finite $u \subseteq \lambda$, let

$$B_u := \{t \in I : M \models \exists x \bigwedge_{i \in u} R(b_i[t], x)\} \text{ and } \mathbf{b}_u = \mathbf{j}(B_u). \tag{6.C}$$

and let

$$\bar{\mathbf{b}} = \langle \mathbf{b}_u : u \in [\lambda]^{<\aleph_0} \rangle. \tag{6.D}$$

First we build an appropriate support for $\bar{\mathbf{b}}$. As before, for each $i, j \in \lambda$ let

$$A_{b_i = b_j} := \{t \in I : b_i[t] = b_j[t]\} \text{ and let } \mathbf{a}_{b_i = b_j} := \mathbf{j}(A_{b_i = b_j}). \tag{6.E}$$

Now for each $\eta \in \mathcal{T}$,

$$\mathbf{a}_{(\exists x)(R(b_i, x) \wedge P_\eta(x))} = \mathbf{j}(\ \{t \in I : M \models \exists x (R(b_i[t], x) \wedge P_\eta(x))\}\ ). \tag{6.F}$$

As $\theta > \aleph_0$, for each $i < \lambda$ and $\eta \in \lim(\mathcal{T})$, define

(6.G) $$\mathbf{a}_{\text{“}(\exists x)(R(b_i,x)\wedge \text{leaf}(x)=\eta)\text{”}} := \bigcap_{k<\omega} \mathbf{a}_{(\exists x)(R(b_i,x)\wedge P_{\eta\restriction k}(x))}.$$

For each $s \in \mathcal{S}$,

$$\mathbf{a}_{Q_s(b_i)} = \mathbf{j}(\ \{t \in I : M \models Q_s(b_i[t])\}\ ).$$

For each $\nu \in \lim(\mathcal{S})$, letting $\nu = \langle s_k : k < \omega\rangle$ (recalling 2.3) so $\nu(k)$ denotes s_k, define

$$\mathbf{a}_{\text{“}Q_\nu(b_i)\text{”}} := \bigcap_{k<\omega} \mathbf{a}_{Q_{\nu(k)}(b_i)}$$

Then for each $i < \lambda$,

(6.H) $$\langle \mathbf{a}_{\text{“}Q_\nu(b_i)\text{”}} : \nu \in \lim(\mathcal{S})\rangle$$

is a partition of $\mathbf{b}_{\{i\}}$. For each finite $u \subseteq \lambda$ let $\mathcal{F}_u$ be the set of all $f \in \mathrm{FI}_{\mu,\theta}(2^\lambda)$ such that the conditions (6.I), (6.J) hold:

(6.I) for $i \in u$, for some $j \leq i$, $\mathbf{x}_f \leq \mathbf{a}_{b_i=b_j}$ and for all $k < j$, $\mathbf{x}_f \cap \mathbf{a}_{b_i=b_k} = 0$.

(6.J) for $i \in u$, for some $\nu \in \lim(\mathcal{S})$, $\mathbf{x}_f \leq \mathbf{a}_{\text{“}Q_\nu(b_i)\text{”}}$.

It follows that for any $f \in \mathcal{F}_{\{i\}}$, if $\nu \in \lim(\mathcal{S})$ is such that $\mathbf{x}_f \leq \mathbf{a}_{\text{“}Q_\nu(b_{\{i\}})\text{”}}$, then for some $\eta \in \lim(\mathcal{T})$, indeed for any η such that $\eta \restriction k \in \nu(k)$ for all $k < \omega$,

(6.K) $$\mathbf{x}_f \leq \mathbf{a}_{\text{“}(\exists x)(R(b_i,x)\wedge \text{leaf}(x)=\eta)\text{”}}.$$

Each $\mathcal{F}_u$ is upward closed, i.e. $f \in \mathcal{F}_u$ and $g \supseteq f$ implies $g \in \mathcal{F}_u$.

For each $u \in [\lambda]^{<\aleph_0}$, by induction on $\zeta < \lambda$, choose a maximal antichain $\langle \mathbf{x}_{f_\epsilon} : \epsilon < \epsilon_*\rangle$ of elements of $\mathfrak{B}$ such that (i) each $\mathbf{x}_{f_\epsilon}$ is either $\leq \mathbf{b}_u$ or $\leq 1 - \mathbf{b}_u$, and (ii) each $f_\epsilon \in \mathcal{F}_u$. Necessarily the construction will stop at an ordinal $< \mu^+$. Renumber this antichain as

$$\langle \mathbf{x}_{f_{u,\zeta}} : \zeta < \mu\rangle.$$

Then

(6.L) $$\{\ \mathbf{x}_{f_{u,\zeta}} : \zeta < \mu, u \in [\lambda]^{<\aleph_0}\ \}$$

is a support of $\bar{\mathbf{b}}$ in the sense of Definition 5.2.

When $u = \{i\}$, we will again write

$$f_{i,\zeta} \text{ to mean } f_{\{i\},\zeta}.$$

Second, we build a multiplicative refinement for $\bar{\mathbf{b}}$.

$$\text{Fix } \alpha < 2^\lambda \text{ such that } \bigcup\{\mathrm{dom}(f_{u,\zeta}) : \zeta < \mu, u \in [\lambda]^{<\aleph_0}\} \subseteq \alpha\ .$$

As $\mu \geq 2^{\aleph_0}$, let

(6.M) $$\langle \eta_\epsilon : \epsilon < \mu\rangle$$

be an enumeration, possibly with repetitions, of all leaves $\eta \in \lim(\mathcal{T})$. Let G be the function from $\{f_{i,\zeta} : i < \lambda, \zeta < \mu\} \times \mu$ to $\{0,1\}$ given by

(6.N) $$G(f_{i,\zeta}, \epsilon) = 1 \text{ iff } \mathbf{x}_{f_{i,\zeta}} \leq \mathbf{a}_{\text{“}Q_\nu(b_i)\text{”}} \text{ and } \eta_\epsilon \restriction k \in \nu(k) \text{ for all } k < \omega.$$

We'll need a new antichain to help us divide the work:

(6.O) let $\bar{\mathbf{c}} = \langle \mathbf{c}_\epsilon : \epsilon < \mu\rangle$ be given by $\mathbf{c}_\epsilon = \mathbf{x}_{\{(\alpha+1,\epsilon)\}}$.

Any element of this antichain will have nonzero intersection with any of the elements from $\mathfrak{B}^+_{\alpha,\theta}$.

For each $i < \lambda$, define

$$\mathbf{b}'_{\{i\}} = \left(\bigcup\{\mathbf{x}_{f_{i,\zeta}} \cap \mathbf{c}_\epsilon : \zeta < \mu \text{ and } G(f_{i,\zeta}, \epsilon) = 1\}\right) \cap \mathbf{b}_{\{i\}}. \tag{6.P}$$

Define

$$\mathbf{b}'_u = \bigcap_{i \in u} \mathbf{b}'_{\{i\}}, \text{ and let } \bar{\mathbf{b}}' = \langle \mathbf{b}'_u : u \in [\lambda]^{<\aleph_0} \rangle.$$

By definition, $\bar{\mathbf{b}}'$ is multiplicative. Again we address the hypotheses of Definition 5.3. For (a), suppose for a contradiction that for some $u \in [\lambda]^{<\aleph_0}$ there were a nonzero

$$\mathbf{c} \leq \mathbf{b}'_u \setminus \mathbf{b}_u.$$

Without loss of generality,

$$\mathbf{c} \leq \mathbf{c}_\epsilon \text{ for some } \epsilon < \mu, \tag{6.Q}$$

and also

$$\mathbf{c} \text{ is either below or disjoint to all elements in } \{\mathbf{x}_{f_{i,\zeta}} : i \in u, \zeta < \mu\}. \tag{6.R}$$

So for each $i \in u$ there is some $\zeta_i < \mu$ with $\mathbf{c} \leq \mathbf{x}_{f_{i,\zeta}}$, and then by (6.Q) and the definition (6.P),

$$\mathbf{c} \leq \bigcap_{i \in u} \mathbf{x}_{f_{i,\zeta_i}} \text{ and } \bigwedge_{i \in u} G(f_{i,\zeta_i}, \epsilon) = 1. \tag{6.S}$$

Now by our hypothesis, $\mathbf{c} \leq \mathbf{b}_{\{i\}}$, meaning

$$\mathbf{c} \leq \mathbf{a}[\exists x R(b_i, x)] \text{ for } i \in u\ . \tag{6.T}$$

And $\mathbf{c} \cap \mathbf{b}_u = \emptyset$ means

$$\mathbf{c} \leq \mathbf{a}[\neg \exists x \bigwedge_{i \in u} R(b_i, x)]. \tag{6.U}$$

Recalling the definition of G and (6.Q), for every $i \in u$,

$$\mathbf{c} \leq \mathbf{x}_{f_{i,\zeta}} \leq \mathbf{a}_{\text{“}(\exists x)(R(b_i,x) \wedge \text{leaf}(x) = \eta_\epsilon)\text{”}}. \tag{6.V}$$

But (6.T), (6.U), and (6.V) together are a contradiction. [Why? (6.U) must be witnessed by full splitting at some finite stage, but (6.V) guarantees that at every finite stage there is a specific piece of the successor partition which is reserved for x.] This contradiction shows that condition (a) of the definition of perfect holds, so $\bar{\mathbf{b}}'$ is indeed a multiplicative refinement of $\bar{\mathbf{b}}$.

For (b), it will suffice to show that for any $\mathbf{c} \in \mathcal{D}_*$ such that $\text{supp}(\mathbf{a}) \subseteq \alpha$, and any finite $u \subseteq \lambda$,

$$\mathbf{c} \cap \bigcap \{\mathbf{b}'_{\{i\}} \cup (1 - \mathbf{b}_{\{i\}}) : i \in u\} > 0.$$

Without loss of generality, we can write $u = v \cup w$ where for each $i \in v$, $\mathbf{c} \leq 1 - \mathbf{b}_{\{i\}}$ and for each $i \in w$, $\mathbf{c} \leq \mathbf{b}_{\{i\}}$. If w is empty we are done, so suppose not, and it will suffice to show that

$$\mathbf{c} \cap \bigcap \{\mathbf{b}'_{\{i\}} : i \in w\} > 0.$$

As $\mathbf{b}_w \in \mathcal{D}_*$, without loss of generality

$$\mathbf{c} \leq \mathbf{b}_w.$$

Recalling Fact B from the beginning of the proof, let k_* be minimal so that $f(k_*) > |w|$, and then

$$\bigcup\{\mathbf{a}[\ (\exists x) \bigwedge_{i\in w} R(b_i, x) \wedge P_\eta(x)\ ]\ :\ \eta \in \mathcal{T}_{k_*}\} = \mathbf{b}_w$$

so, after possibly shrinking $\mathbf{c}$ by taking intersections, we may assume there is $\eta_* \in \mathcal{T}_{k_*}$ such that

$$\mathbf{c} \leq \mathbf{a}[(\exists x) \bigwedge_{i\in w} R(b_i, x) \wedge P_{\eta_*}(x)].$$

Again by Fact B, this implies there is some $\eta \in \lim(\mathcal{T})$ such that $\eta_* \trianglelefteq \eta$ and for all finite k,

$$\mathbf{c} \cap \mathbf{a}[(\exists x) \bigwedge_{i\in w} R(b_i, x) \wedge P_{\eta\restriction k}(x)] > 0$$

so as $\theta > 0$, without loss of generality

$$\mathbf{c} \leq \mathbf{a}_{\text{“}(\exists x)(R(b_i,x)\wedge \text{leaf}(x)=\eta)\text{”}}.$$

Recalling the support (6.L), after possibly shrinking $\mathbf{c}$ by taking intersections, there are ζ_i for each $i \in w$ such that

$$\mathbf{c} \leq \mathbf{x}_{f_{i,\zeta}}$$

i.e. for each $i \in w$,

$$0 < \mathbf{x}_{f_{i,\zeta_i}} \cap \mathbf{c} \leq \mathbf{a}_{\text{“}(\exists x)(R(b_i,x)\wedge \text{leaf}(x)=\eta)\text{”}}.$$

Our choice of partitions in (6.J) and (6.L) means that for each $i \in w$ there is $\nu_i \in \lim(\mathcal{S})$ such that

$$\mathbf{x}_{f_{i,\zeta_i}} \leq \mathbf{a}_{\text{“}Q_{\nu_i}(b_i)\text{''}}.$$

Recalling (6.K), we conclude from these two equations that $\eta \restriction k \in \nu_i(k)$ for all $k < \omega$, for each $i \in w$. So letting ϵ be such that $\eta = \eta_\epsilon$ in the enumeration (6.M), $G(f_{i,\zeta_i}, \epsilon) = 1$ for all $i \in w$. We have shown that

$$\mathbf{c} \cap \bigcap_{i\in w} \mathbf{x}_{f_{i,\zeta_i}} \cap \mathbf{c}_\epsilon > 0$$

and this suffices.

This completes Case 2, and so completes the proof of the Theorem. □

7. Consequences for Keisler's order

THEOREM 7.1. *Let $f \in \mathcal{F}$. In Keisler's order, T_f is strictly above the theory of the random graph.*

PROOF. Recall that $T_{\mathbf{rg}}$ is minimum in Keisler's order among the unstable theories ([**11**], Conclusion 5.3). So as T_f is unstable, $T_{\mathbf{rg}} \trianglelefteq T_f$. By Lemma 3.2, if $\mathcal{D}$ is a regular ultrafilter on λ built from $(\mathcal{D}_0, \mathfrak{B} = \mathfrak{B}^1_{2^\lambda,\aleph_0,\aleph_0}, \mathcal{D}_*)$ where $\mathcal{D}_0$ is any regular good [or so-called excellent] filter on λ and $\mathcal{D}_*$ is any ultrafilter on $\mathfrak{B}$, then $\mathcal{D}$ is not good for T_f. On the other hand, by [**13**] Theorem 3.2 in the case $\mu = \aleph_0$, there is such an ultrafilter which is good for $T_{\mathbf{rg}}$. This shows that $T_{\mathbf{rg}} \triangleleft T_f$. □

We recall the higher analogues of the triangle-free random graph, studied by Hrushovski [**7**]. In particular, he proved that each $T_{n,k}$ is simple unstable with trivial forking for $n > k \geq 2$.

DEFINITION 7.2. Recall that $T_{n,k}$ denotes the generic $(n+1)$-free $(k+1)$-hypergraph, i.e. the model completion of the theory (in a language with a single $(k+1)$-place relation, interpreted as a hyperedge, so symmetric and irreflexive) stating that there do not exist $(n+1)$ distinct elements of which every $(k+1)$ are an edge.

The infinite descending chain in Keisler's order obtained in [**15**] was given by $\cdots T_m \triangleleft T_{n-1} \triangleleft \cdots T_1 \triangleleft T_0$ where T_n is the disjoint union of the theories $T_{k+1,k}$ for $k > 2n+2$.

In the Appendix, Theorem 7 below we update [**15**] Claim 5.1 to allow for the possibility of uncountable θ.

THEOREM ([**15**], Claim 5.1 for possibly uncountable θ, Theorem 7 below). *Suppose that:*

(1) *for integers $2 \leq k < \ell$, and e.g. $\theta = \aleph_1$, $\mu = 2^{\aleph_0}$, $\lambda = \mu^{+\ell}$,*
or just: $(\lambda, k, \mu^+) \rightarrow k+1$ in the sense of [**15**] *Notation 1.2*
(2) $\mathfrak{B} = \mathfrak{B}^1_{2^\lambda,\mu,\theta}$
(3) *$\mathcal{D}_*$ is an ultrafilter on $\mathfrak{B}$*
(4) $T = T_{k+1,k}$

Then $\mathcal{D}_*$ is not (λ, T)-moral.

THEOREM 7.3. *For any finite $k \geq 2$, T_f and $T_{k+1,k}$ are incomparable in Keisler's order. More precisely:*

(1) *let $\mathcal{D}$ be a $(\lambda, \mu, \aleph_0, \aleph_0)$-perfected ultrafilter on λ where*
$$\lambda = \aleph_{k-1}, \ \mu = \aleph_0, \ \theta = \aleph_0.$$
Then $\mathcal{D}$ is not good for T_f, but it is good for $T_{k+1,k}$.
(2) *let $\mathcal{D}_*$ be a $(\lambda, \mu, \aleph_1, \aleph_0)$-perfected ultrafilter on λ where*
$$\mu \geq 2^{\aleph_0} \text{ and } \lambda = \mu^{+n} \text{ for } n > k+1.$$
Then $\mathcal{D}$ is good for T_f, but it is not good for $T_{k+1,k}$.

PROOF. (1) The non-saturation is Lemma 3.2 via separation of variables, and the saturation is [**15**] Theorem 4.1.

(2) The saturation is Theorem 5.5, and the non-saturation is Theorem 7 via separation of variables. □

Appendix: perfect ultrafilters for uncountable θ

In [**14**], we considered *suitable* tuples of cardinals $(\lambda, \mu, \theta, \sigma)$, see Definition 4.1. We defined

"$\mathcal{D}$ is a $(\lambda, \mu, \theta, \sigma)$-ultrafilter on the Boolean algebra $\mathfrak{B}^1_{2^\lambda,\mu,\theta}$"

in the case where $\theta = \sigma = \aleph_0$, and we proved that such ultrafilters did indeed exist.

In this Appendix, we upgrade that definition and existence proof to include the case of uncountable θ. The proof is almost word-for-word the same as that of [**14**] §9, but to eliminate doubt, we have reproduced that proof here with the minor changes. We defined "support" in 5.2 above and "perfect" in 5.3 above.

CONVENTION 9.4. Throughout this section we assume:
$$\lambda \geq \mu^{<\theta} \geq \theta = \mathrm{cof}(\theta) \geq \sigma = \aleph_0.$$

Without loss of generality we may assume $\theta > \sigma$, as the case $\theta = \sigma = \aleph_0$ was the case of [**14**] §9.

OBSERVATION 9.5. Suppose $\alpha < 2^\lambda$ is fixed, D_α is an ultrafilter on $\mathfrak{B}^1_{\alpha,\mu,\theta} \subseteq \mathfrak{B} = \mathfrak{B}^1_{2^\lambda,\mu,\theta}$, and $\langle \mathbf{b}_u : u \in [\lambda]^{<\aleph_0} \rangle$ is a sequence of elements of D_α. Suppose that there exists a multiplicative sequence $\langle \mathbf{b}'_u : u \in [\lambda]^{<\aleph_0} \rangle$ of elements of $\mathfrak{B}^+$ such that

(a) $\mathbf{b}'_u \leq \mathbf{b}_u$ for all $u \in [\lambda]^{<\aleph_0}$,

(b) for every $\mathbf{c} \in \mathfrak{B}^+_{\alpha,\mu,\theta} \cap \mathcal{D}_\alpha$, no intersection of finitely many members of $\{\mathbf{b}'_{\{i\}} \cup (1 - \mathbf{b}_{\{i\}}) : i < \lambda\}$ is disjoint to $\mathbf{c}$.

Then there is a multiplicative sequence $\langle \mathbf{b}''_u : u \in [\lambda]^{<\aleph_0} \rangle$ such that (a), (b) hold with $\mathbf{b}'_u$, $\mathbf{b}'_{\{i\}}$ replaced by $\mathbf{b}''_u$, $\mathbf{b}''_{\{i\}}$ respectively, and such that some support of $\bar{\mathbf{b}}''$ is contained in $\mathfrak{B}_{\alpha+\lambda,\mu,\theta}$.

PROOF. Without loss of generality there is $\mathcal{V}$ of cardinality λ such that some support of $\bar{\mathbf{b}}'$ is contained in $\{\mathbf{x}_f : f \in \mathrm{FI}_{\mu,\theta}(\mathcal{V})\}$. Let π be a permutation of 2^λ which is the identity on α and takes $\mathcal{U}$ into $\alpha + \lambda$. This induces an automorphism ρ of $\mathfrak{B}$ which is the identity on $\mathfrak{B}_{\alpha,\mu,\theta}$, so in particular is the identity on $\mathcal{D}_\alpha$ and thus on $\bar{\mathbf{b}}$. For each $u \in [\lambda]^{<\aleph_0}$, let $\mathbf{b}''_u = \rho(\mathbf{b}'_u)$. Then clearly $\bar{\mathbf{b}}''$ fits the bill. □

THEOREM 9.6 (Existence). *Let $(\lambda, \mu, \theta, \aleph_0)$ be suitable. Let $\mathfrak{B} = \mathfrak{B}^1_{2^\lambda,\mu,\theta}$. Then there exists a $(\lambda, \mu, \theta, \aleph_0)$-perfect ultrafilter on $\mathfrak{B}$.*

PROOF. Begin by letting $\langle \bar{\mathbf{b}}_\delta = \langle \mathbf{b}_{\delta,u} : u \in [\lambda]^{<\aleph_0} \rangle : \delta < 2^\lambda \rangle$ be an enumeration of the monotonic sequences of elements of $\mathfrak{B}^+$, each occurring cofinally often. Let $z : 2^\lambda \to 2^\lambda$ be an increasing continuous function which satisfies: $z(0) \geq 0$ and for all $\beta < 2^\lambda$, $z(\beta) + \lambda = z(\beta+1)$. By induction on $\delta < 2^\lambda$ we will construct $\langle D_\delta : \delta < 2^\lambda \rangle$, an increasing continuous sequence of filters with each D_δ an ultrafilter on $\mathfrak{B}_{z(\delta),\mu,\theta}$, to satisfy:

(*) *if* $\delta = \beta + 1$, if it is the case that

$\langle \mathbf{b}_{\beta,u} : u \in [\lambda]^{<\aleph_0} \rangle$ is a monotonic sequence of elements of $\mathcal{D}_\beta$ and there exists a choice of supp$(\bar{b})$ with $\bigcup\{\mathrm{dom}(f) : \mathbf{x}_f \in \mathrm{supp}(\bar{\mathbf{b}})\} \subseteq \beta$ and there exists a multiplicative sequence

$$\langle \mathbf{b}'_u : u \in [\lambda]^{<\aleph_0} \rangle$$

of elements of $\mathfrak{B}^+$ such that

(a) $\mathbf{b}'_u \leq \mathbf{b}_{\beta,u}$ for all $u \in [\lambda]^{<\aleph_0}$,

(b) for every $\mathbf{c} \in \mathfrak{B}^+_{z(\beta),\mu,\theta} \cap \mathcal{D}_\beta$, no intersection of finitely many members of $\{\mathbf{b}'_{\{i\}} \cup (1 - \mathbf{b}_{\beta,\{i\}}) : i < \lambda\}$ is disjoint to $\mathbf{c}$.

then there is a sequence $\bar{\mathbf{b}}'' = \langle \mathbf{b}''_u : u \in [\lambda]^{<\aleph_0} \rangle$ of elements of $\mathfrak{B}^+$ such that:

(i) $\mathbf{b}''_u \leq \mathbf{b}_{\beta,u}$ for all $u \in [\lambda]^{<\aleph_0}$,

(ii) for every $\mathbf{c} \in \mathfrak{B}^+_{z(\beta),\mu,\theta} \cap \mathcal{D}_\beta$, no intersection of finitely many members of $\{\mathbf{b}''_{\{i\}} \cup (1 - \mathbf{b}_{\beta,\{i\}}) : i < \lambda\}$ is disjoint to $\mathbf{c}$.

(iii) some support of $\bar{\mathbf{b}}''$ is contained in $\mathfrak{B}_{z(\delta),\mu}$, and

(iiv) D_δ is an ultrafilter on $\mathfrak{B}_{z(\delta),\mu,\theta}$ which extends $D_\beta \cup \{\mathbf{b}'_u : u \in [\lambda]^{<\aleph_0}\}$.

The induction may be carried out at limit stages because all of the D_δ are ultrafilters. Suppose $\delta = \beta + 1$. If $\bar{\mathbf{b}}$ satisfies the quoted condition, then let $\bar{\mathbf{b}}''$ be given by Observation 9.5, using $z(\beta)$ here for α there. Then (i), (ii), (iii) are satisfied, so we need to prove that

$$D_\beta \cup \{\mathbf{b}''_u : u \in [\lambda]^{<\aleph_0}\}$$

has the finite intersection property. As D_β is an ultrafilter on $\mathfrak{B}_{z(\beta),\mu,\theta}$, and $\bar{\mathbf{b}}'$ is a multiplicative sequence, it suffices to prove that for any $\mathbf{c} \in \mathcal{D}_\beta$ and any finite $u \subseteq \lambda$,

$$\mathbf{c} \cap \bigcap \{\mathbf{b}''_{\{i\}} : i \in u\} > 0.$$

As $\mathbf{b}_{\{i\}} \in \mathcal{D}_\beta$ for each $i \in u$, we may assume that $\mathbf{c} \cap (1 - \mathbf{b}_{\{i\}}) = 0$ for each $i \in u$. Then we are finished by assumption (ii). This completes the induction. Let $\mathcal{D}_* = \bigcup_{\delta<2^\lambda} D_\delta$.

Let us check that $\mathcal{D}_*$ is indeed a perfect ultrafilter. If $\bar{\mathbf{b}}$ satisfies condition 5.3(A), let $\mathcal{U}$ be as there, and let $\delta = \beta+1$ be an ordinal $< 2^\lambda$ such that $\bar{\mathbf{b}}_\beta = \bar{\mathbf{b}}$ and $\mathcal{U} \subseteq \mathfrak{B}_{\beta,\mu,\theta}$, which is possible as we listed each sequence cofinally often. Then since D_β was an ultrafilter, $\mathcal{D}_* \restriction \mathfrak{B}_{\beta,\mu,\theta} = \mathcal{D}_\beta$ so at stage δ condition (*) of the inductive hypothesis will be activated and we will have ensured that $\bar{\mathbf{b}}$ has a multiplicative refinement in $\mathcal{D}_*$. □

Appendix: non-saturation for $T_{k+1,k}$ and uncountable θ

In this Appendix we update the non-saturation result from [**15**] to allow for possibly uncountable θ. The proof is the same. It has just been slightly rewritten for readability, since it seems a good occasion to call attention to Question 10.9.

THEOREM 10.7 ([**15**], Claim 5.1 for possibly uncountable θ). *Suppose that:*

(1) *for integers* $2 \leq k < \ell$, $\theta \geq \aleph_0$, *and e.g.* $\mu = 2^{\aleph_0}$, $\lambda = \mu^{+\ell}$,
or just: $(\lambda, k, \mu^+) \to k+1$ *in the sense of* [**15**] *Notation 1.2*

(2) $\mathfrak{B} = \mathfrak{B}^1_{2^\lambda,\mu,\theta}$ *[for θ possibly uncountable]*

(3) $\mathcal{D}_*$ *is an ultrafilter on* $\mathfrak{B}$

(4) $T = T_{k+1,k}$

Then $\mathcal{D}_*$ *is not* (λ, T)*-moral.*

PROOF. Fix the objects given in the statement of the theorem. We'll often use the notation $T_{n,k}$ instead of $T_{k+1,k}$, but $n = k+1$ seems necessary in this proof, as we will point out. We will use [**15**] Claim 1.6, which when applied to (λ, k, μ^+) gives:

(10.A) there is a model $M \models T_{n,k}$ such that:

> M has size $\geq \lambda$, and there are λ elements of its domain $\langle b_\alpha : \alpha < \lambda\rangle$ such that if we let $\mathcal{P} = \{w \in [\lambda]^n : (\forall u \in [w]^{k+1})(M \models R(\bar{b}_u)) \}$ denote the indices for near-forbidden configurations,[10] then for any $F : [\lambda]^k \to [\lambda]^{\leq\mu}$ such that $u \subseteq F(u)$ for all $u \in \mathrm{dom}(F)$, there exists $w \in \mathcal{P}$ such that $(\forall v \in [w]^k)(w \not\subseteq F(v))$.

Informally, the conclusion is that for any such F, some near-forbidden configuration escapes the control of its k-element subsets.

The strategy will be to build a possibility pattern that has no multiplicative refinement. Fix a sequence of ordinals $\langle \alpha_w : w \in \mathcal{P}\rangle$, each $< 2^\lambda$, with no repetitions.

---

[10] recall that in $T_{n,k}$ the forbidden configuration is a set of $n+1$ vertices of which every $k+1$ form an edge. So if $\{b_\alpha : \alpha \in w\}$ is near-forbidden, then $\{R(x, \bar{b}_v) : v \in [w]^k\}$ is not a type.

For each $w \in \mathcal{P}$, fix a function $g_w \in \mathrm{FI}_{\mu,\theta}(2^\lambda)$ such that $\mathrm{dom}(g_w) = \{\alpha_w\}$ and $\mathbf{x}_{g_w} = 0 \mod \mathcal{D}_*$.

Let

$$\langle v_\alpha : \alpha < \lambda \rangle \tag{10.B}$$

list $[\lambda]^k$ without repetition.[11] Let $\Omega = [\lambda]^{<\aleph_0}$ and for each $s \in \Omega$, let[12]

$$\mathbf{b}_s = 1_{\mathfrak{B}} - \bigcup\{\mathbf{x}_{g_w} : w \in \mathcal{P} \text{ and } [w]^k \subseteq \{v_\beta : \beta \in s\}\}. \tag{10.C}$$

In order to check that $\bar{\mathbf{b}} = \langle \mathbf{b}_s : s \in \Omega \rangle$ is really a representation of some type in the ultrapower, it will suffice by compactness to argue as follows (since we may always choose the index model to be a λ^+-saturated elementary extension of M). First note that for each finite $s \subseteq \lambda$, each nonzero $\mathbf{c} \in \mathfrak{B}$, $\mathbf{c}$ induces a partition of $\{\mathbf{b}_{s'} : s' \subseteq s\}$ according to whether $\mathbf{c} \cap \mathbf{b}_{s'} = 0$ or $\mathbf{c} \cap \mathbf{b}_{s'} > 0$. By shrinking $\mathbf{c}$ if necessary, we may assume $\mathbf{c}$ induces a partition of $\{\mathbf{b}_{s'} : s' \subseteq s\}$ according to whether $\mathbf{c} \cap \mathbf{b}_{s'} = 0$ or $\mathbf{c} \leq \mathbf{b}_{s'}$. Let $\mathrm{vert}(s) := \bigcup\{v_\beta : \beta \in s\}$ be the set of indices for all elements mentioned in formulas in s. Finally, by shrinking $\mathbf{c}$ if necessary, we may also assume that either $\mathbf{c} \leq \mathbf{x}_{g_w}$ or $\mathbf{c} \leq 1 - \mathbf{x}_{g_w}$ for every $w \in \mathcal{P}$ such that $w \subseteq \mathrm{vert}(s)$. It suffices to show that for each such finite s and nonzero $\mathbf{c}$, we may choose elements $\{b'_\beta : \beta \in \mathrm{vert}(s)\}$ in M such that

$$M \models (\exists x)(\bigwedge_{\beta \in s'} R(x, \bar{b}'_{v_\beta}) \text{ when } \mathbf{c} \leq \mathbf{b}_{s'} \tag{10.D}$$

and

$$M \models \neg(\exists x)(\bigwedge_{\beta \in s'} R(x, \bar{b}'_{v_\beta}) \text{ when } \mathbf{c} \cap \mathbf{b}_{s'} = 0. \tag{10.E}$$

Consider the elements $\{b_\beta : \beta \in \mathrm{vert}(s)\}$. In M, this set may have some edges on it. Informally, what we will do is for each $w \subseteq \mathrm{vert}(s)$ such that $w \in \mathcal{P}$, we remove the edge on $\{b_\beta : \beta \in w\}$ if and only if $\mathbf{c} \leq 1 - \mathbf{x}_{g_w}$. Formally, we choose a set of distinct elements $\{b'_\beta : \beta \in \mathrm{vert}(s)\}$ of M such that $R(b'_{\beta_0}, \ldots, b'_{\beta_k})$ if and only if $\{\beta_0, \ldots, \beta_k\} \in \mathcal{P}$ and $\mathbf{c} \leq \mathbf{x}_{g_w}$.

Let's check that (10.D) and (10.E) are satisfied.

If $\mathbf{c} \cap \mathbf{b}_{s'} = 0$, then there is some w such that $w \subseteq \mathrm{vert}(s)$, $w \in \mathcal{P}$, $[w]^k \subseteq \{v_\beta : \beta \in s\}$ and $\mathbf{c} \leq \mathbf{x}_{g_w}$. [Suppose not. If there were no $w \subseteq \mathrm{vert}(s')$ such that $w \in \mathcal{P}$ and $[w]^k \subseteq \{v_\beta : \beta \in s'\}$, then by definition in (10.C), $\mathbf{b}_{s'} = 1_{\mathfrak{B}}$, so we contradict $\mathbf{c} > 0$. So there must be some such w. Let $w_0, \ldots, w_i$ be a list of all such w. Then again by (10.C), $\bigcup_{j \leq i} \mathbf{x}_{g_{w_j}} = 1 - \mathbf{b}_{s'}$, so it must be that $\mathbf{c} \leq \mathbf{x}_{g_{w_j}}$ for some $j \leq i$.] By construction, $M \models R(\bar{b}'_w)$, so $\{R(x, \bar{b}_{v_\beta}) : \beta \in s'\} \supseteq \{R(x, \bar{b}_{v_\beta} : v_\beta \in [w]^k\}$ is indeed inconsistent.

If $\mathbf{c} \leq \mathbf{b}_{s'}$, then for any w such that $w \subseteq \mathrm{vert}(s)$, $w \in \mathcal{P}$, and $[w]^k \subseteq \{v_\beta : \beta \in s\}$, we must also have (by definition of $\mathbf{b}_{s'}$ in (10.C) that $\mathbf{c} \leq \mathbf{x}_{g_w}$. But in this case we removed the edge on $\{b'_\beta : \beta \in w\}$. Since this is true for all relevant w, $\{R(x, \bar{b}'_{v_\beta}) : \beta \in s'\}$ is indeed consistent.

---

[11]We are aiming at a type in the ultrapower of the form $\{R(x, \bar{a}_{v_\alpha}) : \alpha < \lambda\}$. Since u, v, w are used for sets of indices, we use $s \in \Omega$ for a finite set of formulas in the type.

[12]Note that the condition in (10.C) is set to avoid $\mathbf{x}_{g_w}$ only if all k-element subsets of w occur as v_β for some $\beta \in s$. It's not enough that each element of w occurs in some v_β. For example, in the tetrahedron-free three-hypergraph, if $w = \{1,2,3\}$ and $\{v_\beta : \beta \in s\} = \{\{1,2\},\{2,3\},\{1,3\}\}$, then $\mathbf{b}_s \cap \mathbf{x}_{g_w} = 0$, but not if $\{v_\beta : \beta \in s\} = \{\{1,2\},\{1,3\}\}$.

This shows that $\bar{\mathbf{b}} = \langle \mathbf{b}_s : s \in \Omega \rangle$ is indeed a possibility pattern, and it remains to show it has no multiplicative refinement. Suppose for a contradiction that

$$\bar{\mathbf{b}}' = \langle \mathbf{b}'_s : s \in \Omega \rangle$$

were a multiplicative refinement of $\bar{\mathbf{b}}$, i.e. $\bar{\mathbf{b}}'$ is a sequence of elements of $\mathcal{D}_*$ such that $s_1, s_2 \in \Omega$ implies $\mathbf{b}'_{s_1} \cap \mathbf{b}'_{s_2} = \mathbf{b}'_{s_1 \cap s_2}$ and for each $s \in \Omega$, $\mathbf{b}'_s \leq \mathbf{a}_s$. As each $\mathbf{b}'_{\{\beta\}} \in \mathfrak{B}^+$, we may write $\mathbf{b}'_{\{\beta\}} = \bigcup\{\mathbf{x}_{h_{\beta,i}} : i < i(\beta) \leq \mu\}$ where $\langle h_{\beta,i} : i < i(\beta)\rangle$ is a set of pairwise inconsistent functions from $\mathrm{FI}_{\mu,\theta}(2^\lambda)$. Let $S_\beta = \bigcup\{\mathrm{dom}(h_{\beta,i}) : i < i(\beta)\}$, so $S_\beta \subseteq 2^\lambda$ has cardinality $\leq \mu \cdot \theta = \mu$.

Subclaim 10.8. Let $n = k+1$. If $w \in \mathcal{P}$ then $\alpha_w \in \bigcup\{S_\beta : v_\beta \in [w]^k\}$.

Proof. Let $x = \{\beta : v_\beta \in [w]^k\}$, which is a finite set since (10.B) was without repetitions. As $\bar{\mathbf{b}}'$ is multiplicative, $\mathbf{b}'_x = \bigcap\{\mathbf{b}'_\beta : \beta \in x\}$. Let $f \in \mathrm{FI}_{\mu,\theta}(2^\lambda)$ be such that $\mathbf{x}_f \leq \mathbf{b}'_x$. Thus $\mathbf{x}_f \leq \mathbf{b}'_{\{\beta\}}$ for each $\beta \in x$. Let $g = f \restriction \bigcup\{S_\beta : \beta \in x\}$, noting that g must be nonempty [indeed, if $\mathrm{dom}(f) \cap S_\beta = \emptyset$ for some $\beta \in x$, then necessarily $\mathbf{x}_f \cap (1 - \mathbf{b}'_\beta) > 0$]. Then $\mathbf{x}_g \leq \mathbf{b}'_{\{\beta\}}$ for all $\beta \in x$. This implies that $\mathbf{x}_g \leq \mathbf{b}'_x \leq \mathbf{b}_x$ because $\bar{\mathbf{b}}'$ refines $\bar{\mathbf{b}}$. By definition in (10.C),

$$\mathbf{b}_x = 1_{\mathfrak{B}} - \bigcup\{\mathbf{x}_{g_u} : u \in \mathcal{P} \text{ and } [u]^k \subseteq \{v_\beta : \beta \in x\}\}.$$

So as $[w]^k \subseteq \{v_\beta : \beta \in x\}$, necessarily $\mathbf{x}_g \cap \mathbf{x}_{g_w} = 0_{\mathfrak{B}}$. Since our Boolean algebra $\mathfrak{B}$ was generated freely, it must be that $\mathrm{dom}(g_w) \cap \mathrm{dom}(g) \neq \emptyset$, but $\mathrm{dom}(g_w) = \{\alpha_w\}$. This shows that $\alpha_w \in \bigcup\{S_\beta : v_\beta \in [w]^k\}$ as desired.

This proves Subclaim 10.8. □

Finally, define $F : [\lambda]^k \to [\lambda]^{\leq \mu}$ by: if $v \in [\lambda]^k$ let β be such that $v = v_\beta$, and let

$$F(v) = \bigcup\{w \in [\lambda]^n : w \in \mathcal{P} \text{ and } \alpha_w \in S_\beta\} \cup v.$$

Then $F(v)$ is well defined, $F(v) \subseteq \lambda$, and $|F(v)| \leq \mu$ for $v \in [\lambda]^k$, since (10.B) is without repetition and $|S_\beta| \leq \mu$. Now for all $w \in \mathcal{P}$, there is $v = v_\beta \in [w]^k$ such that $\alpha_w \in S_\beta$. Thus $w \subseteq F(v)$. This shows that for all $w \in \mathcal{P}$,

$$(\exists v \in [w]^k)(w \subseteq F(v)).$$

This is a contradiction to (10.A), so $\bar{\mathbf{b}}$ does not have a multiplicative refinement. Thus, $\mathcal{D}_*$ cannot be moral for $T_{k+1,k}$. This completes the proof. □

Question 10.9. Does Theorem 7 hold for $n > k+1$?

Acknowledgments

We thank the anonymous referee and D. Ulrich for very helpful comments on the manuscript.

References

[1] D. Casey and M. Malliaris, "Notes on cofinality spectrum problems." Notes for the Appalachian Set Theory workshop. ArXiv:1709.02408.

[2] R. Engelking and M. Karłowicz, *Some theorems of set theory and their topological consequences*, Fund. Math. **57** (1965), 275–285, DOI 10.4064/fm-57-3-275-285. MR196693

[3] P. Erdős, A. Hajnal, A. Máté, and R. Rado, *Combinatorial set theory: partition relations for cardinals*, Studies in Logic and the Foundations of Mathematics, vol. 106, North-Holland Publishing Co., Amsterdam, 1984. MR795592

[4] G. Fichtenholz and L. Kantorovitch, "Sur les opérations linéaires dans l'espace des fonctions bornées." Studia Math. 5 (1934), 69–98.
[5] R. Grossberg, J. Iovino, and O. Lessmann, *A primer of simple theories*, Arch. Math. Logic **41** (2002), no. 6, 541–580, DOI 10.1007/s001530100126. MR1923196
[6] F. Hausdorff, "Über zwei Sätze von G. Fichtenholz und L. Kantorovitch." Studia Math. 6 (1936), 18–19.
[7] E. Hrushovski, *Pseudo-finite fields and related structures*, Model theory and applications, Quad. Mat., vol. 11, Aracne, Rome, 2002, pp. 151–212. MR2159717
[8] H. J. Keisler, *Ultraproducts which are not saturated*, J. Symbolic Logic **32** (1967), 23–46, DOI 10.2307/2271240. MR218224
[9] P. Komjáth and S. Shelah, *Two consistency results on set mappings*, J. Symbolic Logic **65** (2000), no. 1, 333–338, DOI 10.2307/2586540. MR1782123
[10] M. E. Malliaris, *Realization of ϕ-types and Keisler's order*, Ann. Pure Appl. Logic **157** (2009), no. 2-3, 220–224, DOI 10.1016/j.apal.2008.09.008. MR2499710
[11] M. E. Malliaris, *Hypergraph sequences as a tool for saturation of ultrapowers*, J. Symbolic Logic **77** (2012), no. 1, 195–223, DOI 10.2178/jsl/1327068699. MR2951637
[12] M. Malliaris and S. Shelah, *A dividing line within simple unstable theories*, Adv. Math. **249** (2013), 250–288, DOI 10.1016/j.aim.2013.08.027. MR3116572
[13] M. Malliaris and S. Shelah. "Saturating the random graph with an independent family of small range." In *Logic Without Borders* (volume in honor of J. Väänänen) ed. A. Hirvonen, J. Kontinen, R. Kossak, A. Villaveces. DeGruyter, 2015.
[14] M. Malliaris and S. Shelah, *Existence of optimal ultrafilters and the fundamental complexity of simple theories*, Adv. Math. **290** (2016), 614–681, DOI 10.1016/j.aim.2015.12.009. MR3451934
[15] M. Malliaris and S. Shelah, *Keisler's order has infinitely many classes*, Israel J. Math. **224** (2018), no. 1, 189–230, DOI 10.1007/s11856-018-1647-7. MR3799754
[16] M. Malliaris and S. Shelah, *A new look at interpretability and saturation*, Ann. Pure Appl. Logic **170** (2019), no. 5, 642–671, DOI 10.1016/j.apal.2019.01.001. MR3926500
[17] S. Shelah, *Classification Theory.* Revised edition. North-Holland, 1990.
[18] D. Ulrich, *Keisler's order is not linear, assuming a supercompact*, J. Symb. Log. **83** (2018), no. 2, 634–641, DOI 10.1017/jsl.2018.1. MR3835081

Department of Mathematics, University of Chicago, 5734 South University Avenue, Chicago, Illinois 60637

Email address: `mem@math.uchicago.edu`

Einstein Institute of Mathematics, Edmond J. Safra Campus, Givat Ram, The Hebrew University of Jerusalem, Jerusalem, 91904, Israel; and Department of Mathematics, Hill Center - Busch Campus, Rutgers, The State University of New Jersey, 110 Frelinghuysen Road, Piscataway, New Jersey 08854-8019

Email address: `shelah@math.huji.ac.il`

URL: `http://shelah.logic.at`

Contemporary Mathematics
Volume **752**, 2020
https://doi.org/10.1090/conm/752/15134

A brief introduction to amenable equivalence relations

Justin Tatch Moore

1. Introduction

The notion of an *amenable equivalence relation* was introduced by Zimmer [**23**] [**24**] in the course of his analysis of orbit equivalence relations in ergodic theory. More recently it played an important role in Monod's striking family of examples of nonamenable groups which do not contain nonabelian free subgroups. Recall that a group is *amenable* if it supports a mean which is translation invariant. Such groups were first studied by von Neumann who observed that amenable groups can not contain nonabelian free subgroups. He famously asked if this was the only obstruction to a group's amenability.

Monod's example[1] can be described as follows. If A is a subring of $\mathbf{R}$, define $H(A)$ to be the group of all piecewise $\mathrm{PSL}_2(A)$ homeomorphisms of the real projective line which fix the point at infinity.

THEOREM 1.1. [**18**] *If A is any dense subring of $\mathbf{R}$, then $H(A)$ is nonamenable. Moreover, if $f, g \in H(\mathbf{R})$, then either $\langle f, g\rangle$ is metabelian or else contains an infinite rank free abelian subgroup. In particular, $H(\mathbf{R})$ does not contain a nonabelian free subgroup.*

Subsequently, Lodha and the author constructed a finitely presented nonamenable subgroup of $H(\mathbf{Z}[1/\sqrt{2}])$ [**16**]. Lodha has since shown that the group of [**16**] is moreover of type F_∞ [**15**].

At least from a group-theoretic perspective, the most novel aspect of [**18**] was the use of Zimmer's notion of an amenable equivalence relation in the proof of the nonamenability of the groups $H(A)$. The purpose of this article is to give a brief survey of the theory of amenable and hyperfinite equivalence relations and illustrate how it can be used to show that certain discrete groups are nonamenable.

The subject matter falls within the broader scope of what is sometimes called *measurable group theory* — the study of groups through the analysis of their action on measure spaces and using analytical tools. This is in contrast with *geometric group theory*, where groups are studied through their actions which preserve an

2010 *Mathematics Subject Classification.* Primary: 43A07; Secondary: 20F65.
Key words and phrases. amenable, equivalence relation, free group, hyperfinite.
The author was supported in part by NSF grant DMS–1262019.

[1]This is a slight and, at least for our purposes, inessential variation of Monod's example from [**18**].

underlying geometry and by employing geometric techniques and concepts. Measurable group theory is closely aligned with the ergodic theory, the dynamics of discrete groups, probability, and descriptive set theory. Further reading can be found in [**7**] and [**12**] and their references. Further background on descriptive set theory can be found in [**11**].

This article is organized as follows. After reviewing some background material and fixing some terminology, we will present the definitions of amenable and hyperfinite equivalence relations in Section 3. This section will culminate with a theorem connecting these two apparently different notions. Section 4 will present several examples of nonamenable equivalence relations. Section 5 will discuss the analogs of the closure properties of amenable groups in the setting of equivalence relations. These played an important role in the isolation of the group in [**16**].

This article does not contain any new results, although Theorem 4.8 below is cast in a more abstract way than in [**9**]. (It is also to the author's knowledge, the first account of this proof in English.) The article's goal is to encourage the reader to pursue further reading in, e.g., [**7**] and [**12**] which contains a much more complete treatment of the subject matter presented here.

2. Preliminaries

Before proceeding, we will fix some terminology. In a few places we will refer to the continuity of extended real valued functions. In this context, the neighborhoods of infinity are the co-bounded sets. Recall that a *Polish space* is a topological space which is separable and completely metrizable. The σ-algebra of Borel sets in a Polish space is said to be a *standard Borel space*. A function f between standard Borel spaces X and Y is *Borel* if preimages of Borel sets are Borel. This is equivalent to the graph of f being a Borel subset of $X \times Y$. It is well known that any two uncountable standard Borel spaces are isomorphic in the sense that there is a bijection f between them such that f and f^{-1} are Borel.

A *Borel measure* on a standard Borel space is a countably additive σ-finite measure defined on its Borel sets. Such a measure extends uniquely to the σ-algebra generated by the Borel sets and the subsets of measure 0 Borel sets. We will generally not distinguish between these measures but note here that *measurable* will always refer to the larger σ-algebra.

In this article we will write (X, μ) is a *measured Polish space* to mean that X is a Polish space and that μ is a Borel measure on X. If in addition the topology on X is generated by the open sets of finite measure, then we say that (X, μ) is *locally finite*. If Γ is a topological group, then we will say that Γ acts continuously on a measured Polish space (X, μ) if:

- the map $(g, x) \mapsto g \cdot x$ is continuous and
- the maps $g \mapsto \mu(g \cdot E)$ are continuous for each measurable $E \subseteq X$.

Notice that this is stronger than the assertion that Γ acts continuously on the metric space X.

We note some useful facts about measured Polish spaces.

FACT 2.1. If (X, μ) is a measured Polish space and $E \subseteq X$ is measurable, then $\mu(E)$ is the supremum of all $\mu(F)$ where F is a closed subset of E.

FACT 2.2. Suppose that (X, μ) is a locally finite measured Polish space. If $E \subseteq X$ is measurable and has positive measure, then for every $\epsilon > 0$ there is an open set $U \subseteq X$ such that $0 < \mu(U) < \infty$ and $\mu(E \cap U) > (1 - \epsilon)\mu(U)$.

We will also need the following proposition.

PROPOSITION 2.3. *Suppose that (X, μ) is a locally finite measured Polish space and Γ is a metrizable group acting continuously on (X, μ). If $E \subseteq X$ is a measurable set of positive measure, then there is an open neighborhood V of the identity of Γ and an $\epsilon > 0$ such that if g is in V, then $\mu((g \cdot E) \cap E) > \epsilon$.*

PROOF. Let $E \subseteq X$ be given and let $U \subseteq X$ be an open set with

$$0 < \frac{3}{4}\mu(U) < \mu(E \cap U) < \mu(U) < \infty.$$

Set $\epsilon = \frac{1}{4}\mu(U)$. Observe by our continuity assumption on the action, we have that for every x in U there is an open $W_x \subseteq U$ containing x and a $\delta_x > 0$ such that if the distance from g to the identity is less than δ_x, then $g \cdot W_x \subseteq U$. Find a $\delta > 0$ such that

$$\mu(\{x \in U : \delta_x \geq \delta\}) > \frac{3}{4}\mu(U)$$

and define $W = \bigcup\{W_x : \delta_x \geq \delta\}$, observing that $\mu(W) > \frac{3}{4}\mu(U)$. In particular, $\mu(E \cap W) > \frac{1}{2}\mu(U)$. By our assumption that Γ acts continuously on (X, μ), there is an open set V containing the identity such that every element of V has distance less than δ to the identity and $\mu(g \cdot (E \cap W)) > \frac{1}{2}\mu(U)$ whenever g is in V. Since $\mu(E \cap U) > \frac{3}{4}\mu(U)$ and since $g \cdot (E \cap W) \subseteq (g \cdot E) \cap U$, it follows that $\mu(E \cap (g \cdot E) \cap U) > \frac{1}{4}\mu(U) = \epsilon$. □

If X is a standard Borel space, an equivalence relation E on X is Borel if it is Borel as a subset of X^2. A Borel equivalence relation is said to be *countable* if every equivalence class is countable. Notice that while this meaning conflicts with the literal interpretation of "countable," there is never a cause for confusion since for an equivalence relation to be countable as a set, it must have a countable underlying set and in the present context one is generally only interested in uncountable standard Borel spaces.

The principal example of a countable Borel equivalence relation is as follows: if G is a countable discrete group acting by Borel automorphisms on a standard Borel space X, then the orbit equivalence relation E^G_X is a countable Borel equivalence relation. That is, (x, y) is in E^G_X if and only if there is a g in G such that $g \cdot x = y$. In fact all countable Borel equivalence relations arise in this way:

THEOREM 2.4. **[6]** *If E is a countable Borel equivalence relation on a standard Borel space, then there is a countable group G and a Borel action of G on X such that $E = E^G_X$.*

The advantage of working with equivalence relations is in part that the notion of a countable Borel equivalence relation is much more flexible than that of a group. For instance while subgroups give rise to subequivalence relations, the converse is not generally true. A more sophisticated use of this generality occurs in the proof of Theorem 1.1: orbit equivalence relations are used to transfer the nonamenability of $\mathrm{PSL}_2(A)$ to the group $H(A)$ even though these groups are quite unrelated from an algebraic perspective.

3. Amenable and hyperfinite equivalence relations

Any action of a countable group on a standard Borel space gives rise to a countable Borel equivalence relation and, conversely, any countable Borel equivalence relation can be generated as the orbit equivalence relation of some group action. The fundamental problem of this subject is to understand the extent to which properties of the group which generated a countable Borel equivalence relation are reflected in properties of the equivalence relation and vice versa.

Our focus in this article will be to develop the properties of equivalence relations which are analogs of the group-theoretic property of *amenability*. Roughly speaking, the notion of an amenable equivalence relation has the property that every action of an amenable group gives rise to an amenable equivalence relation and a group is amenable only when every orbit equivalence relation is amenable.

Now to be more precise. Suppose that (X, μ) is Borel measure on a standard Borel space and E is a countable Borel equivalence relation on X. We say that E is *μ-amenable* if there is a μ-measurable assignment $x \mapsto \nu_x$ such that:

- each ν_x is a finitely additive probability measure on X satisfying that $\nu_x([x]_E) = 1$.
- if $(x, y) \in E$, then $\nu_x = \nu_y$.

By *measurable* we mean that if A is any measurable subset of $X \times X$, then

$$x \mapsto \nu_x(\{y \in X : (x, y) \in A\})$$

is μ-measurable. While we will generally quantify *amenable* with a measure, a Borel equivalence relation is said to be *amenable* if it is μ-amenable with respect to every Borel measure on the underlying standard Borel space.

While it is not apparent from the definition, it is true that every orbit equivalence relation of a countable amenable group acting on standard Borel space is necessarily μ-amenable with respect to any Borel measure μ (this does not require any invariance of μ with respect to the group action). This will follow from Theorem 3.5 below.

Next we turn to a seemingly unrelated notion. A countable Borel equivalence relation E on a standard Borel space X is *hyperfinite* if E is an increasing union of Borel equivalence relations with finite equivalence classes. A good example to keep in mind is that of *eventual equality* on infinite binary sequences: define $x =^* y$ if $x(k) = y(k)$ for all but finitely many k. Notice that this is the union of the equivalence relations $=^n$ defined by $x =^n y$ if $x(k) = y(k)$ for all $k \geq n$. The following theorem gives a powerful criterion for verifying hyperfiniteness.

THEOREM 3.1. [**4**] *Suppose that X is a standard Borel space and $f : X \to X$ is a Borel function which such that $f^{-1}(x)$ is at most countable for each x. The smallest equivalence relation E satisfying that, for each $x \in X$, $(x, f(x)) \in E$ is hyperfinite.*

EXAMPLE 3.2. [**4**] Define an equivalence relation E all infinite binary sequences by xEy if for some m and n, $x(m + i) = y(n + i)$ for all $i > 0$. This equivalence relation is called *tail equivalence* and is generated by the shift map $f : 2^\omega \to 2^\omega$ given by $f(x)(i) = x(i + 1)$.

EXAMPLE 3.3. [**2**] Recall that the real projective line $P^1(\mathbf{R})$ is the collection of all lines in $\mathbf{R}^2$ passing through the origin. Identify an element of $P^1(\mathbf{R})$ with the x-coordinate of its intersection with the line $y = 1$, adopting the convention that the line $y = 0$ becomes identified with ∞. This identification gives $P^1(\mathbf{R})$ a natural compact metric topology — it is in fact homeomorphic to a circle. $\mathrm{PSL}_2(\mathbf{R})$ is the group of fractional linear transformations $t \mapsto \frac{at+b}{ct+d}$ of $P^1(\mathbf{R})$, interpreted as the extended real line. Define a map $\Phi : 2^\omega \to P^1(\mathbf{R})$ by

$$\Phi(\mathtt{1}x) = \phi(x) \qquad\qquad \Phi(\mathtt{0}x) = -\phi(\sim x)$$

($\sim x$ denotes the bitwise complement of x) where ϕ is the unique order preserving map from 2^ω to $[0, \infty]$ which satisfies the equation:

$$\phi(\mathtt{1}x) = 1 + \phi(x) \qquad\qquad \phi(\mathtt{0}x) = \frac{1}{1 + \frac{1}{\phi(x)}}.$$

Informally, the first digit of x determines the sign of $\Phi(x)$ and the remaining digits are used to determine the continued fraction expansion of $|\Phi(x)|$ so that the nth "digit" in this expansion is the length of the nth block of consecutive equal digits in x (after the first digit). Formally, the existence and uniqueness of ϕ can be seen as follows. First, the implicit definition gives a recursive formula for ϕ on on those binary sequences which are eventually 0, noting that $\phi(\mathtt{0})$ is forced to be 0. Moreover ϕ maps these eventually 0 sequences in an order preserving manner to the nonnegative rationals. It is then readily checked that ϕ extends to 2^ω so as to still preserve order and satisfy the implicit equation. The action of $\mathrm{PSL}_2(\mathbf{Z}) = \langle t \mapsto t+1, t \mapsto -1/t\rangle$ on $P^1(\mathbf{R})$ naturally lifts to an action on 2^ω. Specifically, if $a(t) = t + 1$ and $b(t) = -1/t$, then the action of these generators is given by

$$a \cdot (\mathtt{00}x) = \mathtt{0}x \qquad\qquad b \cdot (\mathtt{0}x) = \mathtt{1}x$$
$$a \cdot (\mathtt{01}x) = \mathtt{10}x \qquad\qquad b \cdot (\mathtt{1}x) = \mathtt{0}x$$
$$a \cdot (\mathtt{1}x) = \mathtt{11}x$$

It can be verified that the corresponding orbit equivalence relation on 2^ω is tail equivalence. In particular, this orbit equivalence relation of $\mathrm{PSL}_2(\mathbf{Z})$'s action on $P^1(\mathbf{R})$ is hyperfinite.

The following gives an important characterization of the hyperfinite equivalence relations:

THEOREM 3.4. **[21]** **[22]** *Every Borel action of* $\mathbf{Z}$ *on a standard Borel space generates a hyperfinite orbit equivalence relation. Conversely, every hyperfinite Borel equivalence relation is the orbit equivalence relation of a Borel action of* $\mathbf{Z}$.

While it is not obvious, it turns out that every hyperfinite Borel equivalence relation is in fact μ-amenable with respect to any Borel measure on the underlying space. In fact, a natural weakening captures the notation of μ-amenability exactly. If μ is a Borel measure on X, then we say that E is *μ-hyperfinite* if there is a μ-measure 0 set $Y \subseteq X$ such that the restriction of E to $X \setminus Y$ is hyperfinite.

The following theorem is an amalgamation of several results stated in modern language (see also **[12]**).

THEOREM 3.5. **[2]** **[5]** **[13]** **[19]** **[24]** *Suppose that X is a standard Borel space, E is a countable Borel equivalence relation on X, and μ is a Borel measure on X. The following are equivalent:*

(1) *E is μ-amenable;*
(2) *E is μ-hyperfinite;*

(3) $E = E_X^G$ *for some* μ*-measurable action of an amenable group* G *on* X*;*
(4) $E = E_X^{\mathbf{Z}}$ *for some* μ*-measurable action of* $\mathbf{Z}$ *on* X*.*

It is worth noting that the extent to which the measure 0 sets can be omitted in the previous theorem is a major open problem in descriptive set theory. For instance:

PROBLEM 3.6. [**4**] If E_n $(n < \infty)$ is an increasing sequence of hyperfinite Borel equivalence relations on a standard Borel space, is $\bigcup_{n=0}^{\infty} E_n$ hyperfinite?

PROBLEM 3.7. [**22**] Is every orbit equivalence relation of a Borel action of a countable amenable group acting on a standard Borel space hyperfinite?

In fact it was only relatively recently that a positive solution to Problem 3.7 was proved for the class of abelian groups [**8**]; the strongest result at the time of this writing is [**20**]. Problem 3.7 remains open for the Baumslag-Solitar group $BS(1,2)$ generated by the affine transformations $t \mapsto t+1$ and $t \mapsto 2t$. While not directly related, Marks has recently demonstrated differences between the so-called *Borel context* and *measure-theoretic context* [**17**].

4. Examples

In this section we will consider a number of examples. Perhaps the easiest way to generate nonamenable equivalence relations is through actions of groups which preserve a probability measure.

THEOREM 4.1. [**19**] *Suppose that* G *is a countable group,* (X, μ) *is a standard Borel space equipped with a Borel probability measure, and* E *is the orbit equivalence relation of a measure preserving action of* G *which is free* μ*-a.e.. The equivalence relation* E *is* μ*-amenable if and only if* G *is amenable.*

The following are two typical — but quite different — examples of such actions.

EXAMPLE 4.2. If G is any countably infinite group and (X, μ) is any probability space, then G acts by shift on X^G as follows: $(g \cdot x)(h) = x(g^{-1}h)$. This action preserves the product measure and, unless μ is a point-mass, is free almost everywhere with respect to the product measure.

EXAMPLE 4.3. The action of $\mathrm{SL}_2(\mathbf{Z})$ on the torus $\mathbf{T}^2$ equipped with Lebesgue measure is measure preserving and free λ-a.e.. Since $\mathrm{SL}_2(\mathbf{Z})$ contains the free group on two generators, this orbit equivalence relation is not λ-amenable.

Example 4.3 is in contract with Example 3.3 in which the group $\mathrm{PSL}_2(\mathbf{Z})$ acts on $P^1(\mathbf{R})$, which is homeomorphic to the circle. It is well known that $\mathrm{PSL}_2(\mathbf{Z})$ contains a free subgroup (even one of finite index) and hence is nonamenable. On the other hand, we have seen above that the orbit equivalence relation induced on $P^1(\mathbf{R})$ is just tail equivalence on 2^ω in disguise; in particular it is hyperfinite and hence amenable. Notice that, unlike the action of $\mathrm{SL}_2(\mathbf{Z})$ on the torus, there is no probability measure on $P^1(\mathbf{R})$ which is preserved by the action of $\mathrm{PSL}_2(\mathbf{Z})$.

It turns out, however, that *dense* subgroups of $\mathrm{PSL}_2(\mathbf{R})$ do produce a nonamenable orbit equivalence relation when they act on $P^1(\mathbf{R})$.

THEOREM 4.4. [**9**] *Every nondiscrete subgroup of* $\mathrm{PSL}_2(\mathbf{R})$ *is either solvable or else contains a nondiscrete free subgroup on two generators.*

THEOREM 4.5. [9] *If Γ is a rank 2 free subgroup of a finite dimensional Lie group G and Γ is nondiscrete, then the orbit equivalence relation of Γ's action on G is nonamenable with respect to Haar measure.*

In the case of $\Gamma = \mathrm{PSL}_2(\mathbf{Z}[1/2])$, it is not difficult to exhibit a nondiscrete free subgroup explicitly.

EXAMPLE 4.6. The matrices

$$\alpha = \begin{pmatrix} 1/2 & -4 \\ 1/4 & 0 \end{pmatrix} \qquad \beta = \begin{pmatrix} 1/2 & -1/4 \\ 4 & 0 \end{pmatrix}$$

generate a nondiscrete free subgroup of $\mathrm{PSL}_2(\mathbf{Z}[1/2])$. In order to see this, first observe that the traces of these matrices are $1/2$ and hence both matrices describe *elliptic* transformations of the real projective line (i.e. there are no fixed points). Since any elliptic element of $\mathrm{PSL}_2(\mathbf{R})$ of infinite order generates a nondiscrete subgroup, it suffices to show that the above matrices generate a free group.

Define X to be the set of all rational numbers in $P^1(\mathbf{R}) = \mathbf{R} \cup \{\infty\}$ which can be represented by a fraction with an odd denominator and let Y denote the remaining rational numbers in $P^1(\mathbf{R})$. By the *Ping-Pong Lemma* (see, e.g., [3]), it suffices to show that if $n \neq 0$ is an integer, then $\alpha^n Y \subseteq X$ and $\beta^n X \subseteq Y$. Let X_0 consist of those elements of X which can be represented by a fraction of the form $(4p+2)/q$ where q is odd. Notice that $\alpha(X_0 \cup Y) \subseteq X_0$ and that X_0 and Y are disjoint. It follows that $\alpha^n Y \subseteq X$ whenever n is a nonzero integer. Similarly, $\beta^n X \subseteq Y$.

Theorem 4.4 was generalized considerably by the following result.

THEOREM 4.7. [1] *If Γ is a dense subgroup of a connected semi-simple real Lie group, then Γ contains a dense free subgroup of rank 2.*

The following theorem is a generalization of Theorem 4.5, although the argument closely follows that of [9].

THEOREM 4.8. *Suppose that (X, μ) is a locally finite measured Polish space. If $\Gamma = \langle a, b \rangle$ is a free nondiscrete metrizable group which is acting freely and continuously on (X, μ), then the orbit equivalence relation is not μ-amenable.*

PROOF. Suppose for contradiction that E_X^Γ is μ-amenable and fix a μ-measurable assignment $x \mapsto \nu_x$ such that:

- for each x, ν_x is a finitely additive probability measure supported on the orbit of x;
- if x and y are in the same orbit, then $\nu_x = \nu_y$.

For $u \in \{a, b\}$, define Γ_u to be all those elements of Γ which are representable by a reduced word beginning with u and ending with u^{-1}. Observe that if g is a nonidentity element of Γ, then there is a h in $\{a, ab, ab^{-1}\}$ such that hgh^{-1} is in Γ_a. Thus there is an $h \in \{e, a, ab, ab^{-1}\}$ and a $\Gamma' \subseteq \Gamma$ which accumulates to the identity such that $h\Gamma' h^{-1} \subseteq \Gamma_a$. Since conjugation is continuous, it follows that Γ_a also accumulates to the identity. Furthermore, $b\Gamma_a b^{-1} \subseteq \Gamma_b$ and thus Γ_b accumulates to the identity as well.

Since the action of Γ on X is free, for each $x, y \in X$ which lie in the same orbit, there is a unique $\gamma = \gamma(x, y)$ in Γ such that $x = \gamma \cdot y$. Notice that $\gamma(g \cdot x, y) = g\gamma(x, y)$. Define $\phi : X \to [0, 1]$ by letting $\phi(x) = \nu_x(A_x)$ where A_x is the set of those y in the orbit of x such that the reduced word representing $\gamma(x,y)$ begins with a or a^{-1}. Observe that for any x in X and g in Γ_b, $A_x \cap A_{g\cdot x} = \emptyset$. Hence if x is in X and $\phi(x) > 1/2$, then $\phi(g\cdot x) < 1/2$ whenever g is in Γ_b. Similarly, if $\phi(x) < 1/2$ and g is in Γ_a, then $\phi(g\cdot x) > 1/2$. Furthermore, for all x and $g \in \Gamma_a$, the sets A_x, $A_{bgb^{-1}\cdot x}$ and $A_{b^2gb^{-2}\cdot x}$ are pairwise disjoint and consequently $0 \leq \phi(x) + \phi(bgb^{-1}\cdot x) + \phi(b^2gb^{-2}\cdot x) \leq 1$.

I next claim that $Y = \{x \in X : \phi(x) \neq 1/2\}$ has positive measure with respect to μ. Suppose not. Using our assumption that Γ acts continuously on (X,μ), find an open neighborhood V of the identity such that if g is in V, then Y, $bg^{-1}b^{-1}\cdot Y$, and $b^2g^{-1}b^{-2}\cdot Y$ have total measure less than that of X. Now if x is outside these sets and $g \in V\cap\Gamma_a$, we have that $\phi(x)$, $\phi(bgb^{-1}\cdot x)$, and $\phi(b^2gb^{-2}\cdot x)$ are each $1/2$, contradicting that there sum is at most 1. Thus Y must have positive measure.

Let $Y_a = \{y \in Y : \phi(y) > 1/2\}$ and $Y_b = \{y \in Y : \phi(y) < 1/2\}$. Since Y has positive measure, either Y_a or Y_b have positive measure. If Y_a has positive measure, then by Proposition 2.3 there is a g in Γ_b such that $(g\cdot Y_a)\cap Y_a$ has positive measure and in particular is nonempty. This contradicts our observation that if $\phi(y) > 1/2$ and g is in Γ_b, then $\phi(g\cdot y) < 1/2$. Similarly, if Y_b has positive measure, one obtains a contradiction by finding a g in Γ_a such that $g\cdot Y_b$ intersects Y_b and recalling our observation that if $\phi(y) < 1/2$ and g is in Γ_a, then $\phi(g\cdot y) > 1/2$. It must be, therefore, that the orbit equivalence relation is nonamenable. □

We finish this section with a simple but powerful observation of Monod.

EXAMPLE 4.9. [**18**] If A is a countable dense subring of $\mathbf{R}$, let $H(A)$ denote the group consisting of all orientation preserving homeomorphisms of $P^1(\mathbf{R})$ which fix the point at infinity and which are piecewise $\mathrm{PSL}_2(A)$. Suppose that α is in $\mathrm{PSL}_2(A)$ and that α does not fix ∞. As a fractional linear transformation, the graph of α is a hyperbola. If $r \in \mathbf{R}$ is sufficiently large in magnitude, then $\alpha(t) = t + r$ has two solutions $a < b$; set

$$\alpha_r(t) = \begin{cases} \alpha(t) & \text{if } a < t < b \\ t + r & \text{otherwise.} \end{cases}$$

If r is moreover an integer, then α_r is in $H(A)$. It follows that the restriction of the orbit equivalence relation of $\mathrm{PSL}_2(A)$ to $\mathbf{R}$ coincides with the corresponding restriction of the orbit equivalence relation of $H(A)$'s action on $\mathbf{R}$. By Theorem 3.5 and the results of [**9**] mentioned above, $H(A)$ is nonamenable whenever A is a dense subring of $\mathbf{R}$.

EXAMPLE 4.10. [**16**] Let $P(\mathbf{Z})$ denote the subgroup of $H(\mathbf{Z})$ consisting of those elements which have a continuous derivative. By unpublished work of William Thurston, $P(\mathbf{Z})$ is isomorphic to Richard Thompson's group F. In fact

$$\alpha(t) = t + 1 \qquad\qquad \beta(t) = \begin{cases} t & \text{if } t \leq 0 \\ \frac{t}{1-t} & \text{if } 0 \leq t \leq \frac{1}{2} \\ 3 - \frac{1}{t} & \text{if } \frac{1}{2} \leq t \leq 1 \\ t + 1 & \text{if } 1 \leq t \end{cases}$$

is the standard set of generators with respect to the usual finite presentation of F (see [**16**]). It is not difficult to see that the orbit equivalence relation of $P(\mathbf{Z})$'s action on $P^1(\mathbf{R})$ coincides with that of $\mathrm{PSL}_2(\mathbf{Z})$ except for the point at infinity.

Since $\mathrm{PSL}_2(\mathbf{Z}) \cup \{t \mapsto t+1/2\}$ generates $\mathrm{PSL}_2(\mathbf{Z}[1/2])$, it follows that $\langle t \mapsto t/2, \beta \rangle$ is nonamenable.

The previous example is less relevant to the amenability problem for F, however, than it might initially appear. For instance, Lodha has shown that if Γ is any subgroup of $H(\mathbf{R})$ which is isomorphic to F, then the orbit equivalence relation of Γ's action on $\mathbf{R}$ is λ-amenable where λ is Lebesgue measure [**14**].

5. Closure properties of amenable equivalence relations

One of the most basic facts about amenable groups is that they are closed under taking subgroups, extensions, and directed unions. These operations have their analogs in the setting of countable Borel equivalence relations as well. Notice that if $H \leq G$, then $E^H_X \subseteq E^G_X$ whenever G acts on a standard Borel space. Also, if G is an increasing union of a sequence of subgroups G_n $(n < \infty)$, then $E^G_X = \bigcup_n E^{G_n}_X$. Since the property of being μ-hyperfinite is clearly inherited to subequivalence relations, we have the following corollary of Theorem 3.5.

PROPOSITION 5.1. *If E is a subequivalence relation of a μ-amenable equivalence relation is μ-amenable.*

While there is no natural notion of extension in the setting of equivalence relations, it is easy to formulate what is meant by a product of equivalence relations. It is straightforward to verify the following analog of the closure of the class of amenable groups under taking products.

PROPOSITION 5.2. *Products of μ-amenable equivalence relations are amenable with respect to the corresponding product measure.*

The following is the corresponding analog of the amenability of increasing unions of amenable groups.

THEOREM 5.3. [**5**] [**13**] *Suppose that E_n $(n < \infty)$ is an increasing sequence of countable Borel equivalence relations on a standard Borel space X. If μ is a standard probability measure on X and each E_n is μ-hyperfinite, then $\bigcup_n E_n$ is μ-hyperfinite.*

The power of the closure properties mentioned in this section is that they afford some flexibility which has no analog in the algebraic setting. For instance, while the equivalence relations E_n in the previous theorem are required to be nested, they need not come from a nested sequence of groups. It is also sometimes fruitful to generate equivalence relations with partial homeomorphisms rather than full automorphisms of an underlying space.

EXAMPLE 5.4. Consider the following homeomorphisms of $P^1(\mathbf{R}) = \mathbf{R} \cup \{\infty\}$: $\alpha(t) = t + 1/2$ and $\beta(t) = -1/t$. For $0 < r < \infty$, define α_r to be the restriction of α to $[-r, -1/r]$. Let E be the equivalence relation generated by α and β and E_r be the equivalence relation generated by β and α_r (i.e. E_r is the smallest equivalence relation such that for all t, $(t, t+1/2) \in E_r$ and if additionally $-r \leq t \leq -1/r$, then $(t, -1/t) \in E_r$). Notice that if $0 < r < s < \infty$, then $E_r \subseteq E_s \subseteq E$ and that $E = \bigcup_{r>0} E_r$. Thus there is an r such that $0 < r < \infty$ such that E_r is nonamenable. In fact a more careful analysis reveals that $E_4 = E$, although this is not relevant for the point we wish to illustrate here.

EXAMPLE 5.5. Suppose that α and β are homeomorphisms such that the action of $\langle\alpha, \beta\rangle$ on $\mathbf{R}$ generates a nonamenable equivalence relation. Suppose further that α_n $(n < \infty)$ and β_n $(n < \infty)$ are sequences of homeomorphisms such that for all but countably many t, $\alpha_n(t) = \alpha(t)$ and $\beta_n(t) = \beta(t)$ holds for all but finitely many n. It follows that there exists an n such that the action of $\langle\alpha_n, \beta_n\rangle$ on $\mathbf{R}$ generates a nonamenable equivalence relation. To see this, let X_n denote the set of all t in $\mathbf{R}$ such that for all $k \geq n$, $\alpha_k(t) = \alpha(t)$ and $\beta_k(t) = \beta(t)$. Define E_n to be the equivalence relation generated by the restrictions of α_n and β_n to X_n. It follows that E_n $(n < \infty)$ is an increasing sequence of countable Borel equivalence relations which, off a countable subset of $\mathbf{R}$, unions to the equivalence relation generated by α and β. The claim now follows from Theorem 5.3.

Acknowledgments

I would like to thank Matt Brin and Clinton Conley for their help while preparing this article.

References

[1] E. Breuillard and T. Gelander, *On dense free subgroups of Lie groups*, J. Algebra **261** (2003), no. 2, 448–467, DOI 10.1016/S0021-8693(02)00675-0. MR1966638

[2] A. Connes, J. Feldman, and B. Weiss, *An amenable equivalence relation is generated by a single transformation*, Ergodic Theory Dynamical Systems **1** (1981), no. 4, 431–450 (1982), DOI 10.1017/s014338570000136x. MR662736

[3] P. de la Harpe, *Topics in geometric group theory*, Chicago Lectures in Mathematics, University of Chicago Press, Chicago, IL, 2000. MR1786869

[4] R. Dougherty, S. Jackson, and A. S. Kechris, *The structure of hyperfinite Borel equivalence relations*, Trans. Amer. Math. Soc. **341** (1994), no. 1, 193–225, DOI 10.2307/2154620. MR1149121

[5] H. A. Dye, *On groups of measure preserving transformations. I*, Amer. J. Math. **81** (1959), 119–159, DOI 10.2307/2372852. MR131516

[6] J. Feldman and C. C. Moore, *Ergodic equivalence relations, cohomology, and von Neumann algebras. I*, Trans. Amer. Math. Soc. **234** (1977), no. 2, 289–324, DOI 10.2307/1997924. MR578656

[7] D. Gaboriau, *Orbit equivalence and measured group theory*, Proceedings of the International Congress of Mathematicians. Volume III, Hindustan Book Agency, New Delhi, 2010, pp. 1501–1527. MR2827853

[8] S. Gao and S. Jackson, *Countable abelian group actions and hyperfinite equivalence relations*, Invent. Math. **201** (2015), no. 1, 309–383, DOI 10.1007/s00222-015-0603-y. MR3359054

[9] É. Ghys and Y. Carrière, *Relations d'équivalence moyennables sur les groupes de Lie* (French, with English summary), C. R. Acad. Sci. Paris Sér. I Math. **300** (1985), no. 19, 677–680. MR802650

[10] S. Jackson, A. S. Kechris, and A. Louveau, *Countable Borel equivalence relations*, J. Math. Log. **2** (2002), no. 1, 1–80, DOI 10.1142/S0219061302000138. MR1900547

[11] A. S. Kechris, *Classical descriptive set theory*, Graduate Texts in Mathematics, vol. 156, Springer-Verlag, New York, 1995. MR1321597

[12] A. S. Kechris and B. D. Miller, *Topics in orbit equivalence*, Lecture Notes in Mathematics, vol. 1852, Springer-Verlag, Berlin, 2004. MR2095154

[13] W. Krieger, *On non-singular transformations of a measure space. I, II*, Z. Wahrscheinlichkeitstheorie und Verw. Gebiete **11** (1969), 83–97; ibid. 11 (1969), 98–119, DOI 10.1007/BF00531811. MR240279

[14] Y. Lodha. *Coherent actions by homeomorphisms on the real line or an interval*, Israel J. Math. **235** (2020), no. 1, 183–212. MR4068782

[15] Y. Lodha. A type F_∞ group of piecewise projective homeomorphisms. ArXiv preprint arXiv:1408.3127.

[16] Y. Lodha and J. T. Moore, *A nonamenable finitely presented group of piecewise projective homeomorphisms*, Groups Geom. Dyn. **10** (2016), no. 1, 177–200, DOI 10.4171/GGD/347. MR3460335

[17] A. S. Marks, *A determinacy approach to Borel combinatorics*, J. Amer. Math. Soc. **29** (2016), no. 2, 579–600, DOI 10.1090/jams/836. MR3454384

[18] N. Monod, *Groups of piecewise projective homeomorphisms*, Proc. Natl. Acad. Sci. USA **110** (2013), no. 12, 4524–4527, DOI 10.1073/pnas.1218426110. MR3047655

[19] D. S. Ornstein and B. Weiss, *Ergodic theory of amenable group actions. I. The Rohlin lemma*, Bull. Amer. Math. Soc. (N.S.) **2** (1980), no. 1, 161–164, DOI 10.1090/S0273-0979-1980-14702-3. MR551753

[20] B. Seward and S. Schneider. Locally nilpotent groups and hyperfinite equivalence relations. ArXiv preprint arXiv:1308.5853, 2013.

[21] T. A. Slaman and J. R. Steel, *Definable functions on degrees*, Cabal Seminar 81–85, Lecture Notes in Math., vol. 1333, Springer, Berlin, 1988, pp. 37–55, DOI 10.1007/BFb0084969. MR960895

[22] B. Weiss, *Measurable dynamics*, Conference in modern analysis and probability (New Haven, Conn., 1982), Contemp. Math., vol. 26, Amer. Math. Soc., Providence, RI, 1984, pp. 395–421, DOI 10.1090/conm/026/737417. MR737417

[23] R. J. Zimmer, *Amenable ergodic group actions and an application to Poisson boundaries of random walks*, J. Functional Analysis **27** (1978), no. 3, 350–372, DOI 10.1016/0022-1236(78)90013-7. MR0473096

[24] R. J. Zimmer, *Hyperfinite factors and amenable ergodic actions*, Invent. Math. **41** (1977), no. 1, 23–31, DOI 10.1007/BF01390162. MR470692

Department of Mathematics, Cornell University, Ithaca, New York 14853–4201
Email address: `justin@math.cornell.edu`

Contemporary Mathematics
Volume **752**, 2020
https://doi.org/10.1090/conm/752/15135

Singularizing cardinals

Dima Sinapova

ABSTRACT. We survey methods and applications of singularizing cardinals. The main ways of singularizing a cardinal are by Priky forcing, Namba forcing, and Woodin tower. We survey what properties of these forcings, especially Prikry type posets, can be abstractly formalized without making specific assumptions on the forcing. Then we describe some applications, focusing on square properties. These show some limitations on ways to get failure of square properties together with violating the singular cardinal hypothesis. We also discuss what happens to cofinalities when collapsing cardinals.

1. Introduction

Singularizing cardinals has a key role in infinitary combinatorics. For example, obtaining models of the failure of the singular cardinal hypothesis (SCH) usually is done by adding a lot of Cohen subsets to a large cardinal, and then singularizing it. On the other hand, it is hard to combine failure of SCH with compactness properties at the first successor of a singular. Compactness is the phenomenon that if a given property holds at every substructure of smaller cardinality of some object, then it holds for the whole object. This motivates the following question.

QUESTION 1. Suppose that $V \subset W$ are transitive models of set theory, κ is regular in V and a singular cardinal in W.

(1) How close is W to a Prikry type forcing extension?
(2) What can we say about compactness properties at $(\kappa^+)^W$? In which cases do we have square sequences at κ in W?

Let us motivate the question. Prikry forcing is one of the standard ways to singularize cardinals: given a measurable cardinal κ and a normal measure U on it, Prikry forcing with respect to U adds a generic cofinal sequence $\langle \alpha_n \mid n < \omega \rangle$ through κ, with the following characterization of genericity (†):

$$\forall A \in U, \text{ for all large } n, \alpha_n \in A.$$

For a complete definition of Prikry forcing, see, for example, Jech [**8**], page 400. A similar characterization of genericity holds in the uncountable cofinality case, when adding a Magidor sequence ([**10**]).

2010 *Mathematics Subject Classification.* Primary 03E35, 03E55, 03E04.

The author was partially supported by the National Science Foundation, grants DMS-1362485 and Career-1454945.

On the other hand, suppose V is a core model, i.e. a canonical inner model with L-like properties and rich enough structure to have large cardinals. Then if $W \supset V$ is an outer model, such that a V-regular cardinal κ is singularized in W, then there will be Prikry/Magidor sequences in W.

In the next section we will survey some results which approximate this phenomenon (†), but without making any assumptions about forcing extensions. It turns out that, under some reasonable assumptions, in the situation of the above question, there are cofinal sequences through κ in W, that can meet many club subsets on a tail end. Here meeting clubs takes the place of meeting measure one sets, since we do not want to make any assumptions about W being a forcing extension. We use the term "pseudo Prikry" to describe these sequences because their properties approximate the genericity of a Prikry sequence, without referring to any measure.

Another way of singularizing cardinals is by a stationary Woodin tower forcing (see [**8**], page 653). A third way to singularize cardinals is by Namba forcing (see [**8**], page 561). In the basic case, forcing with Namba over a ground model V makes $(\aleph_2)^V$ have cofinality ω, while preserving $(\aleph_1)^V$. In the generic extension both $(\aleph_2)^V$ and $(\aleph_3)^V$ are collapsed.

Let us now motivate part two of Question 1. A typical example of compactness is failure of square. The square principle was isolated by Jensen in his fine structure analysis of L. Square principles hold in models that sufficiently resemble L. On the other hand they are at odds with reflection properties and fail above large cardinals. For more on square and reflection, see [**2**]. There is tension between failure of SCH and failure of the weaker square properties. The reason is that it is difficult to avoid weaker square principles at singular cardinals. A detailed history of this tension is given in the introduction of Neeman [**12**]. We will discuss applications of singularizing to square in section 3.

We will use the following notation: $\mathrm{cof}(\tau)$ and $\mathrm{cof}(<\tau)$ denote points of cofinality τ and less than τ, respectively. Similarly, we write $\mathrm{cof}^W(\tau)$ to denote the points of cofinality τ in W. For a set C, $\lim(C)$ is the set of all limit points of C and $\mathrm{nacc}(C)$ is the non-limit points of C. Also, we say that cardinal is collapsed to mean that it is no longer a cardinal in the outer model. Otherwise, it is preserved.

2. Pseudo Prikry sequences and more

By arguments in Gitik [**5**] and independently in Džamonja-Shelah [**4**], if $V \subset W$, and κ is an inaccessible cardinal in V and a singular cardinal in W and $(\kappa^+)^V$ is preserved, then for every family $\langle C_\alpha \mid \alpha < \kappa^+\rangle \in V$ of clubs in κ^+, there is a sequence $\langle \delta_n \mid n < \omega\rangle$ cofinal in κ, such that for each α, for all large n, $\delta_n \in C_\alpha$. Moreover, for every $\lambda < \kappa$, it can be arranged that the cofinality of the δ_n's is above λ.

In [**11**] this theorem was generalized:

THEOREM 2.1. *Suppose $V \subset W$ and $\kappa < \nu$ are cardinals in V, such that, ν is regular in V,*

- $(\omega_1)^W < \kappa$, $\mathrm{cf}^W(\kappa) = \omega$,
- *for all V-regular cardinals, τ, $\kappa \leq \tau \leq \nu$, $\mathrm{cf}^W \tau = \omega$, and*
- $\mu := (\nu^+)^V$ *remains a cardinals in W.*

Suppose that $\langle D_\alpha \mid \alpha < \mu\rangle \in V$ *is a sequence of club subsets of* ν. *Let* $\lambda < \kappa$. *Then in* W *there is a sequence* $\langle \delta_n \mid n < \omega\rangle$ *cofinal in* ν, *with each* $\mathrm{cf}^W(\delta_n) \geq \lambda$, *such that for every* $\alpha < \mu$, *for all large* n, $\delta_n \in D_\alpha$.

The proof relies on the following fact due to Shelah, whose proof can be found in [**13**], Theorem 2.14.

FACT 2.2. Suppose that $\nu > \kappa > \aleph_1$ are cardinals, where ν is regular. Then there is a sequence $\langle S_\delta \mid \delta < \nu\rangle$ of stationary subsets of ν^+, such that $\bigcup_{\delta<\nu} S_\delta = \nu^+ \cap \mathrm{cof}(<\kappa)$ and each S_δ carries a partial square sequence $\langle C^\delta_\alpha \mid \alpha \in S_\delta \cap \mathrm{Lim}\rangle$. More precisely:

- each C^δ_α is a club subset of α with $o.t.(C^\delta_\alpha) < \kappa$, and
- if $\beta \in \lim(C^\delta_\alpha)$, then $\beta \in S_\delta$ and $C^\delta_\alpha \cap \beta = C^\delta_\beta$.

Recall the characterization of genericity for Prikry forcing with respect to a normal measure U: a sequence $\langle \alpha_n \mid n < \omega\rangle$ is generic if and only if for all measure one sets $A \in U$, for all large n, $\alpha_n \in U$. The above theorem provides an abstraction of that notion. Since we do not want to assume anything about a forcing or measures, we are looking at sequences meeting club subsets instead on a tail end. We call such sequences *pseudo Prikry.* Below we give a similar abstraction of supercompact Prikry. Supercompact Prikry with respect to a normal measure U on $\mathcal{P}_\kappa(\kappa^{+m})$ adds a generic ω-sequence through $\mathcal{P}_\kappa(\kappa^{+m})$. And actually, $\langle x_n \mid n < \omega\rangle$ is a geniric sequence if and only if for every $A \in U$, for all large n, $x_n \in A$. The next theorem abstracts this notion, by considering club subsets of $\mathcal{P}_\kappa(\kappa^{+m})$:

Below $V \subseteq W$ are two transitive models of set theory. When we use some set theoretic terminology or notation like "a regular cardinal", δ^+ etc. we shall mean it in the sense of V, unless otherwise stated. Also, for a set of ordinals x and a regular cardinal $\tau \leq \sup(x)$, we define the characteristic function $\chi_x(\tau) := \sup(x \cap \tau)$.

THEOREM 2.3. ([**11**]) *Suppose that* κ *is a regular cardinal in* V, *and* $m < \omega$ *is such that in* W $\kappa, \kappa^+, \dots \kappa^{+m}$ *all have cofinality* ω, $(\omega_1)^W < \kappa$, *and* $(\kappa^+)^W = \kappa^{+m+1}$. *In* V *let* $\langle D_\alpha \mid \alpha < \kappa^{+m+1}\rangle$ *be a sequence of clubs of* $P_\kappa(\kappa^{+m})$. *Then in* W *there is a sequence* $\langle x_n \mid n < \omega\rangle$ *of elements of* $P_\kappa(\kappa^{+m})$, *such that for every* $\alpha < \kappa^{+m+1}$ *for large enough* $n < \omega$, $x_n \in D_\alpha$.

Moreover, if $\lambda < \kappa$ *is an uncountable regular cardinal in* W, *we can assume that for* $n < \omega$ *and* $k \leq m$, $\mathrm{cf}^W(\chi_{x_n}(\kappa^{+k})) \geq \lambda$.

In [**11**] it was asked whether the above can be generalized to infinite gaps. It turns out that the answer is yes. Moti Gitik in [**6**] showed the existence of a pseudo Prikry sequence for meeting clubs in $\mathcal{P}_\kappa(\tau)$, in the case of $V \subset W$, $V \models \kappa < \tau$ are both regular, and in W, $\tau = \bigcup_n z_n$, where each $z_n \in (\mathcal{P}_\kappa(\tau))^V$ (and some mild additional cardinal arithmetic assumptions). This generalizes Theorem 2.3. Below is a summary of his results.

THEOREM 2.4. *(Gitik* [**6**]*) Suppose that* $V \subset W$ *are transitive models of set theory, and* κ *is regular uncountable in* V, *such that* $\mu = (\tau^+)^V = (\kappa^+)^W$ *and* $(\dagger) W \models \tau = \bigcup_n x_n$, *each* $x_n \in \mathcal{P}_\kappa(\tau)$. *Assume also that* $V \models \tau^{<\tau} = \tau$ *and* $\tau \geq ((2^\omega)^+)^W$.

Then for every family $\langle C_\alpha \mid \alpha < \mu\rangle \in V$ *of clubs subsets of* $\mathcal{P}_\kappa(\tau)$, *there is a pseudo Prikry sequence.*

REMARK 1. (1) The assumption that τ is regular in V is essential. If τ is singular of cofinality less than κ, Gitik showed that this always will fail, just like it does in the model in [**7**].

(2) Note that the assumption above (†) implies that all V-regular cardinals in the interval $[\kappa, \tau]$ are singularized to have countable cofinality in W. This for example, will not be the case in a Namba type forcing extension.

(3) If $\tau < \kappa^{+\omega_1}$, then the converse also holds, as shown below.

LEMMA 2.5. *(Gitik [**6**]) Suppose that $V \subset W$ are transitive models of set theory. Suppose that κ is a regular cardinal in V, and for some countable δ, $\mu := (\kappa^+)^W < (\kappa^{+\omega_1})^V$, and also that every V-regular τ in the interval $[\kappa, \mu)$ is singularized to have cofinality ω in W. Then for every $\delta < \omega_1$, such that $\kappa^{+\delta} < \mu$, there is $\langle x_n \mid n < \omega \rangle \in W$, such that every $x_n \in (\mathcal{P}_\kappa(\kappa^{+\delta}))^V$, such that $(\kappa^{+\delta})^V = \bigcup_n x_n$.*

PROOF. By induction on δ. If $\delta = 0$, then it is clear. Suppose that $\delta = \eta + 1$ and the conclusion holds for η. I.e. $(\kappa^{+\eta})^V = \bigcup_n x_n$ for some $x_n \in (\mathcal{P}_\kappa(\kappa^{+\eta}))^V$. In W, let $(\kappa^{+\delta})^V = \sup_n \alpha_n$, with each $|\alpha_n|^V = \kappa^{+\eta}$. Suppose that this is witnessed by bijections $g_n : \kappa^{+\eta} \to \alpha_n$. Let $y_{n,m} = g_n{}''x_m$. Then each $y_{n,m} \in (\mathcal{P}_\kappa(\kappa^{+\delta}))^V$, and $(\kappa^{+\delta})^V = \bigcup_{n,m} y_{n,m}$.

Next, suppose that δ is limit. I.e. $\delta = \sup_n \delta_n$ and by induction, for each n there is a sequence $\langle x^n_k \mid k < \omega \rangle$ of elements in $(\mathcal{P}_\kappa(\kappa^{+\delta_n}))^V$, such that $(\kappa^{+\delta_n})^V = \bigcup_k x^n_k$. Let $y_m = \bigcup_{n,k<m} x^n_k$, which is in $(\mathcal{P}_\kappa(\kappa^{+\delta}))^V$. Then $\bigcup_m y_m = \bigcup_{n,k<\omega} x^n_k = (\kappa^{+\delta})^V$. □

In particular if every V-regular cardinal between $[\kappa, \kappa^{+n}]$ is singularized to countable cofinality, then $(\kappa^{+n})^V$ can be written as the union of ω many elements in $(\mathcal{P}_\kappa(\kappa^{+n}))^V$. Same if every V-regular cardinal between $[\kappa, \kappa^{+\omega}]$ is singularized to countable cofinality.

The existence of pseudo Prikry sequences was generalized to the case for uncountable cofinalities by Lambie-Hanson [**9**]. In that paper he also generalized some of the above theorems to the case when the new successor of κ was a successor of a singular in V, and to the case of existence of *diagonal* pseudo-Prikry sequences.

In the above results, the recurring assumption is that regular cardinals in some interval $[\kappa, \nu]$, ν regular, are singularized to the same cofinality. By a theorem of Shelah, the last such cardinal (i.e. ν) will always have the same cofinality as its cardinality:

THEOREM 2.6. *(Shelah) Suppose that $V \subset W$ are transitive models of set theory, ν is a regular cardinal in V, which is collapsed in W, so that $(\nu^+)^V$ is preserved. Then $W \models \mathrm{cf}(\nu) = \mathrm{cf}(|\nu|)$.*

PROOF. Let $\mu := (\nu^+)^V$, and let $\langle A_\alpha \mid \alpha < \mu \rangle \in V$ be a family of almost disjoint unbounded subsets of ν. I.e. for every $\beta < \mu$, there are $\delta_\alpha < \kappa$, for all $\alpha < \beta$, such that $\langle A_\alpha \setminus \delta_\alpha \mid \alpha < \beta \rangle$ are pairwise disjoint. Such a sequence is called *an almost disjoint sequence* and it always exists for a regular cardinal ν (for a proof, see [**14**], page 440).

Let $\kappa := |\nu|^W$ and let $g : \kappa \to \nu$ be a bijection. Suppose for contradiction that in W, $\mathrm{cf}(\nu) \neq \mathrm{cf}(\kappa)$. Then for all $\alpha < \mu$, there is $\tau_\alpha < \kappa$, such that $g''\tau_\alpha$ is unbounded in A_α. Since μ remains a regular cardinal in W, the map $\alpha \mapsto \tau_\alpha$ is constant on an unbounded set. Let $I \subset \mu$ be unbounded, and $\tau < \kappa$ be such that for all $\alpha \in I$, $\tau = \tau_\alpha$. Let $\beta < \mu$, be such that $W \models |I \cap \beta| > \tau$. We can choose such a point since in W, $|I| = \kappa^+$. Let $\langle \delta_\alpha \mid \alpha < \beta \rangle$ be such that setting $B_\alpha := A_\alpha \setminus \delta_\alpha$, we have that $\langle B_\alpha \mid \alpha < \beta \rangle$ are pairwise disjoint.

But then since $g''\tau$ is unbounded in each B_α for $\alpha \in I$, we have that $\langle g''\tau \cap B_\alpha \mid \alpha \in I \cap \beta \rangle$ are more than $|\tau|$ many pairwise disjoint non empty subsets of $g''\tau$. Contradiction. $\square$

It is an old open problem whether this is true when ν is singular in V. In that case there is not always an almost disjoint sequence of unbounded subsets of ν. We will discuss this more in the last section. It is also open whether cardinals in a finite interval $[\kappa, \kappa^{+n}]$ can be singularized to nonuniform cofinalities.

3. Consequences on square properties

In this section we survey some application of the above theorems on singularizing cardinals and square properties. This is especially useful if we want to understand how much compactness can hold in the presence of failure the singular cardinal hypothesis (SCH). Recall that SCH holds at a singular strong limit cardinal κ if $2^\kappa = \kappa^+$. More generally, for a singular κ, SCH at κ is the statement that if $2^{\mathrm{cf}(\kappa)} < \kappa$, then $\kappa^{\mathrm{cf}(\kappa)} = \kappa^+$. The standard way of violating SCH involves singularizing a cardinal. So, we are especially interested in what square sequences are added then.

Cummings and Schimmerling showed in [**3**] that after Prikry forcing at κ, $\square_{\kappa,\omega}$ holds in the generic extension.

Let us recall some definitions. Square at κ, $\square_\kappa$, states that there exists a sequence $\langle C_\alpha \mid \alpha < \kappa^+ \rangle$ such that each C_α is a club subset of α, $o.t.(C_\alpha) \leq \kappa$, and if $\delta \in \lim C_\alpha$, then $C_\alpha \cap \delta = C_\delta$. There are various weakenings allowing multiple guesses for the clubs at each point. More precisely, the principle $\square_{\kappa,\lambda}$ states that there is a sequence $\langle \mathcal{C}_\alpha \mid \alpha < \kappa^+ \rangle$, such that:

(1) $1 \leq |\mathcal{C}_\alpha| \leq \lambda$,
(2) every $C \in \mathcal{C}_\alpha$ is a club in α, $o.t.(C) \leq \kappa$,
(3) if $C \in \mathcal{C}_\alpha$ and $\delta \in \lim(C)$, then $C \cap \delta \in \mathcal{C}_\delta$.

Note that $\square_{\kappa,1}$ is just the usual square at κ, and $\square_{\kappa,\kappa}$ is weak square at κ, $\square^*_\kappa$. By the results in Gitik [**5**] and independently in Džamonja-Shelah [**4**], it was implicit that if $V \subset W$ are transitive class models of ZFC such that κ is an inaccessible cardinal in V, singular of countable cofinality in W, and $(\kappa^+)^V = (\kappa^+)^W$, then $W \models \square_{\kappa,\omega}$. It turns out that results from Cummings-Schimmerling and Gitik/Dzamonja/Shelah can be generalized to show $\square_{\kappa,\omega}$ even if we collapse some cardinals above κ:

THEOREM 3.1. *([**11**]) Suppose $V \subset W$ and $\kappa < \nu$ are such that, ν is a regular cardinal in V, κ is a singular cardinal in W of countable cofinality, for all V-regular cardinals τ, with $\kappa \leq \tau \leq \nu$, $W \models \mathrm{cf}(\tau) = \omega$, and $(\nu^+)^V = (\kappa^+)^W$. Then in W, $\square_{\kappa,\omega}$ holds.*

PROOF. Work in V. Denote $\mu := \nu^+$. Let χ be some big enough cardinal, and $<_\chi$ be a well-ordering of H_χ. For $\alpha < \mu$ with $\mathrm{cf}(\alpha) < \kappa$, let $\langle M^\alpha_\delta \mid \delta < \nu \rangle$ be a continuous $\subset$-increasing sequence of elementary submodels of $\langle H_\chi, \in, <_\chi ... \rangle$, such that:

(1) $\alpha, \kappa \in M^\alpha_0$,
(2) for each $\delta < \nu$, $|M^\alpha_\delta| < \nu$,

(3) for each $\delta < \nu$, $M^\alpha_\delta \cap \nu \in \nu$.

CLAIM 3.2. For all $\alpha < \mu$, $\delta < \nu$ with $\mathrm{cf}(\delta) > \omega$, if $M^\alpha_\delta \cap \alpha$ is cofinal in α, then $M^\alpha_\delta \cap \alpha$ is ω-closed.

PROOF. Otherwise, let $\langle \beta_i \mid i < \omega \rangle$ be an increasing sequence of points in $M^\alpha_\delta \cap \alpha$, such that $\beta := \sup_i \beta_i \notin M^\alpha_\delta \cap \alpha$. Let β^* be the least ordinal in $M^\alpha_\delta \cap \alpha$ above β. If $\lambda := \mathrm{cf}(\beta^*) < \nu$, then $\lambda \in M^\alpha_\delta \cap \nu \in \nu$. Then $\lambda + 1 \subset M^\alpha_\delta$, and so M^α_δ is cofinal in β^*, which is a contradiction.

It follows that the cofinality of β^* must be ν. Say $M^\alpha_\delta \models h : \nu \to \beta^*$ is cofinal. For every i, let $\beta^*_i \in \mathrm{ran}(h) \setminus \beta_i$. The order type of $\mathrm{ran}(h) \cap M^\alpha_\delta$ is $\nu \cap M^\alpha_\delta$. And $\mathrm{cf}(\nu \cap M^\alpha_\delta) = \mathrm{cf}(\delta) > \omega$. So there is something in $\mathrm{ran}(h) \cap M^\alpha_\delta$ above β. Contradiction with the choice of β^*. ☐

For $\alpha < \mu$ define $D_\alpha := \{M^\alpha_\delta \cap \nu \mid \delta < \nu\}$. We claim that this is club in ν. D_α is closed by construction since the sequence of models is continuous. It is unbounded, since $\nu \subset \bigcup_\delta M^\alpha_\delta$. Note that this also implies that $\alpha \subset \bigcup_\delta M^\alpha_\delta$, since the $<_\chi$-least function in H_χ from ν onto α is in each M^α_δ.

By Theorem 2.1 there is a sequence $\langle \lambda_i \mid i < \omega \rangle \in W$, such that for all large i, $\lambda_i \in D_\alpha$ and for all i, $\mathrm{cf}^W(\lambda_i) > \omega$. For $\alpha < \mu$ with $\mathrm{cf}^V(\alpha) < \kappa$, define $\mathcal{C}_\alpha := \{\overline{M^\beta_\delta \cap \alpha} \mid \alpha \leq \beta < \mu, \delta < \nu, M^\beta_\delta \text{ is cofinal in } \beta \text{ and } \alpha, (\exists i) M^\beta_\delta \cap \nu = \lambda_i\}$. Here $\overline{M^\beta_\delta \cap \alpha}$ denotes the closure of $M^\beta_\delta \cap \alpha$. Otherwise, if $\mathrm{cf}^V(\alpha) = \kappa$, and so $\mathrm{cf}^W(\alpha) = \omega$, we let $\mathcal{C}_\alpha$ be a singleton of some ω-sequence cofinal in α.

CLAIM 3.3. $1 \leq |\mathcal{C}_\alpha| \leq \omega$.

PROOF. Each $\mathcal{C}_\alpha$ is nonempty because for all large i, $\lambda_i \in D_\alpha$, and so $\overline{M^\alpha_\delta} \in \mathcal{C}_\alpha$. Now suppose $\beta, \beta', \delta, \delta'$ are such that:

- $M^\beta_\delta \cap \nu = M^{\beta'}_{\delta'} \cap \nu = \lambda_i$,
- $M^\beta_\delta, M^{\beta'}_{\delta'}$ are both cofinal in α.

We claim that $M^\beta_\delta \cap \alpha = M^{\beta'}_{\delta'} \cap \alpha$.

CASE 1. If $\mathrm{cf}(\alpha) = \omega$, then by Claim 3.2, $\alpha \in M^\beta_\delta, M^{\beta'}_{\delta'}$. By elementarity, the $\prec_\chi$-least function from ν onto α is in $M^\beta_\delta \cap M^{\beta'}_{\delta'}$, and the result follows.

CASE 2. Suppose that $\mathrm{cf}(\alpha) > \omega$. Since $\mathrm{cf}(\delta) = \mathrm{cf}(\delta') = \mathrm{cf}(\lambda_i) > \omega$, then by Claim 3.2, $M^\beta_\delta \cap \alpha, M^{\beta'}_{\delta'} \cap \alpha$ are both ω-clubs in α, and so $M^\beta_\delta \cap M^{\beta'}_{\delta'} \cap \alpha$ is an ω-club in α. By the argument in case 1, for any $\eta \in M^\beta_\delta \cap M^{\beta'}_{\delta'} \cap \alpha$, $M^\beta_\delta \cap \eta = M^{\beta'}_{\delta'} \cap \eta$, and so $M^\beta_\delta \cap \alpha = M^{\beta'}_{\delta'} \cap \alpha$.

It follows that the number of possible elements in $\mathcal{C}_\alpha$ is at most the number of possible intersection with ν, which is ω, by definition of $\mathcal{C}_\alpha$. ☐

Next we use $\langle \mathcal{C}_\alpha \mid \alpha < \mu \rangle$ to obtain a $\square_{\kappa,\omega}$ sequence. In W fix is a sequence $\langle F_\beta \mid \beta < \nu \rangle$, such that each F_β is a club in β of order type at most κ, and for $\delta \in \lim(F_\beta)$, $F_\beta \cap \delta = F_\delta$. Enumerate $\mathcal{C}_\alpha := \{C^\alpha_n \mid n < \omega\}$ and $C^\alpha_n = \{\gamma^{\alpha,n}_\xi \mid \xi < \nu^\alpha_n\}$, where $\nu^\alpha_n = o.t.(C^\alpha_n) < \nu$. Define $E^\alpha_n := \{\gamma^{\alpha,n}_\xi \mid \xi \in F_{\nu^\alpha_n}\}$ and let $\mathcal{E}_\alpha = \{E^\alpha_n \mid n < \omega\}$. $\langle \mathcal{E}_\alpha \mid \alpha < \mu \rangle$ is coherent because both $\langle F_\beta \mid \beta < \nu \rangle$ and $\langle \mathcal{C}_\alpha \mid \alpha < \mu \rangle$ are coherent. Then $\langle \mathcal{E}_\alpha \mid \alpha < \mu \rangle$ is a $\square_{\kappa,\omega}$ sequence. ☐

On the other hand, Gitik and Sharon [7] showed that failure of weak square at a singular is consistent with the failure of SCH. They constructed a forcing where κ is singularized, $(\kappa^{+\omega+1})^V$ becomes the successor of κ, and weak square of κ fails in the generic extension. Later, Neeman [12] showed that failure of SCH at κ is consistent even with the tree property at κ^+. The tree property at λ states that every tree of height λ and levels of size less than λ has a cofinal branch, and in the case of $\lambda = \kappa^+$, it implies failure of weak square at κ.

So the assumption that $\mu := (\kappa^+)^W$ was a successor of a V-regular cardinal is necessary. What about when μ is inaccessible in V? In that case, by standard arguments one can usually define a weak square sequence.

PROPOSITION 3.4. *Suppose $V \subset W$ and $\kappa < \mu$ are cardinals in both V, W, such that, $V \models \mu$ is inaccessible, κ is regular; and $W \models \mathrm{cf}(\kappa) = \omega, \kappa^+ = \mu$, and for all V-regular cardinals τ, with $\kappa \leq \tau < \mu$, $W \models \mathrm{cf}(\tau) = \omega$. Then in W, $\square^*_\kappa$ holds.*

PROOF. For $\alpha < \mu$ with $\mathrm{cf}^V(\alpha) < \kappa$, let $\mathcal{C}_\alpha$ be the set of all clubs $C \subset \alpha$, such that $C \in V$, $o.t.(C) < \kappa$. Since μ is inaccessible, $|\mathcal{C}_\alpha| < \mu$. And for $\alpha < \mu$ with $\mathrm{cf}^V(\alpha) = \kappa$, and so $\mathrm{cf}^W(\alpha) = \omega$, let $\mathcal{C}_\alpha$ be a singleton containing some ω-cofinal sequence through α (in W). Then $\langle \mathcal{C}_\alpha \mid \alpha < \mu \rangle$ is a weak square sequence at κ. $\square$

Woodin asked whether one can violate SCH at $\aleph_\omega$ and at the same time have failure of weak square at $\aleph_\omega$. As pointed out above, Gitik and Sharon showed that this is possible for some singular cardinal. Then they bought down their result to $\aleph_{\omega^2}$:

THEOREM 3.5. *(Gitik-Sharon [7]) Suppose that in V, κ is supercompact, $\mu := \kappa^{+\omega+1}$. Then there is a forcing extension W, where $\kappa = \aleph_{\omega^2}$, $\mu = \aleph_{\omega^2+1}$, SCH fails at κ and weak square at κ also fails.*

It is still open whether the above can be achieved for $\aleph_\omega$, and answer Woodin's original question. By the results discussed above, if one starts with a regular (and possibly large) cardinal κ in the ground model, and singularizes it, then the new successor of κ should have been a successor of a singular in the ground model. More precisely:

COROLLARY 3.6. *Suppose $V \subset W$, κ is a regular cardinal in V, a singular cardinal in W, $\mu = (\kappa^+)^W$, all V-regular cardinals in the interval $[\kappa, \mu)$ are have cofinality ω in W, and $\neg\square^*_\kappa$ in W. Then μ must be the successor of a singular cardinal in V.*

4. Open problems

Many of the theorems on pseudo Prikry sequences do generalize for uncountable cofinalities. But it remains open whether applications on square properties as in Theorem 3.1 hold for uncountable cofinalities. For example, in the simplest case, one can ask:

QUESTION 2. Suppose that $V \subset W$, κ is inaccessible in V and singular in W with $\mathrm{cf}^W(\kappa) = \tau > \omega$, $(\kappa^+)^V = (\kappa^+)^W$. Does $\square_{\kappa,\tau}$ hold in W?

The main obstacle here would be to deal with the points whose V-cofinality is κ when trying to build a square sequence. In the countable cofinality case, one can take an ω sequence, and since there are no limit points, that suffices. We note that in the uncountable case, using the methods of Theorem 3.1, one can build a partial $\square_{\kappa,\mathrm{cf}(\kappa)}$ sequence in the outer model, over the set of points of V-cofinality less than κ.

QUESTION 3. Suppose $V \subset W$ are models of set theory. If κ is regular in V, and in W, κ is a singular cardinal of cofinality ω, $(\kappa^{+n+2})^V = (\kappa^+)^W$, then what is the cofinality of $(\kappa^+)^V, ..., (\kappa^{+n})^V$ in W?

By Shelah, we know the cofinality of the last cardinal collapsed, $(\kappa^{n+1})^V$, will be ω in W. And with supercompact Prikry forcing we can arrange $V \subset W$, where all of the cardinals $(\kappa^+)^V, ..., (\kappa^{+n})^V$ will have countable cofinality in W. So the question is, is this always the case? By lemma 2.5, this is equivalent to asking whether κ^{+n+1} is the union of a sequence $\langle x_n \mid n < \omega\rangle$, where each $x_n \in (\mathcal{P}_\kappa(\kappa^{+n+1}))^V$.

Next, we turn to the problem mentioned at the end of Section 2.

QUESTION 4. Can we have models $V \subset W$, such that $(\aleph_{\omega+1})^V = (\aleph_2)^W$?

Let us first note that if such a pair $V \subset W$ exist, CH must fail in W, since setting $\nu := (\aleph_\omega)^V$, we have $(\aleph_2)^W = \mu \leq \nu^\omega = \aleph_1^\omega = 2^\omega$, as computed in W. Moreover, if in V there is an almost disjoint sequence of unbounded subsets of $\aleph_\omega$, then the same proof as in Shelah's theorem will work to give a negative answer. Note that the existence of such a sequence follows from square. So a positive answer would require large cardinals. By Cummings [**1**] a positive answer would imply that there is *a bad scale* in V at $\aleph_\omega$ and that W cannot be a $(\aleph_{\omega+1})^V$-c.c. extension of V. More generally, he showed that if $V \subset W$ are such that $\aleph_{\omega+1}^V = \aleph_{n+1}^W$ for some $n > 0$, then in V there is a stationary $S \subset \aleph_{\omega+1}$ of bad points of cofinality ω_n. By pcf facts almost every point in $\aleph_{\omega+1}$ of cofinality $\geq \omega_4$ is good ([**15**]). It follows that $\aleph_{\omega+1}$ cannot be turned into $\aleph_5$ or more. The following remain open:

QUESTION 5.
(1) Can we have a stationary set of bad points of cofinality ω_2, or of cofinality ω_3?
(2) Is it true that almost all points cofinality $\geq \omega_4$ are approachable (which is a strengthening of good points)?

We finish with the following conjecture by Moti Gitik [**6**]:

CONJECTURE 1. *Suppose that $V \subset W$ have the same cardinals on a final segment, and κ is a regular cardinal in V, singular with cofinality ω in W, and $(\aleph_1)^V = (\aleph_1)^W$. Then for some intermediate model $V \subset V' \subset W$ with $V' \models \mathrm{cf}(\kappa) = \omega$, V' is a forcing extension of V by either Prikry, Namba, or Woodin tower forcing.*

References

[1] James Cummings, *Collapsing successors of singulars*, Proc. Amer. Math. Soc. **125** (1997), no. 9, 2703–2709, DOI 10.1090/S0002-9939-97-03995-6. MR1416080

[2] James Cummings, Matthew Foreman, and Menachem Magidor, *Squares, scales and stationary reflection*, J. Math. Log. **1** (2001), no. 1, 35–98, DOI 10.1142/S021906130100003X. MR1838355

[3] James Cummings and Ernest Schimmerling, *Indexed squares*, Israel J. Math. **131** (2002), 61–99, DOI 10.1007/BF02785851. MR1942302

[4] Mirna Džamonja and Saharon Shelah, *On squares, outside guessing of clubs and $I_{<f}[\lambda]$*, Fund. Math. **148** (1995), no. 2, 165–198, DOI 10.4064/fm-148-2-165-198. MR1360144

[5] Moti Gitik, *Some results on the nonstationary ideal. II*, Israel J. Math. **99** (1997), 175–188, DOI 10.1007/BF02760681. MR1469092

[6] Moti Gitik, *A note on sequences witnessing singularity, following Magidor and Sinapova*, MLQ Math. Log. Q. **64** (2018), no. 3, 249–253, DOI 10.1002/malq.201700067. MR3845131

[7] Moti Gitik and Assaf Sharon, *On SCH and the approachability property*, Proc. Amer. Math. Soc. **136** (2008), no. 1, 311–320, DOI 10.1090/S0002-9939-07-08716-3. MR2350418

[8] Thomas Jech, *Set theory*, Springer Monographs in Mathematics, Springer-Verlag, Berlin, 2003. The third millennium edition, revised and expanded. MR1940513

[9] Chris Lambie-Hanson, *Pseudo-Prikry sequences*, Proc. Amer. Math. Soc. **146** (2018), no. 11, 4905–4920, DOI 10.1090/proc/13996. MR3856157

[10] Menachem Magidor, *Changing cofinality of cardinals*, Fund. Math. **99** (1978), no. 1, 61–71, DOI 10.4064/fm-99-1-61-71. MR465868

[11] Menachem Magidor and Dima Sinapova, *Singular cardinals and square properties*, Proc. Amer. Math. Soc. **145** (2017), no. 11, 4971–4980, DOI 10.1090/proc/13650. MR3692010

[12] Itay Neeman, *Aronszajn trees and failure of the singular cardinal hypothesis*, J. Math. Log. **9** (2009), no. 1, 139–157, DOI 10.1142/S021906130900080X. MR2665784

[13] Saharon Shelah, *Cardinal arithmetic*, Oxford Logic Guides, vol. 29, The Clarendon Press, Oxford University Press, New York, 1994. Oxford Science Publications. MR1318912

[14] Saharon Shelah, *Proper forcing*, Lecture Notes in Mathematics, vol. 940, Springer-Verlag, Berlin-New York, 1982. MR675955

[15] Assaf Sharon and Matteo Viale, *Some consequences of reflection on the approachability ideal*, Trans. Amer. Math. Soc. **362** (2010), no. 8, 4201–4212, DOI 10.1090/S0002-9947-10-04976-7. MR2608402

Department of Mathematics, Statistics, and Computer Science, University of Illinois at Chicago, Chicago, Illinois 60613

Email address: `sinapova@uic.edu`

Contemporary Mathematics
Volume **752**, 2020
https://doi.org/10.1090/conm/752/15136

A brief account of recent developments in inner model theory

Nam Trang

The goal of this survey paper is to give an overview of recent developments in inner model theory. We discuss several most important questions in the field and relevant work (partially) answering these questions roughly from the early 2000's to present. Most of the major questions are still open. These questions drive future research directions in inner model theory. We mention some of the most prominent projects pertaining to these.[1]

Inner model theory starts with Gödel's work on *the constructible universe* L in 1938. L is the first example of a "canonical inner model". There is no official definition of "canonical inner models", but in the case of L, it is canonical because it is the minimal transitive proper class model of ZFC and it satisfies various benchmark properties that one would want canonical inner models to satisfy, such as *the Generalized Continuum Hypothesis* (GCH), $\diamondsuit$ and $\Box$ (the latter two are due to Jensen in his seminal work [**8**]). L, however, does not admit very large large cardinals. For instance, Scott, [**31**], has shown that L cannot contain a measurable cardinal.

A main theme in inner model theory after Gödel's work and Scott's result is the search for canonical inner models that can accommodate large cardinals beyond those accommodated by L. This is sometimes known as the *Gödel program.* An immediate target is a measurable cardinal. Models of the form $L[U]$ that satisfy "U is a κ-complete, normal, nonprincipal measure on a cardinal κ" have been extensively studied. Most notably, it can be shown that $L[U]$ satisfies the GCH (this is due to Silver [**32**]), $\diamondsuit, \Box$. Kunen, in [**11**], develops the influential theory of M-ultrapowers (based on previous work by H. Gaifman) and uses it to prove fundamental results on the canonicity of $L[U]$. In particular, he shows that $L[U]$ is unique internally (namely U is the unique normal measure on κ in $L[U]$) and for $i \in \{0,1\}$, if U_i is a filter on κ such that $L[U_i] \vDash$ "U_i is a normal measure on κ", then $L[U_0] = L[U_1]$ and $U_0 \cap L[U_0] = U_1 \cap L[U_1]$. At the heart of Kunen's analysis is the comparability of two models of the form $L[U]$, i.e. for $i \in \{0,1\}$, if $M_i = L[U_i] \vDash$ "U_i is a normal measure on κ_i", then either M_0 is an iterated

2010 *Mathematics Subject Classification.* Primary 03E15, 03E45, 03E60.

The author was supported through grant DMS-1849295.

[1]This paper is not meant to be a comprehensive encyclopedia of the subject. We apologize for any omissions of recent works that are not mentioned here.

ultrapower of M_1 or M_1 is an iterated ultrapower of M_0 (via the measure U_i and its images).

Mitchell, in **[14]** and **[15]**, generalizes Kunen's analysis to models of the form $L[\vec{U}]$ where $\vec{U}$ is a sequence of measures. Mitchell's definition of *an extender* (see Section 4 for a definition) then plays a key role in subsequent searches for canonical inner models of larger large cardinals such as strong cardinals (cf. **[5]**) and Woodin cardinals and beyond (cf. **[13]**, **[17]**).[2] These models are typically called *fine-structural extender models* (*premice*) and are of the form $L[\vec{E}]$ where $\vec{E}$ is a nice sequence of extenders (cf. **[41]**, **[51]** for the precise meaning of "nice"). The analysis of these models relies on generalizing Jensen's fine structure in **[8]**[3], and proving *iterability* (iterable premice are called *mice*) and *comparison* for countable mice. To produce these models, one typically performs a *backgrounded construction* to produce a sequence $(\mathcal{N}_\xi : \xi \leq \Omega)$ of models (each of which is a fine-structural extender model) and show that some such $\mathcal{N}_\xi$ satisfies the desired large cardinal axiom. Iterability takes the form: suppose for some $\xi \leq \Omega$, $\pi : M \to \mathcal{N}_\xi$ is such that M is countable, transitive, π is sufficiently elementary, then M is sufficiently iterable, that is, there is an *iteration strategy* Λ for M acting on (normal) trees of length $\leq \omega_1$; we say that Λ witnesses M is $\omega_1 + 1$-iterable. We say that $\mathcal{N}_\xi$ is *countably iterable*. This forms of iterability is sufficient to prove *comparison*: if M and N are countable, transitive and embeddable into some $\mathcal{N}_\xi$ and $\mathcal{N}_\zeta$ respectively, letting Σ be an $\omega_1 + 1$-iteration strategy for M and Λ an $\omega_1 + 1$-iteration strategy for N, then there are countable (normal) trees $\mathcal{T}$ according to Σ with last model $\mathcal{M}^\mathcal{T}$ and $\mathcal{U}$ according to Λ with last model $\mathcal{M}^\mathcal{U}$, such that either $\mathcal{M}^\mathcal{T}$ is an *initial segment* of $\mathcal{M}^\mathcal{U}$ (denoted $\mathcal{M}^\mathcal{T} \trianglelefteq \mathcal{M}^\mathcal{U}$) or vice versa.

Inner models in the papers discussed above are typically constructed via a procedure called *the full backgrounded construction*. This means the extenders giving rise to the model $L[\vec{E}]$ come from extenders in V (and these in turn exist because one assumes an appropriate large cardinal hypothesis holds in V). One could also ask whether (canonical) inner models of large cardinals exist without assuming large cardinals exist in V. In which case, *partial backgrounded constructions* or *K^c-constructions* are used. Some remarkable results in computing lower bound consistency strength have been obtained using this kind of constructions, for instance **[36]** and **[9]** show Con(ZFC + there is a presaturated ideal on ω_1) implies Con(ZFC + there is a Woodin cardinal)[4], **[7]** shows *the Proper Forcing Axiom* (PFA) implies the existence of inner models with very large cardinals and many others.

Finally, the discovery that large cardinals and determinacy are intimately connected opens up other venues where some of the most important and exciting research in modern inner model theory is being done. More precisely, the research along this line belongs to an emerging field called *descriptive inner model theory* (DIMT). DIMT uses tools from both *descriptive set theory* (DST) and inner model theory (IMT) to study and deepen the connection between canonical models of large cardinals and canonical models of *the axiom of determinacy* (AD). One of the first significant developments in DIMT comes in the 1980's with works of Martin,

[2]It has been observed that linear iterations are no longer sufficient for the analysis of models at the level of Woodin cardinals. Iteration trees are what one needs.

[3]There are two main such generalizations: the Mitchell-Steel fine structure in **[13]** and Jensen's $*$-fine structure, cf. **[51]**. These works define exactly what "nice" means.

[4]The converse had been obtained by S. Shelah.

Steel, Woodin and others (cf [**12**] and [**41**]); their work, for instance, shows that one can construct models of AD (e.g. they showed AD holds in $L(\mathbb{R})$) assuming large cardinal axioms (those that involve the crucial notion of Woodin cardinals) and conversely, one can recover models of large cardinal axioms from models of AD. The key to uncover these connections is to analyze structure theory of models of AD, and in particular their HOD. The work of Steel and Woodin mentioned above shows that HOD of $L(\mathbb{R})$ is fine-structural (in particular, it satisfies $\mathsf{GCH}, \Box, \diamondsuit$) but it is not an extender model; rather, HOD of $L(\mathbb{R})$ is what's called a *strategic extender model*, a model constructed from a predicate coding a nice extender sequence and a predicate coding a (partial) iteration strategy of its own initial segments. Much of DIMT aims to extend this analysis to larger AD^+ models, studying various kinds of strategic extender models and developing ways of constructing them. For instance, the conjecture that "PFA has the exact consistency strength as that of a supercompact cardinal" is one of the most longstanding and important open problems in set theory. Techniques recently developed in DIMT enable us to make significant progress in calibrating the consistency strength lower bound for PFA (and many other theories). See for instance[**37**], [**22**], [**44**]. Some of the main ingredients that come into these theorems and their applications include the HOD analysis and the core model induction techniques. We will discuss these ongoing developments in more details.

To us, the following are the most fundamental questions and projects facing inner model theorists in the 21-st century. We will address progress and partial answers to these questions in this survey paper.

(1) Prove iterability of short-extender models.
(2) Does HOD of natural AD^+ models admit fine structure? More generally, analyze the (extender, hybrid) mice in AD^+ models.
(3) Compute lower bound consistency strength of PFA and other strong theories.
(4) Develop a fine structure theory of long-extender models and understand canonical models at the level of supercompact cardinals and above.

1. Iterability

Proving iterability is arguably the most important problem in inner model theory. Suppose a backgrounded construction has produced models $(\mathcal{N}_\xi : \xi \leq \alpha)$. One typically is interested in continuing the construction, i.e. produce model $\mathcal{N}_{\alpha+1}$. By definition, $\mathcal{N}_{\alpha+1} = \mathcal{J}_1(\mathcal{M}_\alpha)$, where $\mathcal{M}_\alpha$ is the (fine-structural) core of $\mathcal{N}_\alpha$. To prove that $\mathcal{M}_\alpha$ exists, i.e. $\mathcal{M}_\alpha$ is solid and universal, we need to show that $\mathcal{N}_\alpha$ is countably iterable. For *fully backgrounded constructions*, that is extenders used in constructing the models $\mathcal{N}_\alpha$'s come from V-extenders, then the strongest iterability results are proved in [**17**]. Roughly, [**17**] shows that assuming there is a Woodin cardinal which is a limit of Woodin cardinals, then there is ξ such that $\mathcal{N}_\xi$ has a Woodin cardinal which is a limit of Woodin cardinals and for every $\xi' \leq \xi$, $\mathcal{N}_{\xi'}$ is countably iterable. For partially backgrounded constructions, or K^c-constructions, similar results are obtained only for a much more modest large cardinal strength. The strongest such result is in [**1**], where $\mathcal{N}_\xi$ satisfies "there is an inaccessible cardinal λ which is a limit of Woodin cardinals and $<$-λ-strong cardinals". This cardinal strength is typically called *non-domestic*. Improving iterability results proved in [**17**] for full-backgrounded constructions and in [**1**] for the K^c-construction is the main challenge going forward for inner model theorists.

Results such as that proved in [**1**] are used to obtain consistency lower bound for various theories. One examples was by Jensen-Steel and Shelah as mentioned above. In the proof of Theorem 1.1 (1), the crucial notion of *the core model* K is used. The existence of the core model K is given by sufficient iterability of K^c, which in turn follows from an anti-large cardinal assumption; for example, in Theorem 1.1 (1), the anti-large cardinal assumption is "there is no ZFC model with a Woodin cardinal".[5] In many applications, K is used (instead of K^c) because K exhibits many nice properties such as: K is invariant under set generic extensions (for appropriate forcings) while K^c is dependent on which universe it is defined (this is used significantly in the proof of Theorem 1.1 and similar applications), one can prove various useful covering properties involving K such as K computes successors of many cardinals correctly (see more below) etc.

Another example concerns consistency lower bound of PFA: using [**1**], [**7**] shows that Con(PFA) implies Con(there is an inaccessible cardinal λ which is a limit of Woodin cardinals and $<$-λ-strong cardinals). We summarize some of the main theorems along this line below.

THEOREM 1.1. (1) *(Jensen-Steel [**9**], Shelah) Con(*ZFC+ *there is a presaturated ideal on* ω_1*)* $\Leftrightarrow$ *Con(*ZFC+ *there is a Woodin cardinal).*

(2) *(Jensen-Schimmerling-Schindler-Steel [**7**]) Con(*PFA*) implies Con(*ZFC + *there is a non-domestic premouse).*

In both of the theorems above, the paradigm is roughly as follows: assume the desired large cardinal hypothesis fails to hold. Then K^c is sufficiently iterable. Then there is some appropriate regular cardinal κ such that $\mathrm{cof}((\kappa^+)^{K^c}) \geq \kappa$; in (1), K exists and one can even show $(\kappa^+)^K = \kappa^+$ for every singular cardinal κ. This contradicts the original assumption (PFA or the existence of a presaturated ideal on ω_1).

We will come back to the question concerning the consistency strength of PFA later when we discuss the core model induction and results improving (2) above.

The theories for extender models developed in [**13**] (Mitchell-Steel fine structure) and [**51**] (Jensen fine structure) allow one to construct extender models in the region of superstrong cardinals assuming appropriate large cardinals in V and countably substructures of V_δ (for an appropriate δ) are sufficiently iterable. For instance, Neeman and Steel, [**19**], have shown that assuming there is a cardinal δ which is Woodin, and SBH_δ holds,[6] then PFA implies the existence of extender models of a Π^2_1 subcompact cardinal, which slightly stronger than a superstrong. Another example (to be discussed in more details later) is assuming there are enough iterable (hod) mice, one can analyze HOD of AD^+ models below a superstrong cardinal (cf [**42**]).

2. Hybrid extender models and hod mice

As mentioned above, one of the most significant developments in inner model theory is through the work of Martin and Steel [**12**], and Woodin [**41**] in the 1980's

[5] L is the core model if $0^\sharp$ does not exist.

[6] SBH_δ states that countable $H \prec V_\delta$, the transitive collapse of H has an $\omega_1 + 1$-iteration strategy for a certain class of normal, non-overlapping trees using only short extenders.

that established the connections between large cardinals and determinacy. One way of formalizing this connection is through *the Mouse Set Conjecture* (MSC), which states that, assuming the AD or a more technical version of it (AD^+),[7] then whenever a real x is ordinal definable from a real y, then x belongs to a canonical model of large cardinal (mouse) over y. MSC conjectures that the most complicated form of definability can be captured by canonical structures of large cardinals. An early instance of this is well-known theorem of Shoenfield that every Δ^1_2 real is in the Gödel's constructible universe L. Another instance is a theorem of Woodin's that in the minimal class model containing all the reals, $L(\mathbb{R})$, if AD holds, then MSC holds. However, the full MSC is open and is one of the main open problems in DIMT. Instances of MSC have been proved in determinacy models constructed by their sets of reals, i.e. those of the form $L(\wp(\mathbb{R}))$, (much larger than $L(\mathbb{R})$) and these proofs typically obtain canonical models of large cardinals (mice) that capture the relevant ordinal definable real by analyzing HOD of the determinacy models. Hence, the key link between these two kinds of structures (models of large cardinals and models of determinacy) is the HOD of the determinacy models. The problem of analyzing HOD of AD^+ models has gradually grown into a central problem of inner model theory and spurs the development of descriptive inner model theory.

In the following, by models of AD^+, we mean models of AD^+ of the form $L(\wp(\mathbb{R}))$. One main reason for the importance of the HOD analysis is that it provides insights into the relationship between canonical inner models of large cardinals (pure extender models) and models of AD^+ (as alluded to earlier). Unlike HOD of ZFC models which are more or less intractable from the point of view of inner model theory, HOD of AD^+ models in some sense are very well-behaved and code up the AD^+ models in a canonical way; hence understanding HOD of such models provide more insights into the models themselves. We will give a more detailed discussion of the HOD analysis shortly. Another reason why studying HOD of AD^+ models is important is because the techniques developed in this study can be used to tackle a variety of problems, even those outside inner model theory. One of the first applications of the HOD analysis is Steel's proof that in $L(\mathbb{R})$, AD implies all regular cardinals below Θ are measurable (see [**41**]); here Θ is the supremum of ordinals α such that there is a surjection from $\mathbb{R}$ onto α. This result is not known to be proved by any other method (from traditional descriptive set theory or anywhere else). Many other applications of the HOD analysis have been found, including Neeman's proof that ω_1 is $<$-Θ-supercompact in $L(\mathbb{R})$ (cf. [**18**]), the positive proof of Woodin's conjecture on the uniqueness of natural models of $\mathsf{AD}+\omega_1$ is $\mathbb{R}$-supercompact (cf. [**20**]), and the construction and analysis of Varsovian models (cf. [**26**]).

Under *the Axiom of Choice* (AC), Θ is the successor of the Continuum. Under AD, Θ is very large, e.g. Θ is a limit of measurable cardinals and more. Steel, [**35**], is the first person to prove that HOD up to Θ of $L(\mathbb{R})$ (assuming AD) is a fine-structural extender model[8]; Woodin, unpublished but see [**40**], extends Steel's work and shows that full HOD of $L(\mathbb{R})$ is a fine-structural model, albeit a hybrid

[7] AD^+ is $\mathsf{AD} + \mathsf{DC}_{\mathbb{R}}+$ "every set of reals has an ∞-Borel code" + Ordinal Determinacy. AD^+ is arguably a completion of AD in that it completes the structural analysis of Suslin cardinals. In particular, it implies the set of Suslin cardinals is closed below Θ, which is not known to follow from AD. It is not known if AD implies AD^+. However, AD^+ holds in all known models of AD constructed in practice. In $L(\mathbb{R})$, one can show AD implies AD^+.

[8] More precisely, Steel shows that V^{HOD}_Θ is the universe of an extender model.

one in that it is constructed from an extender sequence and a (partial) strategy of HOD on trees based on HOD$|\Theta$. This is an example of a Σ-mouse for some iteration strategy Σ, which in turns is a special class of *hybrid mice*. For a strategy Σ of a model M that has nice condensation properties such as *hull condensation* or *branch condensation*, one can construct a fine-structural model $\mathcal{P}$ containing M as an element and $\mathcal{P}$ is constructible from a predicate coding an extender sequence and a predicate coding $\Sigma \cap \mathcal{P}$. The general fine structure of such objects has been investigated by various people including Woodin, Steel, Schindler, Sargsyan, Schlutzenberg, Trang and is written in somewhat more generality in [**30**] and [**29**].[9] These papers provide partial answers to the following important problem. An operator F is a function typically defined on (a cone over) some H_κ; F is nice if it has sufficient condensation properties, and determines itself on generic extensions. See for example [**28**], [**29**] for a more precise definition. Examples of nice operators are: $rud : x \mapsto rud(x)$, $\sharp : x \mapsto \sharp(x)$. Given a nice operator F, one typically is interested in analyzing the scales pattern in models such as $L^F(\mathbb{R})$ (i.e. in this model, given a pointclass Γ, where is the first place in the constructible hierarchy of the model one can show every set in Γ has a scale) much like Steel's analysis of scales in $L(\mathbb{R})$ [**34**].

PROBLEM 2.1. *Given a nice operator F, develop fine structure theory of hybrid extender models of the form $L[\mathbb{E}, F]$, prove that iterable $L[\mathbb{E}, F]$ models with large cardinals exist (we call these F-mice), and analyze the scales pattern in F-mice over $\mathbb{R}$.*

Problem 2.1 is an integral part of developing the theory of hybrid mice suitable for core model induction applications. As mentioned above, partial answers to Problem 2.1 are given in [**30**]. The answers in [**30**] are partial in the sense that the authors assume additionally that F, in some sense, is fine-structural and has some smallness properties. This is enough for applications that aim to construct models up to LSA and a bit beyond (from large cardinals or strong combinatorial principles) but seems hardly complete. Here LSA, which stands for "the Largest Suslin Axiom", is the theory "$\mathsf{AD}^+ + \Theta = \theta_{\alpha+1} + \theta_\alpha$ is the largest Suslin cardinal" (see the next section for a more detailed discussion of the Solovay sequence $(\theta_\alpha)_\alpha$). LSA is a very strong determinacy principle isolated by W.H. Woodin in the 1990's; LSA implies the existence of many models of $\mathsf{AD}_\mathbb{R} + \Theta$ is regular and much more. It was recently shown to be consistent relative to large cardinals (cf. [**24**]).

There are various non-fine-structural operators that are of interest. [**29**] develops the theory of F-mice without assuming F is fine-structural by carefully formulating the notion of hybrid premouse and F's condensation properties. However, the theory developed there is not general enough to cover some important non-fine-structural operators, like the C_Γ operator from the AD^+ world, where $C_\Gamma(x)$ is the largest countable set in Γ containing x.

[9][**30**] fully resolves fine-structural issues related to feeding in branch information of Σ and is used in the development of the fine structure of least-branch hod mice, cf. [**42**]. Furthermore, [**29**] defines a general class of operators F such that F-mice carries fine structure similar to that of extender models. Σ-mice for Σ a strategy with nice condensation properties is an example of F-mice.

Woodin proves that *the Mouse Set Conjecture* (MSC)[10] holds in $L(\mathbb{R})$, see [**33**]. Sargsyan has generalized the aforementioned analyses to much larger AD^+ models, see [**21**] and [**24**]. A central notion in the proof of MSC and the analysis of HOD is the notion of *hodmice* (developed by G. Sargsyan, cf. [**21**], which built on and generalized earlier unpublished work of Woodin). Hod mice are a type of strategic extender models. The extender sequence allows hod mice to satisfy some large cardinal theory and the strategies allow them to generate models of determinacy. Unlike pure extender models and Σ-mice (here we mean F-mice, where the strategy Σ is coded by an operator F), there are many basic structural questions that are still open for hod mice, for example, the extent of condensation and whether combinatorial principles such as $\Box$ hold in hod mice.

The approach taken by Sargsyan in [**21**] and [**24**] in analyzing HOD is called *rigid layering*. This approach to defining hod mice has many advantages, one of which is that constructing such hod mice in practice (see for example [**22**], [**44**], [**24**]) is relatively easy. The complexity of the rigidly layered hod mice constructed matches that with the complexity of the determinacy pointclasses they generate fairly closely (i.e. the hierarchy does not grow too quickly) However, such hod mice may not satisfty some of the fine structural properties extender models satisfy such as condensation and $\Box$[11] due to the non-uniformity in its definition, e.g. rigidly layered hod mice are too extender-biased, in other words, extenders seem to be indexed more often than branches of iteration trees on the model. Further, it is not clear how to define rigidly layered hod mice beyond those defined in [**24**]; it appears that the complexity in defining rigidly layered hod mice increases along with the complexity of the underlying AD^+ theory.

Steel, in [**42**], has provided an alternative approach to defining hod mice and analyzing HOD. His approach is called the *least branch* approach. This approach has the advantage that we no longer need to "layer"; instead, strategies and extenders are fed into the models in a uniform way. This ensures that the fine structure of least-branch hod mice is similar to that of extender models. See, for example, [**42**] and [**43**] for a proof of solidity and condensation for least branch hod mice. Provided we have enough hod mice in our AD^+ universe, more precisely, *Hod Pair Capturing* (HPC) holds (cf. [**42**]), where HPC is the alternative version of *Generation of Pointclasses* in the context of least-branch hod mice, which says that for any $A \subseteq \mathbb{R}$ such that A is Suslin co-Suslin, there is a least-branch hod pair $(\mathcal{P}, \Sigma)$ such that $A <_\omega Code(\Sigma)$,[12] then HOD (up to Θ) is the diret limit of least-branch hod mice. The definition of HOD (up to Θ) as a least branch hod mouse is uniform and does not depend on the complexity of the underlying theory. The theory of least-branch hod mice works for AD^+ models roughly up to the level of superstrong cardinals. The fine structure of least branch hod mice mirrors that of extender

[10] MSC is the statement: assuming there are no iterable mice with a superstrong cardinal, for any reals x, y such that x is $OD(y)$, there is a countable mouse $\mathcal{M}$ over y such that $\mathcal{M}$ is sound, $\rho_\omega(\mathcal{M}) = \omega$, and $x \in \mathcal{M}$.

[11] We know such hod mice fail to satisfy instances of condensation extender models satisfy. Regarding $\Box$, [**24**] provides a partial answer for hod mice in the minimal model of the theory LSA, which is $\mathsf{AD}^+ + \Theta = \theta_{\alpha+1} + \theta_\alpha$ is the largest Suslin cardinal.

[12] $Code(\Sigma)$ is a canonical coding of the strategy Σ by a set of reals.

models. The results in [**42**] and [**43**] immediately imply HOD satisfies $\mathsf{GCH} + \diamondsuit$.[13] and move us a step closer to characterizing $\square$ in HOD (under HPC). However, constructing least branch hod mice in core model induction applications is a challenge. It is worthwhile to develop methods for constructing least branch hod mice that can be applied in a variety of applications.

In summary, we believe that it is very important studying both the rigidly layered hierarchy and the least-branch hierarchy as the two approaches complement each other in improving our understanding of the structure theory of hod mice under AD^+. The most important problems for those studying least-branch hod mice to tackle are the following (see Section 4 for the definition of a (long) extender).

PROBLEM 2.2 (Hod Pair Capturing). *Assume* AD^+. *Show that* HPC *is true assuming no mice with a long extender exist.*

The hypothesis "no mice with long extenders" seems necessary here. The theory of "mice" that we are familiar with (i.e. Mitchell-Steel mice and Jensen mice) are called short-extender mice; mice with short-extenders can satisfy large cardinals up to superstrong and subcompact cardinals, but not much more. Long-extender mice can satisfy cardinals in the supercompact region, but our understanding of long-extender mice is limited and their basic theory is still being developed.[14]

In the rigidly layered hierarchy, we can state the corresponding problem to the above. Resolving this is a key step in proving MSC and would allow us to compute HOD (of AD^+ models beyond those considered in [**24**]). Generation of Pointclasses (GP) was first introduced in [**21**] for the purpose of showing HOD is the direct limit of hod mice; Steel (cf. [**42**]) observes that one can simply require the hod pairs are Wadge cofinal in the Suslin coSuslin sets for the HOD analysis. For all intended purposes, the differences between HPC and GP are cosmetic, though as mentioned above, there are difficulties in defining rigidly layered hod mice beyond those defined in [**24**].

PROBLEM 2.3 (Generation of Pointclasses). *Prove that* AD^+ *implies Generation of Pointclasses (*GP*), assuming no mice with a long extender exist. Here* GP *is the statement for any sufficiently closed pointclass* Γ *such that every* $A \in \Gamma$ *is Suslin co-Suslin and* Γ *does not contain all Suslin co-Suslin sets, there is a (rigidly layered) hod pair* $(\mathcal{P}, \Sigma)$ *such that* $(\mathcal{P}, \Sigma)$ *generates* Γ.[15]

3. The core model induction and its applications

Another way of exploring the determinacy/large cardinal connection is via the Core Model Induction (CMI). The core model induction is pioneered by Woodin and developed further by Steel, Schindler and others. CMI draws strength from strong set-theoretic principles such as PFA to inductively construct canonical models of determinacy and those of large cardinals in a locked-step process by combining core model techniques (for constructing the core model K) with descriptive set theory,

[13]In the $\mathsf{AD}^+ + \neg\mathsf{AD}_{\mathbb{R}} + V = L(\wp(\mathbb{R}))$ case, it is still open whether HOD up to Θ is a direct limit of least-branch hod mice. However, Steel's analysis shows that HOD up to the largest Suslin cardinal is such a direct limit; and this suffices to show HOD satisfies $\mathsf{GCH} + \diamondsuit$.

[14]At the point this paper is written, the theory of long extenders has been developed at the level of κ being κ^{+n}-supercompact for all $n < \omega$.

[15]Roughly, $(\mathcal{P}, \Sigma)$ generates Γ means that every $A \in \Gamma$ is in the derived model of $\mathcal{P}$ generated by Σ.

in particular the scales analysis in $L(\mathbb{R})$ and its generalizations (cf. [**34**], [**38**], [**30**]). CMI is the only known systematic method for computing lower bound consistency strength of strong theories extending ZFC and it is hoped that it will allow one to compute the exact strength of important theories such as PFA.

The Solovay sequence is the sequence $\vec{\theta} = (\theta_\alpha : \alpha \leq \Omega)$ such that $\vec{\theta}$ is continuous (i.e. range of $\vec{\theta}$ is closed), sup $\vec{\theta} = \Theta$, θ_0 is the supremum of ordinals α such that there is an OD surjection from $\mathbb{R}$ onto α, for $\alpha < \Omega$, $\theta_{\alpha+1}$ is the supremum of ordinals α such that there is a $OD(A)$ surjection from $\mathbb{R}$ onto α, where A is a set of reals of Wadge rank θ_α. $L(\mathbb{R})$ is the minimal (transitive, proper class) model of AD containing all reals. It can be shown that in $L(\mathbb{R})$, assuming AD, $\mathsf{AD}^+ + \Theta = \theta_0$ holds. Here are some AD^+ theories listed in increasing strength: AD^+, $\mathsf{AD}^+ + \Theta > \theta_0$, $\mathsf{AD}_\mathbb{R}$, $\mathsf{AD}_\mathbb{R} + \mathsf{DC}$, $\mathsf{AD}_\mathbb{R} + \Theta$ is regular, LSA. LSA is a very strong determinacy principle, first isolated by Woodin in [**48**]. [**48**] conjectures that LSA has strength roughly in the region of a supercompact cardinal. However, [**24**] shows that LSA is consistent relative to the existence of a Woodin cardinal which is a limit of Woodin cardinals, which is much weaker. However, LSA is still much stronger than $\mathsf{AD}_\mathbb{R} + \mathsf{DC}$, which is roughly the lower bound consistency strength of PFA obtained by [**7**] via core model (K^c) techniques in [**1**].

Many of the early work in CMI typically obtains $\mathsf{AD}^{L(\mathbb{R})}$ as lower bound of various theories (cf. [**28**], [**3**], [**39**], [**37**]). Ketchersid, using ideas of Woodin, is the first to have constructed inner models cf $\mathsf{AD}^+ + \Theta > \theta_0$, [**10**]. [**10**] contains a wealth of techniques for constructing mice generating models of $\mathsf{AD}^+ + \Theta > \theta_0$ and a bit beyond.[16] They are, however, insufficient to construct models of $\mathsf{AD}_\mathbb{R} + \Theta$ is regular, LSA and beyond. To construct models of these stronger theories, there are many ingredients that come into play. One needs to analyze HOD of such theories. Sargsyan (cf. [**21**], [**24**]) has solved Problem 2.3 for models of AD^+ up to LSA, computed HOD of such models, and developed various techniques for constructing hod mice in practice. One needs to develop the theory of F-mice for complicated operators F, the scales analysis in F-mice over $\mathbb{R}$, and various techniques for constructing new scales and pointclasses of determinacy. Again, this has been done by various people ([**28**], [**30**],[**29**], [**46**]).

We discuss below several problems involving determining consistency strength of important set-theoretic principles. The arguably most important one is.

QUESTION 3.1. What is the consistency strength of PFA?

As mentioned above, several authors have used the core model induction to continually improve the lower bound consistency of PFA. [**37**], [**25**], [**44**], and [**24**] show that the strength of PFA is at least that of LSA. The strongest theorem along this direction is.

THEOREM 3.2 (Sargsyan and Trang [**24**]). *Assume* PFA. *Then there is an inner model M containing $\mathbb{R} \cup ORD$ such that $M \models$* LSA.

However, as mentioned above, LSA is shown to be consistent relative to a Woodin limit of Woodin cardinals and the consensus amongst set theorists is that PFA should be as strong as a supercompact cardinal (it is well-known that the

[16] These results and techniques can help us construct models of $\mathsf{AD}_\mathbb{R}$; however, they do not give lower bound strengths beyond those obtained by K^c-constructions as in [**1**], which is roughly $\mathsf{AD}_\mathbb{R} + \mathsf{DC}$.

upper bound for PFA is a supercompact, cf [**2**]). A complete answer to Question 3.1 is the holy grail of inner model theory. From the point of view of descriptive inner model theory, to completely solve this problem, one needs to understand HOD of AD^+ models; in particular, one needs to resolve Problems 2.2 and/or 2.3 and their generalizations to the long-extender regions. Futhermore, at the level of $\mathsf{AD}_{\mathbb{R}} + \Theta$ is regular and beyond, one needs to do a significant amount of work to construct hod-like objects to eventually generate HOD of the AD^+ models; this is where the bulk of the construction is and seems to be hypothesis-dependent, i.e. the construction methods seem to depend on the types of strong hypotheses one assumes as well as the target AD^+ theory one is trying to construct models for. For example, [**24**], from PFA, combines techniques for constructing K^c from core model theory and techniques for analyzing HOD of AD^+ models, introduces a strategic K^c construction, whose outcome is a hod-like object $\mathcal{M}$ that generates HOD of an LSA model (and hence the LSA model itself).

It has been observed (independently) by I. Neeman and the author that

THEOREM 3.3. *Assume* PFA + *there is a Woodin cardinal. Then there are inner models that satisfy* ZFC + *there is a Woodin cardinal which is a limit of Woodin cardinals.*

The proof of the above theorem will appear in an upcoming paper; the basic argument is: suppose not. Let δ be a Woodin cardinal. Then the fully backgrounded construction in V_δ described in [**17**] will produce an extender model $\mathcal{N}$ of height δ. Let $\mathcal{N}^+$ be the stack of mice over $\mathcal{N}$ (as defined in [**7**] and [**19**]), then [**19**] shows that $\mathrm{cof}(o(\mathcal{N}^+)) \geq \delta$; this contradicts PFA by a standard argument, using the fact that $\square(\delta^+)$ and $\square(\delta)$ both fail.[17]

So with an extra, mild large cardinal assumption, we can improve the lower bound of Theorem 3.2 significantly.[18] Unfortunately, the method of Theorem 3.3 does not seem to generalize; this is because the proof of the theorem is highly dependent on the construction in [**17**], which has no known improvement. So it seems the right approach to systematically study the universe of sets and its canonical structures in the presence of strong hypotheses like PFA is continue to generalize the approach of [**24**].

In closing, we list some sample problems that seem within reach of current core model induction techniques. Below $\mathsf{MM}(\kappa)$ is the Martin's Maximum for posets of size $\leq \kappa$ and c is the size of the continuum.

QUESTION 3.4.
(a) What is the consistency strength of $\neg\square_\kappa$ for some singular strong limit κ?
(b) What is the consistency strength of $\neg\square_\kappa + \neg\square(\kappa)$ for $\kappa \geq \aleph_3$ such that $\kappa^\omega = \kappa$ and $2^{<\kappa} = \kappa$?
(c) What is the consistency strength of $\mathsf{MM}(c^+)$?
(d) Is the consistency strength of $\mathsf{MM}(c)$ the same as that of $\mathsf{AD}_{\mathbb{R}} + \Theta$ is regular?
(e) What is the consistency strength of $\mathsf{ZF} + \mathsf{DC} + \omega_1$ is supercompact?

[17] $\square(\alpha)$ says that there is a sequence of $(C_\beta : \beta < \alpha)$ such that C_β is a club subset of β, $C_\beta \cap \gamma = C_\gamma$ for every limit point γ of C_β, and there is no "thread" through the sequence, i.e. there is no club C in α such that for any limit point β of C, $C \cap \beta = C_\beta$.

[18] Note that unlike Neeman-Steel theorem [**19**, Theorem 1.2], which assumes SBH_δ, which is a very strong iterability assumption, no assumption on iterability is made here.

(f) What is the consistency strength of the theory ZF+ every cardinal is singular?
(g) What is the consistency strength of the failure of the Unique Branch Hypothesis (UBH) for nice trees?

The first four questions concern consistency strength of certain fragments of MM. Some significant refinements of techniques in **[37]**, **[44]**, **[22]**, and **[24]** may be needed to make progress on these questions. For example, regarding (a), **[37]** proves AD holds in $L(\mathbb{R})$ and D. Adolf has recently improved on this result to construct the minimal model of "$\mathsf{AD}_{\mathbb{R}} + \Theta$ is regular". Regarding (b), we believe the hypothesis is very strong. If we set $\kappa = \omega_2$ in the hypothesis, **[4]** shows that it is weak. The next two concern developing core model induction methods in the choiceless (or limited choice) context. See **[45]** and **[3]**. Concerning the last question, UBH for nice trees is the statement that every nice iteration tree on V has at most one cofinal well-founded branch, where a nice iteration tree is normal, nonoverlapping, and every extender E used in the tree has $length(E) = strength(E) =$ an inaccessible cardinal in the model in which E is chosen from. The motivation for this question is the belief that UBH for nice trees should be true, cf **[23]** and **[27]**. Note that Woodin, **[49]**, has constructed nonoverlapping iteration trees on V with distinct cofinal wellfounded branches from a very large cardinal assumption. However, the trees constructed in **[49]** are not nice.

4. Inner model theory at the level of supercompact cardinals

Let M and N be transitive models of (a fragment of) ZFC and $j : M \to N$ be an elementary embedding with critical point κ. Let $\lambda > \kappa$ (λ typically is a limit ordinal). A (κ, λ)-extender $E = E_j$ (over M or M-extender) derived from j is $(E_a : a \in [\lambda]^{<\omega})$,[19] where E_a is a (nonprincipal, countably complete) measure defined as follows: letting κ_a be the least γ such that $j(\gamma) \geq max(a) + 1$, for each $A \subseteq [\kappa_a]^{|a|}$ in M,

$$A \in E_a \Leftrightarrow a \in j(A).$$

A (κ, λ)-extender E is *short* if all measures in E concentrate on κ (i.e. $\kappa_a = \kappa$). Otherwise, E is *long*. For $\gamma < \lambda$, $E|\gamma = (E_a : a \in [\gamma]^{<\omega})$ is an initial segment of E. As mentioned above, fine structure theory has been fully developed for short-extender models (cf. **[13]** and **[51]**).

Developing fine structure for long-extender models proves to be very challenging. There are several problems with generalizing fine structure theory for short extenders to long ones. One main problem is the *Failure of the Initial Segment Condition.* The Initial Segment Condition, which roughly states that initial segments of an extender F on the extender sequence of a (fine-structural) model are on the sequence, is a staple of inner model theory in the region of short extenders. This condition fails in fine-structural extender models having cardinals just past superstrong cardinals. This and the fact that long generators may be moved in iteration trees further complicate comparison arguments for mice at this level. The basic theory for extender models satisfying the existence of a cardinal κ which is

[19] One can define abstractly the notion of an extender by specifying the properties the E_a's must have and how they relate to one another. See e.g. **[41]** for such a definition.

κ^{+n}-supercompact for some $n < \omega$ has been developed by Woodin ([**47**]) and Neeman, Steel ([**16**]).[20] Beyond this level of complexity, various attempts have been made with no significant progress. It is worth noting that Woodin (in [**49**]) has generalized the Martin-Steel theory of coarse extender models to the levels of supercompact cardinals. These are called *suitable extender models.* Woodin also shows that an extender model (not necessarily fine-structural) of a (fully) supercompact cardinal absorbs all large cardinals from V (these are called weak extender models, as defined below, cf Definition 4.4). This remarkable theorem suggests that it is enough to develop fine structure theory for models of a supercompact cardinal. Recent work of Woodin (see [**50**] for a more detailed discussion) suggests that the right approach for developing such a theory is via going into the strategic-extender hierarchy[21] and again the connection with AD^+ and HOD of models of AD^+ emerges at this level of large cardinal.

[**6**] defines *the Ultrapower Axiom* (UA), which states that suppose M_0, M_1 are two ultrapowers of the universe V, with $j_0 : V \to M_0$, $j_1 : V \to M_1$ be the respective ultrapower maps, then there are internal ultrapower maps $i_0 : M_0 \to N$, $i_1 : M_1 \to N$ such that $i_0 \circ j_0 = i_1 \circ j_1$. This axiom was isolated by Goldberg from Woodin's previous work on weak comparisons. UA is known to hold in all fine-structural extender models,[22] has many consequences that suggests the universe in which UA holds resembles that of a fine-structural extender model (e.g. UA implies the normal fine κ-complete ultrafilters on $\wp_\kappa(\lambda)$ are well-ordered by the (generalized) Mitchell order, GCH holds above a supercompact cardinal κ), and permits the development of a semi-fine structural theory of large cardinals in the region of supercompact cardinals. This work suggests UA must be consistent with supercompacts. This in turn suggests that inner model theory must reach the levels of many supercompact cardinals since (at this stage) it is hard to conceive how one can build models satisfying UA which do not come from fine-structural models. Most importantly, UA suggests non-strategic extender hierarchy cannot reach the infinite levels of supercompact (cf. [**6**, Conjecture 8.3]).

Finally, one way of formalizing the development discussed above is the axiom $V =$ *Ultimate-L*. In the following, a set $A \subseteq \mathbb{R}$ is universally Baire if for each γ, there are trees T_γ, U_γ on $\omega \times Ord$ such that $A = p[T_\gamma] = \mathbb{R}\backslash p[U_\gamma]$, and furthermore, for any forcing extension $V[g]$ for V such that the forcing has size $< \gamma$, in $V[g]$, $p[T_\gamma] = \mathbb{R}\backslash p[U_\gamma]$. In the presence of a proper class of Woodin cardinals, if A is universally Baire, then $L(A, \mathbb{R}) \vDash \mathsf{AD}^+$.

DEFINITION 4.1 ($V =$ Ultimate-L, Woodin [**50**]).
(1) There is a proper class of Woodin cardinals.
(2) For each Σ_2-sentence Φ, if Φ holds in V then there exists a universally Baire set $A \subseteq \mathbb{R}$ such that
$$(HOD \cap V_\Theta)^{L(A,\mathbb{R})} \vDash \Phi.$$

[20] There are many difficulties in developing inner model theory at the level of supercompactness. See, for example, [**50**] for more details.

[21] Models in the strategic-extender hierarchy are constructed from a sequence of partial extenders and partial strategies of their own initial segments, much like the hod mice discussed above. Woodin also suggests that at this level of complexity, the strategic-extender hierarchy has to be different from the layered hierarchy in [**21**] and [**24**].

[22] This is the only known proof of consistency of UA. Forcing UA holds (along with a large cardinal) seems very difficult. See [**6**, Proposition 3.4].

There are many variations of the version of the axiom stated above (cf. [**50**]). One main motivation for this axiom is that $V = Ultimate - L$, when supplemented by large cardinals, is expected to axiomatize the structure of canonical inner models of a supercompact cardinal (if they exist). Another important feature of this axiom is that it suggests the universe looks like a strategic extender model, which is the HOD of $L(A, \mathbb{R})$ where A is a universally Baire set. The following is a version of Woodin's *Ultimate-L Conjecture.* We need a few definitions.

DEFINITION 4.2. A sequence $N = (N_\alpha : \alpha \in Ord)$ is *weakly Σ_2-definable* if there is a formula $\varphi(x)$ such that:

(1) For all $\beta < \eta_1 < \eta_2 < \eta_3$, if $(N_\varphi)^{V_{\eta_1}}|\beta = (N_\varphi)^{V_{\eta_3}}|\beta$, then
$$(N_\varphi)^{V_{\eta_1}}|\beta = (N_\varphi)^{V_{\eta_2}}|\beta = (N_\varphi)^{V_{\eta_3}}|\beta.$$
(2) For all $\beta \in Ord$, $N|\beta = (N_\varphi)^{V_\eta}|\beta$ for all sufficiently large η.

Here for $\gamma \in Ord$, $(N_\varphi)^{V_\gamma} = \{a : V_\gamma \vDash \varphi[a]\}$.

DEFINITION 4.3. Suppose that $N \subseteq V$ is an inner model and $N \vDash \mathsf{ZFC}$. Then N is *weakly Σ_2-definable* if the sequence
$$(N \cap V_\alpha : \alpha \in Ord)$$
is weakly Σ_2-definable.

Inner models of V which are Σ_2-definable are weakly Σ_2-definable. For instance, HOD is Σ_2-definable, hence it is weakly Σ_2-definable. The increasing enumeration of supercompact cardinals is weakly Σ_2-definable.

DEFINITION 4.4. A transitive class $N \vDash \mathsf{ZFC}$ is a *weak extender model for δ is supercompact* if for every $\gamma > \delta$, there exists a δ-complete normal fine measure U on $\wp_\delta(\gamma)$ such that

(1) $N \cap \wp_\delta(\gamma) \in N$,
(2) $U \cap N \in N$.

By the remarkable [**50**, The Universality Theorem], weak extender models for δ is supercompact absorb all large cardinal properties given by extenders or elementary embeddings in the universe.

CONJECTURE 4.5 (Woodin, [**50**]). *Suppose that δ is an extendible cardinal. Then there exists a weak extender model N for the supercompactness of δ such that:*

(1) *N is weakly Σ_2-definable and $N \subset HOD$,*
(2) *$N \vDash$ "$V =$ Ultimate-L".*

Proving this conjecture or its variants would show in a decisive fashion that the strategic-extender hierarchy is the correct hierarchy for the canonical universe of sets.

Acknowledgment

The author would like to thank the NSF for its generous support through grant DMS-1849295. The author would also like to thank the referee for several helpful comments on an older version of the paper.

References

[1] Alessandro Andretta, Itay Neeman, and John Steel, *The domestic levels of K^c are iterable*, Israel J. Math. **125** (2001), 157–201, DOI 10.1007/BF02773379. MR1853810

[2] James E. Baumgartner, *Iterated forcing*, Surveys in set theory, London Math. Soc. Lecture Note Ser., vol. 87, Cambridge Univ. Press, Cambridge, 1983, pp. 1–59, DOI 10.1017/CBO9780511758867.002. MR823775

[3] Daniel Busche and Ralf Schindler, *The strength of choiceless patterns of singular and weakly compact cardinals*, Ann. Pure Appl. Logic **159** (2009), no. 1-2, 198–248, DOI 10.1016/j.apal.2008.12.001. MR2523718

[4] Andrés Eduardo Caicedo, Paul Larson, Grigor Sargsyan, Ralf Schindler, John Steel, and Martin Zeman, *Square principles in $\mathbb{P}_{\max}$ extensions*, Israel J. Math. **217** (2017), no. 1, 231–261, DOI 10.1007/s11856-017-1444-8. MR3625110

[5] A. Dodd and R. Jensen, *The core model*, Ann. Math. Logic **20** (1981), no. 1, 43–75, DOI 10.1016/0003-4843(81)90011-5. MR611394

[6] G. Goldberg, *The ultrapower axiom*, (2018), Same Volume.

[7] Ronald Jensen, Ernest Schimmerling, Ralf Schindler, and John Steel, *Stacking mice*, J. Symbolic Logic **74** (2009), no. 1, 315–335, DOI 10.2178/jsl/1231082314. MR2499432

[8] R. Björn Jensen, *The fine structure of the constructible hierarchy*, Ann. Math. Logic **4** (1972), 229–308; erratum, ibid. 4 (1972), 443, DOI 10.1016/0003-4843(72)90001-0. With a section by Jack Silver. MR0309729

[9] Ronald Jensen and John Steel, *K without the measurable*, J. Symbolic Logic **78** (2013), no. 3, 708–734. MR3135495

[10] Richard O'Neal Ketchersid, *Toward AD(R) from the continuum hypothesis and an omega(1)-dense ideal*, ProQuest LLC, Ann Arbor, MI, 2000. Thesis (Ph.D.)–University of California, Berkeley. MR2701033

[11] Kenneth Kunen, *Some applications of iterated ultrapowers in set theory*, Ann. Math. Logic **1** (1970), 179–227, DOI 10.1016/0003-4843(70)90013-6. MR0277346

[12] Donald A. Martin and John R. Steel, *A proof of projective determinacy*, J. Amer. Math. Soc. **2** (1989), no. 1, 71–125, DOI 10.2307/1990913. MR955605

[13] William J. Mitchell and John R. Steel, *Fine structure and iteration trees*, Lecture Notes in Logic, vol. 3, Springer-Verlag, Berlin, 1994. MR1300637

[14] William J. Mitchell, *Sets constructible from sequences of ultrafilters*, J. Symbolic Logic **39** (1974), 57–66, DOI 10.2307/2272343. MR344123

[15] William J. Mitchell, *Sets constructed from sequences of measures: revisited*, J. Symbolic Logic **48** (1983), no. 3, 600–609, DOI 10.2307/2273452. MR716621

[16] I. Neeman and J.R. Steel, *Fine structure of plus-one premice*, (2014), handwritten notes, available at math.berkeley.edu/∼steel/papers/Publications.html.

[17] Itay Neeman, *Inner models in the region of a Woodin limit of Woodin cardinals*, Ann. Pure Appl. Logic **116** (2002), no. 1-3, 67–155, DOI 10.1016/S0168-0072(01)00103-8. MR1900902

[18] Itay Neeman, *Inner models and ultrafilters in $L(\mathbb{R})$*, Bull. Symbolic Logic **13** (2007), no. 1, 31–53. MR2300902

[19] Itay Neeman and John Steel, *Equiconsistencies at subcompact cardinals*, Arch. Math. Logic **55** (2016), no. 1-2, 207–238, DOI 10.1007/s00153-015-0465-4. MR3453585

[20] Daniel Rodríguez and Nam Trang, *$L(\mathbb{R}, \mu)$ is unique*, Adv. Math. **324** (2018), 355–393, DOI 10.1016/j.aim.2017.11.010. MR3733890

[21] Grigor Sargsyan, *Hod mice and the mouse set conjecture*, Mem. Amer. Math. Soc. **236** (2015), no. 1111, viii+172, DOI 10.1090/memo/1111. MR3362806

[22] Grigor Sargsyan, *Covering with universally Baire operators*, Adv. Math. **268** (2015), 603–665, DOI 10.1016/j.aim.2014.09.016. MR3276606

[23] Grigor Sargsyan and Nam Trang, *Non-tame mice from tame failures of the unique branch hypothesis*, Canad. J. Math. **66** (2014), no. 4, 903–923, DOI 10.4153/CJM-2013-036-x. MR3224270

[24] ———, *The largest suslin axiom, submitted*, 2017, available at math.unt.edu/∼ntrang.

[25] Grigor Sargsyan, *Nontame mouse from the failure of square at a singular strong limit cardinal*, J. Math. Log. **14** (2014), no. 1, 1450003, 47, DOI 10.1142/S0219061314500032. MR3225589

[26] Grigor Sargsyan and Ralf Schindler, *Varsovian models I*, J. Symb. Log. **83** (2018), no. 2, 496–528, DOI 10.1017/jsl.2018.5. MR3835075

[27] Grigor Sargsyan and Nam Trang, *Tame failures of the unique branch hypothesis and models of* $\mathsf{AD}_{\mathbb{R}} + \Theta$ *is regular*, J. Math. Log. **16** (2016), no. 2, 1650007, 31, DOI 10.1142/S0219061316500070. MR3580892

[28] R. Schindler and J. R. Steel, *The core model induction*, available at http://www.math.uni-muenster.de/logik/Personen/rds/, 2013.

[29] Farmer Schlutzenberg and Nam Trang, *The fine structure of operator mice*, arXiv preprint arXiv:1604.00083 (2016).

[30] ———, *Scales in hybrid mice over* $\mathbb{R}$, arXiv preprint arXiv:1210.7258 (2016).

[31] Dana Scott, *Measurable cardinals and constructible sets*, Mathematical Logic In The 20th Century, World Scientific, 2003, pp. 407–410.

[32] Jack Silver, *The consistency of the* GCH *with the existence of a measurable cardinal*, Axiomatic Set Theory (Proc. Sympos. Pure Math., Vol. XIII, Part I, Univ. California, Los Angeles, Calif., 1967), Amer. Math. Soc., Providence, R.I., 1971, pp. 391–395. MR0278937

[33] J. R. Steel, *A theorem of Woodin on mouse sets, available at http://math.berkeley.edu/~steel/.*

[34] John R. Steel, *Scales in* $L(\mathbf{R})$, Cabal seminar 79–81, Lecture Notes in Math., vol. 1019, Springer, Berlin, 1983, pp. 107–156, DOI 10.1007/BFb0071699. MR730592

[35] John R. Steel, $\mathrm{HOD}^{L(\mathbf{R})}$ *is a core model below* Θ, Bull. Symbolic Logic **1** (1995), no. 1, 75–84, DOI 10.2307/420947. MR1324625

[36] John R. Steel, *The core model iterability problem*, Lecture Notes in Logic, vol. 8, Springer-Verlag, Berlin, 1996. MR1480175

[37] John R. Steel, *PFA implies* $\mathrm{AD}^{L(\mathbb{R})}$, J. Symbolic Logic **70** (2005), no. 4, 1255–1296, DOI 10.2178/jsl/1129642125. MR2194247

[38] J. R. Steel, *Scales in* $K(\mathbb{R})$ *at the end of a weak gap*, J. Symbolic Logic **73** (2008), no. 2, 369–390, DOI 10.2178/jsl/1208359049. MR2414454

[39] John Steel and Stuart Zoble, *Determinacy from strong reflection*, Trans. Amer. Math. Soc. **366** (2014), no. 8, 4443–4490, DOI 10.1090/S0002-9947-2013-06058-8. MR3206466

[40] John R Steel and W Hugh Woodin, *HOD as a core model*, Ordinal definability and recursion theory: Cabal Seminar, vol. 3, 2012.

[41] John R. Steel, *An outline of inner model theory*, Handbook of set theory. Vols. 1, 2, 3, Springer, Dordrecht, 2010, pp. 1595–1684, DOI 10.1007/978-1-4020-5764-9_20. MR2768698

[42] ———, *Normalizing iteration trees and comparing iteration strategies*, (2018), available at math.berkeley.edu/~steel/papers/Publications.html.

[43] J.R. Steel and N. Trang, *Condensation for mouse pairs*, (2018), submitted, available at math.unt.edu/~ntrang.

[44] Nam Trang, PFA *and guessing models*, Israel J. Math. **215** (2016), no. 2, 607–667, DOI 10.1007/s11856-016-1390-x. MR3552291

[45] N. Trang and T. Wilson, *Supercompact measures on* $\mathcal{P}_{\omega_1}(\mathcal{P}(\mathbb{R}))$, (2016), submitted, available at math.unt.edu/~ntrang.

[46] Trevor M. Wilson, *The envelope of a pointclass under a local determinacy hypothesis*, Ann. Pure Appl. Logic **166** (2015), no. 10, 991–1018, DOI 10.1016/j.apal.2015.05.003. MR3356616

[47] H.W. Woodin, *The fine structure of suitable extender models I*, (2014).

[48] W. Hugh Woodin, *The axiom of determinacy, forcing axioms, and the nonstationary ideal*, De Gruyter Series in Logic and its Applications, vol. 1, Walter de Gruyter & Co., Berlin, 1999. MR1713438

[49] W. Hugh Woodin, *Suitable extender models I*, J. Math. Log. **10** (2010), no. 1-2, 101–339, DOI 10.1142/S021906131000095X. MR2802084

[50] W. Hugh Woodin, *In search of Ultimate-L: the 19th Midrasha Mathematicae Lectures*, Bull. Symb. Log. **23** (2017), no. 1, 1–109, DOI 10.1017/bsl.2016.34. MR3632568

[51] Martin Zeman, *Inner models and large cardinals*, De Gruyter Series in Logic and its Applications, vol. 5, Walter de Gruyter & Co., Berlin, 2002. MR1876087

Department of Mathematics, University of North Texas, Denton, Texas
Email address: `Nam.Trang@unt.edu`

Contemporary Mathematics
Volume **752**, 2020
https://doi.org/10.1090/conm/752/15137

A descriptive construction of trees and Stallings' theorem

Anush Tserunyan

ABSTRACT. We give a descriptive construction of trees for multi-ended graphs, which yields yet another proof of Stallings' theorem on ends of groups. Even though our proof is, in principle, not very different from already existing proofs and it draws ideas from **[Krö10]** it is written in a way that easily adapts to the setting of countable Borel equivalence relations, leading to a free decomposition result and a sufficient condition for treeability.

CONTENTS

1. Introduction

Stallings' theorem on ends of groups equates (as most impressive results in geometric group theory do) a geometric property of the Cayley graph of a finitely generated (f.g.) group with a structural/algebraic characterization of the group. Call a f.g. group Γ *multi-ended* if for some (equivalently, any) finite generating subset $F \subseteq \Gamma$, the Cayley graph G induced by F has more than one *end*, that is: there is a finite set of edges of G, removing which results in at least two infinite connected components. We say that a group Γ *splits* over its subgroup $\Delta < \Gamma$ if either $\Gamma = *_\Delta K$ (HNN-extension) or $\Gamma = K *_\Delta \Lambda$ (amalgamated product). The following theorem was proven in [**Sta68**] for torsion-free groups and in [**Sta71**] for general f.g. groups.

THEOREM 1.1 (Stallings 1968–71). *A f.g. group Γ is multi-ended if and only if it splits over a finite subgroup $\Delta < \Gamma$.*

There are many proofs of this theorem and in this paper, we give yet another one, which, however, is adaptable to the Borel context when working with countable

2010 *Mathematics Subject Classification.* Primary 20E08, 37A20, 03E15.

The author's research was partially supported by NSF Grant DMS-1501036.

Borel equivalence relations on standard Borel spaces (see [**JKL02**] for a survey on the subject).

The main statement we prove is the following structural result, which implies Stallings' theorem via Bass–Serre theory:

THEOREM 1.2. *If a group Γ admits a transitive action on a connected multi-ended graph G with all vertex-stabilizers being finite, then it also admits an action on a tree with finite edge-stabilizers and without fixed points.*

The latter implies a slight strengthening of Stallings' theorem:

THEOREM 1.3. *For a f.g. group Γ, the following are equivalent:*

(1) Γ *is multi-ended.*
(2) Γ *admits a transitive action on a multi-ended graph G with all vertex-stabilizers being finite.*
(3) Γ *splits over a finite subgroup Δ.*

Our proof of Theorem 1.2 is based on Krön's slick construction of a nested family of cuts in G that is invariant under the action of Γ [**Krö10**, Theorem 3.3], however our construction of the tree on this family is different. It is this construction that adapts to countable Borel equivalence relations, yielding Theorem 1.4.

For the theory of countable Borel equivalence relations on standard Borel spaces we refer to [**JKL02**] and we recall the standard definitions of this context (e.g. of Borel graph and graphing) in Section 4.

For a graph G on a set X, a *cut* is an infinite set $C \subseteq X$ contained in a single connected component Y of G such that $Y \setminus C$ is also infinite but there are only finitely many edges of G between C and $Y \setminus C$. If G is a locally countable Borel graph on a standard Borel space X, then the set $\mathcal{C}_G$ of all cuts is also naturally a standard Borel space. Call a set $\mathcal{C} \subseteq \mathcal{C}_G$ *G-complete* if it contains at least one cut from every connected component of G.

The collection $\mathcal{C}_G$ admits G-complete Borel subsets $\mathcal{C} \subseteq \mathcal{C}_G$ with certain desired properties (being nested, self-dual, chain-vanishing, and meeting every G-connected component) and we temporarily call such $\mathcal{C}$ *good.* Good collections are readily available, e.g. $\hat{\mathcal{C}}'_G$, as defined below right before Proposition 3.20, or any Borel maximal non-nested subset of the set of thin cuts, see Remark 4.3. On a collection $\mathcal{C}$, we define a Borel binary relation $\sim_{\mathcal{C}}$ (Definition 2.17), which turns out to be an equivalence relation if $\mathcal{C}$ is good.

THEOREM 1.4 (Stallings for equivalence relations). *Let E be a countable Borel equivalence relation on a standard Borel space X and let G be a multi-ended Borel graphing of E. For any G-complete good Borel collection $\mathcal{C} \subseteq \mathcal{C}_G$, there is a treeable subequivalence relation E_T and a Borel equivalence relation $E_{\mathcal{C}} \leqslant_B \sim_{\mathcal{C}}$ such that $E = E_T * E_{\mathcal{C}}$.*

Here, by $E = E_T * E_{\mathcal{C}}$, we mean that E is the free product of E_T and $E_{\mathcal{C}}$ as introduced in [**Gab00**, Subsection IV-B], and $E_{\mathcal{C}} \leqslant_B \sim_{\mathcal{C}}$ means that $E_{\mathcal{C}}$ is *Borel reducible* to $\sim_{\mathcal{C}}$, i.e. there is a Borel map $\pi : X \to \mathcal{C}$ such that for any $x, y \in X$,

$$xE_{\mathcal{C}}y \iff \pi(x) \sim_{\mathcal{C}} \pi(y).$$

The precise statement of Theorem 1.4 is given in Theorem 4.1.

Theorem 1.4 can be used to prove treeability of some equivalence relations E via isolating suitable Borel collections $\mathcal{C}$, to which the following applies:

COROLLARY 1.5 (Condition for treeability). *Let E be a countable Borel equivalence relation on a standard Borel space X. If E admits a multi-ended Borel graphing G and a G-complete good Borel collection $\mathcal{C} \subseteq \mathcal{C}_G$ such that the equivalence relation $\sim_{\mathcal{C}}$ is treeable (in particular, if it is smooth or hyperfinite), then E is treeable.*

Comparison with other results and proofs. Our proof of Theorem 1.2 is not, in principle, too different from other proofs existing in the literature, e.g., [**Dun17**], [**DD89**], and [**Krö10**], in the sense that it uses some of the common ideas involved in constructions of trees such as nested sets (see Subsection 2.C) and thin cuts (see Subsection 3.A), as well as the equivalence relation in Definition 2.17.

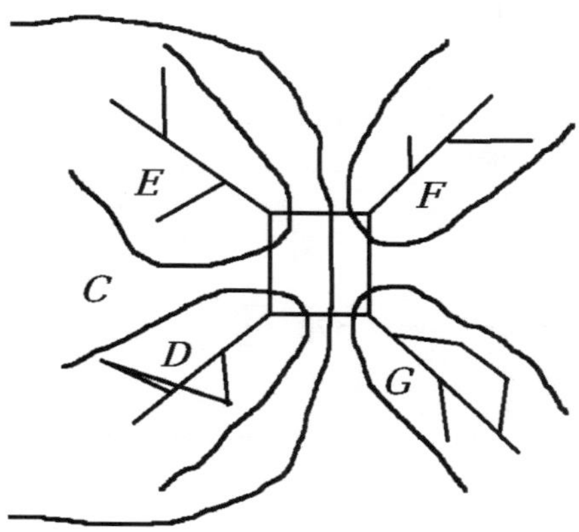

FIGURE 1. Nested collection $\mathcal{C} := \{C, D, E, F, G\}$ of thin cuts such that $C \notin \mathcal{C}(B)$ for every $\mathcal{C}$-block B.

The shortest proof of a version of Theorem 1.2 that the current author is aware of is presented in [**Krö10**], featuring a slick construction of a nested collection of cuts invariant under the action of the group [**Krö10**, Sections 2 and 3] and a simpler construction of a tree using blocks [**Krö10**, Section 4]. Trying to understand [**Krö10**, Section 4] is what initiated the present research because there seem to be issues in the very definition of the tree. More precisely, the existence of a block B_C for each $C \in \mathcal{C}$ claimed in [**Krö10**, Lemma 4.1] is false and Fig. 1 depicts a counterexample. This invalidates the claim of [**Krö10**, Theorem 4.2] that the defined graph $T(\mathcal{C})$ is a tree, namely, it can be disconnected. Also, even if [**Krö10**, Lemma 4.1] was true, the proof of [**Krö10**, Theorem 4.2] (both parts: acyclicity and connectedness) seems oversimplified and incomplete.

The incorrectness of some parts of [**Krö10**] is also mentioned in [**HK18**, Footnote 1 on page 4].

Thus, the proof of Theorem 1.2 given in the current paper is the simplest one known to the present author: it combines the construction of the non-nested collection given in [**Krö10**, Sections 2 and 3] and a construction of a tree that adapts to the setting of countable Borel equivalence relations. To keep the present paper self-contained, [**Krö10**, Sections 2 and 3] are rewritten in Subsection 3.B in the local terminology.

There are a number of other related results and proofs revolving around similar ideas appear, see, for example, [**Dun79, DD89, Dun82, DK15, EP14, Krö01,**

Möl92a, Möl92b, HK18]. We refer the reader to [**Krö10**, Section 5] for a concise description and comparison of some of these results.

As for Theorem 1.4, a similar free decomposition result was proven in [**Ghy95**] via different methods. The most relevant statement in the latter paper is that if a graphing of a countable Borel equivalence relation E has infinitely-many ends in each connected component then the equivalence relation decomposes into a free product of a nontrivial hyperfinite subequivalence relation and some other equivalence relation E'. It does not, however, give a direct insight into the nature/structure of E_1 as Theorem 1.4 does with $E_{\mathcal{C}}$, and the proof is significantly longer.

Ends of Borel graphs have also been extensively studied, from a different angle, in [**Mil04**], [**Mil09**], and [**HM09**].

Organization. Section 2 describes a construction of a tree on an abstract collection of sets satisfying certain properties. This is then applied to collections of cuts in multi-ended graphs in Section 3, yielding Theorems 1.2 and 1.3. Finally, Section 4 is where we discuss the context of countable Borel equivalence relation and prove Theorem 1.4.

2. Constructing a tree on a collection of sets

Fix an ambient set $X \neq \emptyset$ and, henceforth, by a *set* we will mean a subset of X. Let $\mathcal{S}$ always denote a collection of **nonempty** subsets of X.

For a set $A \subseteq X$, we denote by A^c the complement of A within X and for $i \in \{-1, 1\}$, we put

$$A^i := \begin{cases} A & \text{if } i = 1 \\ A^c & \text{if } i = -1. \end{cases}$$

2.A. Orthogonality, domination, and basis.

DEFINITION 2.1. Call sets A, B *orthogonal* and write $A \perp B$ if $A \cap B = \emptyset$ and $A \neq B^c$. On the other hand, say that A (resp., *exactly*) *dominates* B if $A \supseteq B^i$ for some $i \in \{-1, 1\}$ (resp., $A \supseteq B$). We say that a collection $\mathcal{A}$ of sets (resp., *exactly*) *dominates* B if some set in $\mathcal{A}$ (resp., exactly) dominates B.

OBSERVATION 2.2. For sets A, B, if $A \perp B$, then neither dominates the other. In particular, $A^i \not\perp B^j$ for any $(i, j) \in \{-1, 1\}^2$ other than $(i, j) = (1, 1)$.

DEFINITION 2.3. A collection $\mathcal{B}$ of sets is called *orthogonal* if any two distinct sets in $\mathcal{B}$ are orthogonal. We say that $\mathcal{B}$ *dominates* a collection of sets $\mathcal{S}$ if every set in $\mathcal{S}$ is dominated by some set in $\mathcal{B}$. We call $\mathcal{B} \subseteq \mathcal{S}$ a *basis* for $\mathcal{S}$ if it is orthogonal and dominates $\mathcal{S}$.

Note that a basis can be infinite as well as finite, and we give concrete examples of both in Remark 3.4 for collections of cuts in a graph.

Our global goal is to define a special tree on the set $\mathfrak{B}(\mathcal{S})$ of all bases for $\mathcal{S}$. However, at this point, even the existence of a basis is not clear and our local goal is to prove it under some hypotheses on $\mathcal{S}$.

We continue by recording some easy properties of bases.

OBSERVATION 2.4. For a collection of sets $\mathcal{S}$ and $\mathcal{B}_0 \subseteq \mathcal{B}_1 \subseteq \mathcal{S}$, if both $\mathcal{B}_0, \mathcal{B}_1$ are bases for $\mathcal{S}$, then $\mathcal{B}_0 = \mathcal{B}_1$.

PROOF. Immediate from Observation 2.2 □

LEMMA 2.5. *For a collection $\mathcal{S}$ of sets with a basis $\mathcal{B} \subseteq \mathcal{S}$, if a set $A \in \mathcal{S}$ is orthogonal to a set $B \in \mathcal{B}$, then A is exactly dominated by $\mathcal{B}$.*

PROOF. Since $\mathcal{B}$ dominates A, there is $C \in \mathcal{B}$ with $C \supseteq A^i$, for some $i \in \{-1, 1\}$, and $C \neq B$, by Observation 2.2. If $i = -1$, then $B \perp A$ gives $B \subseteq A^c \subseteq C$, contradicting $B \perp C$, so $i = 1$. □

2.B. Maximally orthogonal sets. Here, we define a tool for building bases.

DEFINITION 2.6. For a collection $\mathcal{S}$ of sets and sets $A, B \in \mathcal{S}$, say that B is *$\mathcal{S}$-maximally orthogonal to A* if it is an inclusion-maximal set in $\mathcal{S}$ that is orthogonal to A, i.e. $B \perp A$ and for any $C \in \mathcal{S}$, $C \perp A$ and $C \supseteq B$ implies $C = B$. Put

$$[A]_{\mathcal{S}} := \{A\} \cup \{B \in \mathcal{S} : B \ \mathcal{S}\text{-maximally orthogonal to } A\}.$$

LEMMA 2.7. *For a collection $\mathcal{S}$ with a basis $\mathcal{B} \subseteq \mathcal{S}$, for any $A \in \mathcal{B}$, $[A]_{\mathcal{S}} \subseteq \mathcal{B}$.*

PROOF. Fix $B \in [A]_{\mathcal{S}} \setminus \{A\}$, so $B \perp A$. By Lemma 2.5, there is $C \in \mathcal{B}$ distinct from A such that $C \supseteq B$. But $C \perp A$, so by the maximality of B, $C = B$. □

We now introduce a condition on $\mathcal{S}$, which ensures that $[A]_{\mathcal{S}}$ contains enough sets.

DEFINITION 2.8. A sequence $(A_n)_{n \in \mathbb{N}}$ of sets is called a *chain* if it is strictly monotone (i.e. either $\subsetneq$-increasing or $\subsetneq$-decreasing). Call the chain $(A_n^c)_{n \in \mathbb{N}}$ the *dual* of $(A_n)_{n \in \mathbb{N}}$. For a collection $\mathcal{S}$ of sets, call a decreasing (resp. increasing) chain $(A_n)_{n \in \mathbb{N}}$ *$\mathcal{S}$-vanishing* if there is no $B \in \mathcal{S}$ such that $A_n \supseteq B$ (resp. $A_n \cap B = \emptyset$) for every $n \in \mathbb{N}$. A collection of sets $\mathcal{S}$ is said to be *chain-vanishing* if every chain in it is $\mathcal{S}$-vanishing.

In Example 2.12 below, we provide an instance of a chain-vanishing collection of sets.

LEMMA 2.9. *For a chain-vanishing collection $\mathcal{S}$ and any $A, B \in \mathcal{S}$, if $A \perp B$, then there is $C \in [A]_{\mathcal{S}}$ with $C \supseteq B$.*

PROOF. Otherwise, we contradict the chain-vanishing property by recursively building an increasing chain $(B_n)_{n \in \mathbb{N}}$ with $B_0 = B$ and $B_n \perp A$ for every $n \in \mathbb{N}$. Indeed, assume $(B_i)_{i \leqslant n}$ is already defined and $B_n \perp A$. We know that $B_n \notin [A]_{\mathcal{S}}$, so it is not maximal in $\mathcal{S}$ among the sets orthogonal to A, and thus, there is $B_{n+1} \in \mathcal{S}$ with $B_{n+1} \supsetneq B_n$ and $B_{n+1} \perp A$. □

2.C. Nested sets.

TERMINOLOGY 2.10. For sets A, B, the sets $A^i \cap B^j$, $i, j = \pm 1$, are called the *corners* of (A, B).

DEFINITION 2.11. Sets A and B are called *nested* if they have an empty corner. A collection of sets $\mathcal{S}$ is called *nested* if any two sets in $\mathcal{S}$ are nested.

EXAMPLE 2.12. Let $\mathcal{T} := (X, E)$ be tree on X and let $\mathcal{S}$ be the collection of all subsets of X that are connected components of the graphs obtained from $\mathcal{T}$ by removing a single edge, i.e.

$$\mathcal{S} := \{C \subsetneq X : C \text{ is a connected component of } \mathcal{T} - \{e\} \text{ for some } e \in E\}.$$

Clearly, $\mathcal{S}$ is nested for any graph. The acyclicity of $\mathcal{T}$ guarantees that $\mathcal{S} \neq \emptyset$, and the connectedness implies that $\mathcal{S}$ is self-dual[1] and chain-vanishing.

Using orthogonality and domination, we rephrase nestedness as a (nonexclusive) alternative.

OBSERVATION 2.13 (Alternative for nested sets). For any nested sets A, B, either A dominates B or $A \perp B^j$, for some $j \in \{-1, 1\}$.

PROOF. $A \supseteq B^i$ is the same as $A^c \cap B^i = \emptyset$. Thus, if A doesn't dominate B, then, by nestedness, it must be that $A \cap B^j = \emptyset$ for some $j \in \{-1, 1\}$. □

For a set $A \in \mathcal{S}$, if A belongs to a basis then Lemma 2.7 implies that $[A]_{\mathcal{S}}$ is an orthogonal family. However, we still haven't shown that every $A \in \mathcal{S}$ belongs to a basis. In fact, we show the orthogonality of $[A]_{\mathcal{S}}$ first in order to show the existence of a basis.

LEMMA 2.14. *For any nested collection of sets $\mathcal{S}$ and $A \in \mathcal{S}$, $[A]_{\mathcal{S}}$ is an orthogonal family.*

PROOF. Because $A \perp B$ and $A \perp C$, $B^c \cap C^c \supseteq A \neq \emptyset$, which, in particular, implies that $B^c \neq C$. Also, by the maximality of B and C, $B \not\subseteq C$ and $C \not\subseteq B$, or equivalently, $B \cap C^c \neq \emptyset$ and $B^c \cap C \neq \emptyset$. Thus, by nestedness, it must be that $B \cap C = \emptyset$, so $B \perp C$. □

PROPOSITION 2.15. *For any nested, self-dual, chain-vanishing collection $\mathcal{S}$ of sets and any $A \in \mathcal{S}$, $[A]_{\mathcal{S}}$ is a basis for $\mathcal{S}$.*

PROOF. By Lemma 2.14, we only need to show that $[A]_{\mathcal{S}}$ dominates $\mathcal{S}$, so take $B \in \mathcal{S}$. By Observation 2.13, either A dominates B, in which case we are done, or $B^j \perp A$ for some $j \in \{-1, 1\}$. In the latter case, $B^j \in \mathcal{S}$ by self-duality, and by the chain-vanishing property (Lemma 2.9), there is $C \in [A]_{\mathcal{S}}$ with $C \supseteq B^j$. □

COROLLARY 2.16. *For any nested, self-dual, chain-vanishing collection $\mathcal{S}$ of sets, the bases for $\mathcal{S}$ are precisely the sets of the form $[A]_{\mathcal{S}}$ for $A \in \mathcal{B}$. In particular, for any $A, B \in \mathcal{B}$, $[A]_{\mathcal{S}} = [B]_{\mathcal{S}}$, and two distinct bases are disjoint.*

PROOF. Immediate from Lemma 2.7, Proposition 2.15, and Observation 2.4. □

DEFINITION 2.17. For $A, B \in \mathcal{S}$, we write $A \sim_{\mathcal{S}} B$ if $A \in [B]_{\mathcal{S}}$.

COROLLARY 2.18. *For any nested, self-dual, chain-vanishing collection $\mathcal{S}$ of sets, $\sim_{\mathcal{S}}$ is an equivalence relation on $\mathcal{S}$.*

2.D. The graph $\mathcal{T}_{\mathcal{S}}$. Throughout this subsection, let $\mathcal{S}$ be a nested, self-dual, chain-vanishing collection of sets. We define an undirected graph $\mathcal{T}_{\mathcal{S}}$ on $\mathcal{S}/\sim_{\mathcal{S}}$ by putting an edge $\big([A]_{\mathcal{S}}, [A^c]_{\mathcal{S}}\big)$ whenever both $A, A^c \in \mathcal{S}$. Note that this is an undirected (i.e. symmetric) graph and all of the graph terminology used below is in the sense of undirected graphs.

OBSERVATION 2.19. $\mathcal{T}_{\mathcal{S}}$ has no loops or multi-edges, i.e. for any $A, B \in \mathcal{S}$,

(a) $[A^c]_{\mathcal{S}} \neq [A]_{\mathcal{S}}$;
(b) $[A]_{\mathcal{S}} = [B]_{\mathcal{S}}$ and $[A^c]_{\mathcal{S}} = [B^c]_{\mathcal{S}}$ implies $A = B$.

[1] A collection of sets is *self-dual* if it is closed under complements.

PROOF. (a) follows from the fact that $A \not\perp A^c$ and (b) from Observation 2.2. □

DEFINITION 2.20. Let $\mathcal{T}_\mathcal{S}$ be as above and let $\lambda \in \mathbb{N} \cup \{\infty\}$. We say that a (finite or infinite) sequence $(A_n)_{n<\lambda} \subseteq \mathcal{S}$ *represents* a path if $[A_0]_\mathcal{S}, [A_1]_\mathcal{S}, [A_2]_\mathcal{S}, \dots$ is a path in $\mathcal{T}_\mathcal{S}$ and $A_{n-1} \sim_\mathcal{S} A_n^c$ for each $1 \leqslant n < \lambda$.

LEMMA 2.21. *A path in $\mathcal{T}_\mathcal{S}$ represented by $(A_n)_{n<\lambda} \subseteq \mathcal{S}$, $\lambda \geqslant 3$, has no backtracking*[2] *if and only if $A_{n-1}^c \neq A_n$ for each $1 \leqslant n < \lambda$.*

PROOF. For each $2 \leqslant n < \lambda$, because $A_{n-1}^c \sim_\mathcal{S} A_{n-2}$, we have

$$[A_{n-2}]_\mathcal{S} = [A_n]_\mathcal{S} \iff A_{n-1}^c \sim_\mathcal{S} A_n.$$

But we also have $A_{n-1} \sim_\mathcal{S} A_n^c$, so (b) of Observation 2.19 gives

$$A_{n-1}^c \sim_\mathcal{S} A_n \iff A_{n-1}^c = A_n.$$ □

PROPOSITION 2.22. *Let $\mathcal{S}$ be a nested, chain-vanishing collection of sets and let $(A_n)_{n<\lambda} \subseteq \mathcal{S}$, $\lambda \in \mathbb{N} \cup \{\infty\}$, represent a path in $\mathcal{T}_\mathcal{S}$ with no backtracking. Then $(A_n)_{n<\lambda}$ is strictly increasing.*

PROOF. By Lemma 2.21, $A_{n-1} \neq A_n^c$, but we also have $A_{n-1} \sim_\mathcal{S} A_n^c$, so $A_{n-1} \perp A_n^c$, or equivalently, $A_{n-1} \subsetneq A_n$. □

COROLLARY 2.23. *$\mathcal{T}_\mathcal{S}$ is acyclic.*

PROOF. Let $n \geqslant 2$ and let $(A_i)_{i \leqslant n}$ represent a path in $\mathcal{T}_\mathcal{S}$ with no backtracking. Assuming that $[A_0]_\mathcal{S} = [A_n]_\mathcal{S}$, the sequence $(A_n, A_1, A_2, \dots, A_n)$ still represents a path (the same one). But now Proposition 2.22 implies $A_n \subsetneq A_n$, a contradiction. □

THEOREM 2.24. *For any self-dual, nested, chain-vanishing collection of sets $\mathcal{S}$, the graph $\mathcal{T}_\mathcal{S}$ is a tree.*

PROOF. By the previous proposition, we only need to show connectedness, so fix distinct $[A]_\mathcal{S}, [B]_\mathcal{S} \in \mathcal{S}/\sim_\mathcal{S}$. By nestedness, we have $A^i \perp B^j$ for some $i, j \in \{-1, 1\}$, and by the definition of $\mathcal{T}_\mathcal{S}$, it is enough to show that $[A^i]_\mathcal{S}$ and $[B^j]_\mathcal{S}$ are connected, so we may assume without loss of generality that $A \perp B$ to begin with.

Suppose towards a contradiction that there is no path connecting $[A]_\mathcal{S}$ and $[B]_\mathcal{S}$.

CLAIM. There is an infinite sequence $(A_n)_{n \in \mathbb{N}} \subseteq \mathcal{S}$ representing a path in $\mathcal{T}_\mathcal{S}$ and such that $A_0 = A$ and $A_n \perp B$ for all $n \in \mathbb{N}$.

PROOF OF CLAIM. Putting $A_0 := A$ we assume by induction that $(A_i)_{i \leqslant n}$, $n \geqslant 0$, represents a path in $\mathcal{T}_\mathcal{S}$ and $A_n \perp B$. By Lemma 2.9, there is $C \in [A_n]_\mathcal{S}$ with $C \supseteq B$. Thus, $A_{n+1} := C^c$ is disjoint from B and $(A_i)_{i \leqslant n+1}$ represents a path in $\mathcal{T}_\mathcal{S}$. Because we assume that there is no path between $[A]_\mathcal{S}$ and $[B]_\mathcal{S}$, $[A_{n+1}]_\mathcal{S} \neq [B]_\mathcal{S}$, in particular, $A_{n+1} \neq B$, hence, $A_{n+1} \perp B$. ⊠

[2] A (finite or infinite) path $(v_n)_{n<\lambda}$ in an undirected graph is said to have *backtracking* if $v_{n-2} = v_n$ for some $2 \leqslant n < \lambda$.

For each $n \in \mathbb{N}$, $A_n \perp B \perp A_{n+1}$ implies $A_n^c \supseteq B$ and $A_{n+1} \cap B = \emptyset$, so $A_n^c \neq A_{n+1}$. Therefore, by Lemma 2.21, the path $[A_0]_{\mathcal{S}}, [A_1]_{\mathcal{S}}, [A_2]_{\mathcal{S}}, \ldots$ has no backtracking, so Proposition 2.22 implies that $(A_n)_{n \in \mathbb{N}}$ is an increasing chain, contradicting the chain-vanishing property. □

We now apply this theorem to an action $\Gamma \curvearrowright^{\alpha} X$ of a group Γ. The latter naturally induces an action of Γ on $\mathscr{P}(X)$ and, for $A \in \mathscr{P}(X)$, we denote by $\mathrm{Stab}_{\alpha}(A) \leqslant \Gamma$ the (setwise) stabilizer of A, omitting the subscript when the action is clear from the context.

COROLLARY 2.25. *Let $\Gamma \curvearrowright^{\alpha} X$ be an action of a group Γ on a set X and let $\mathcal{S} \subseteq \mathscr{P}(X)$ be invariant under the action α. If $\mathcal{S}$ is nested, self-dual, and chain-vanishing, then there is a tree $\mathcal{T}_{\mathcal{S}}$ of cardinality at most $|\mathcal{S}|$, on which Γ acts such that the set of all (directed) edge-stabilizers is exactly $\{\mathrm{Stab}_{\alpha}(A) : A \in \mathcal{S}\}$. Moreover, if the action α is such that for any $A \in \mathcal{S}$, there is $\gamma \in \Gamma$ with $\gamma \cdot_{\alpha} A \neq A$ and $\gamma \cdot_{\alpha} A \cap A \neq \emptyset$, then the action $\Gamma \curvearrowright \mathcal{T}_{\mathcal{S}}$ has no fixed points.*

PROOF. Clearly, the action $\Gamma \curvearrowright \mathcal{S}$ respects complements, orthogonality, and containment, and hence also the equivalence relation $\sim_{\mathcal{S}}$. This naturally induces an action $\Gamma \curvearrowright \mathcal{S}/\sim_{\mathcal{S}}$ as well as on the set of edges of $\mathcal{T}_{\mathcal{S}}$; in other words, Γ acts on $\mathcal{T}_{\mathcal{S}}$.

For the edge-stabilizers, for $A \in \mathcal{S}$, every $g \in \mathrm{Stab}(A)$ fixes the edge $\big([A]_{\mathcal{S}}, [A^c]_{\mathcal{S}}\big)$. Conversely, if $\gamma \in \Gamma$ fixes the edge $\big([A]_{\mathcal{S}}, [A^c]_{\mathcal{S}}\big)$, i.e. $\big([A]_{\mathcal{S}}, [A^c]_{\mathcal{S}}\big) = \big([\gamma A]_{\mathcal{S}}, [(\gamma A)^c]_{\mathcal{S}}\big)$, then (b) of Observation 2.19 applied to A and γA implies that $A = \gamma A$, and hence, $\gamma \in \mathrm{Stab}(A)$.

Now assuming the hypothesis of the "moreover" part, let $[A]_{\mathcal{S}}$ be a vertex of $\mathcal{T}_{\mathcal{S}}$ and let $\gamma \in \Gamma$ be as in this hypothesis. Then $\gamma \cdot [A]_{\mathcal{S}} \neq [A]_{\mathcal{S}}$ because otherwise, $\gamma \cdot_{\alpha} A \in [A]_{\mathcal{S}}$, so either $\gamma \cdot_{\alpha} A = A$ or $\gamma \cdot_{\alpha} A \perp A$, a contradiction. □

3. Application to ends of graphs

Throughout this section, by a *graph* on a set X we mean an irreflexive symmetric subset of X^2. Let G be a connected graph on X.

NOTATION 3.1. For sets $A, B \subseteq X$, let $\delta_G(A, B)$ denote the set of edges incident (in G) to both A and B. We write $\delta_G A$ to denote $\delta_G(A, A^c)$ and call this set the *edge-boundary* (or *coboundary*) of A. Below, we omit writing the subscript G, unless it is not clear from the context.

DEFINITION 3.2. A *cut* in a connected graph G is a subset $A \subseteq X$ such that A and A^c are infinite and δA is finite.

We recall that a connected graph G is said to *have more than one end* if it admits a cut; we also call such a graph *multi-ended*. Henceforth, we assume that G has more than one end and we let $\mathcal{C}(G)$ (or just $\mathcal{C}$) denote the set of all cuts of G.

Letting $\mathrm{Aut}(G)$ denote the group of automorphisms of G, call a set of vertices or edges *invariant*, if it is invariant under (i.e. setwise fixed by) the natural action of $\mathrm{Aut}(G)$. As the action of $\mathrm{Aut}(G)$ on X naturally induces an action on $\mathscr{P}(X)$, we also call a collection $\mathcal{S} \subseteq \mathscr{P}(X)$ *invariant* if it is closed under this action of $\mathrm{Aut}(G)$.

Our goal is to find an invariant, nested, self-dual, chain-vanishing subcollection of $\mathcal{C}$ and we do this in two stages: first, we isolate an invariant, self-dual, chain-vanishing collection of cuts, and then, we restrict it further subcollection that is also nested.

3.A. Thin and neat cuts. As our first restriction, we take the collection $\hat{\mathcal{C}}$ of all *thin* cuts, which, by definition, are those cuts C that minimize $|\delta C|$, i.e.

$$|\delta C| = k_0 := \min_{A \in \mathcal{C}} |\delta A|.$$

Below we show that $\hat{\mathcal{C}}$ is chain-vanishing.

Call a set $A \subseteq X$ *G-connected* (or just *connected*) if the induced subgraph $G|_A$ is connected. Call a cut A *neat* if both A and A^c are connected.

LEMMA 3.3. *For any cut C, $G|_C$ has at most $|\delta(C)|$-many connected components; in fact, letting a and b denote the numbers of finite and infinite connected components, respectively,*

$$a < |\delta(C)| \text{ and } 1 \leqslant b \leqslant \frac{|\delta(C)| - a}{k_0}.$$

In particular, if C is thin, then it is connected. Thus, thin cuts are neat.

PROOF. The main observation is that for each connected component $A \subseteq C$ of $G|_C$, $\delta A \subseteq \delta C$, so δC is the (disjoint) union of the δA with A ranging over the connected components of $G|_C$. If A is infinite, then it is a cut, so $|\delta A| \geqslant k_0$, and hence the inequality above. Finally, the neatness of thin cuts follows from their closedness under complements. □

For $k \geqslant 1$, let $\mathcal{C}_k$ denote the collection of neat cuts, whose edge-boundary has exactly k elements. By Lemma 3.3, $\mathcal{C}_{k_0} = \hat{\mathcal{C}}$.

REMARK 3.4. It is worth emphasizing that bases for $\hat{\mathcal{C}}$ can be infinite as well as finite, see Figs. 2 and 3.

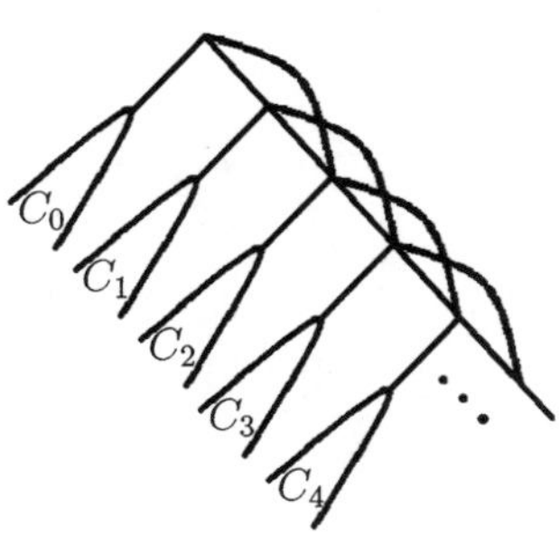

FIGURE 2. An infinite basis $\{C_0, C_1, C_2, \dots\}$ for $\hat{\mathcal{C}} = \mathcal{C}_{k_0}$ with $k_0 = 1$.

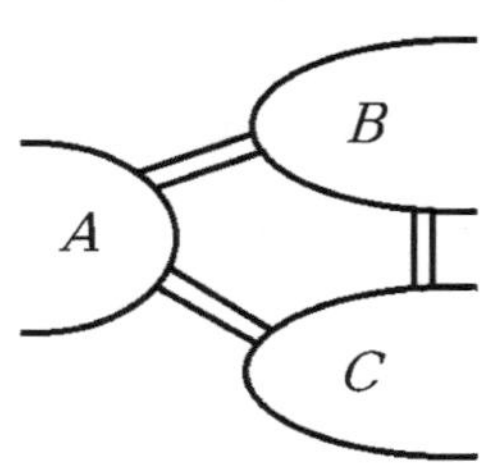

FIGURE 3. A finite basis $\{A, B, C\}$ for $\hat{\mathcal{C}} = \mathcal{C}_{k_0}$ with $k_0 = 2$.

PROPOSITION 3.5. *For any $k \geqslant 1$, $\mathcal{C}_k$ is self-dual and chain-vanishing. In particular, $\hat{\mathcal{C}}$ is self-dual and chain-vanishing.*

PROOF. Each $\mathcal{C}_k$ is self-dual by definition, so it is enough to prove that there are no proper decreasing chains. Assuming towards a contradiction that $(C_n)_{n \in \mathbb{N}} \subseteq \mathcal{C}_k$ is a strictly decreasing chain with $C_\infty := \bigcap_{n \in \mathbb{N}} C_n \neq \emptyset$, we recursively build a sequence $(e_m)_{m\in\mathbb{N}}$ of pairwise distinct edges of G such that for each $m \in \mathbb{N}$, $e_m \in \delta C_n$ for all large enough $n \in \mathbb{N}$. Granted such a sequence, we get $n \in \mathbb{N}$ with $\delta C_n \supseteq \{e_0, e_1, \dots, e_k\}$, contradicting $|\delta C_n| = k$.

Suppose by induction that a desired sequence $(e_m)_{m<\ell}$ of length $\ell \geqslant 0$ has already been constructed. Thus, there is n large enough such that $\delta C_n \supseteq \{e_m\}_{m<\ell}$. By the strictness of our chain, $C_n \setminus C_\infty \neq \emptyset$. Because C_n is connected, there is a path connecting the set $C_n \setminus C_\infty$ to C_∞ (i.e. a vertex in one to a vertex in the other). Consequently, there exists $e_\ell \in \delta(C_n \setminus C_\infty, C_\infty)$; in particular, $e_\ell \notin \delta C_n \supseteq \{e_m\}_{m<\ell}$. Let $\{u, v\} = e_\ell$ with $u \in C_n \setminus C_\infty$ and $v \in C_\infty$, and let $m > n$ be large enough such that $u \notin C_m$. Because $v \in C_m$ and $u \in C_m^c$, $e_\ell \in \delta C_m$, concluding the recursive construction. □

3.B. Minimizing the degree of non-nestedness. This subsection is almost entirely taken from [**Krö10**, Sections 2 and 3] and simply rewritten here in our terminology for the sake of keeping the paper self-contained.

Note that for any $k \geqslant 1$, $\mathcal{C}_k$ is invariant, so, by Proposition 3.5, we could take any $\mathcal{C}_k$ as our first restriction, as long as it is nonempty. Now consider the non-nestedness graph $\mathcal{N}$ on $\mathscr{P}(X)$, i.e.

$$\{A, B\} \in \mathcal{N} :\Leftrightarrow A \text{ and } B \text{ are not nested.}$$

The nested subcollections of $\mathcal{C}_k$ are exactly the $\mathcal{N}$-independent[3] ones, and our goal is to find one that is invariant. Proposition 3.11 below implies that $\mathcal{N}_k := \mathcal{N}|_{\mathcal{C}_k}$ is locally finite, and we take as a candidate the subcollection $\mathcal{C}_k' \subseteq \mathcal{C}_k$ of all vertices with minimum $\mathcal{N}_k$-degree. This is clearly invariant, but may not be $\mathcal{N}$-independent in general. However, when k is minimum (i.e. $k = k_0$, so $\mathcal{C}_k = \hat{\mathcal{C}}$), $\mathcal{C}_k'$ turns out to indeed be $\mathcal{N}$-independent. The requirement of k being minimum is used through Lemma 3.17 below and is essential for the argument.

3.B(i). *The local finiteness of $\mathcal{N}_k$.*

Definition 3.6. Call a set of edges $F \subseteq G$ an *edge-cut* if $F = \delta A$ for some cut A.

Observation 3.7. A cut A is neat if and only if δA is a minimal edge-cut, i.e. no proper subset of δA is an edge-cut.

Lemma 3.8. *For every $k \geqslant 1$, each edge $e \in G$ belongs to only finitely many minimal edge-cuts of size k.*

Proof. We prove this by induction on k. The base case $k = 1$ is obvious, so assume the statement is true for $k \geqslant 1$. If e does not belong to any minimal edge-cut of size $k + 1$, we are done, so suppose it does. Let u, v be the vertices incident to e.

Claim. There is a path P in $G - \{e\}$ connecting u to v; in particular, $G \setminus \{e\}$ is connected.

Proof of Claim. By Observation 3.7 and because $k + 1 \geqslant 2$, there is a neat cut A with $\delta A \supsetneq \{e\}$ and $u \in A$. Let $e' := \{u', v'\}$ be another edge in δA with $u' \in A$. Because A is G-connected, there is a path P_A in $G|_A$ connecting u to u'. Similarly, there is a path P_{A^c} in $G|_{A^c}$ connecting v' and v. Thus, the path $P_A{}^\frown e'{}^\frown P_{A^c}$ lies in $G \setminus \{e\}$ and connects u to v. ⊠

[3]For a graph G on a set V, a set $I \subseteq V$ is called *G-independent* if there is no G-edge between the elements of I, i.e. $I^2 \cap G = \emptyset$.

Every minimal edge-cut of size $k+1$ containing e becomes a minimal edge-cut of size k in $G-\{e\}$ and must contain at least one edge lying on the path P. But by the induction hypothesis applied to $G-\{e\}$, every edge lying on P belongs to only finitely many minimal edge-cuts of size k in $G-\{e\}$, so the fact that P is finite concludes the proof. □

NOTATION 3.9. For a set $A \subseteq X$, let $N_k^G(A)$ (or just $N_k(A)$) denote the set of all cuts in $\mathcal{C}_k$ that are not nested with A; call it the *k-neighborhood* of A.

NOTATION 3.10. For a set $A \subseteq X$, let $\partial_G A$ (or just ∂A) denote the set of vertices in A that are adjacent to vertices in A^c (equivalently, incident to edges in δA); call ∂A the *vertex-boundary* (or just *boundary*) of A.

PROPOSITION 3.11. *For every $k \geqslant 1$ and set $A \subseteq X$ with finite edge-boundary, $N_k(A)$ is finite.*

PROOF. First, we relate the sets in $N_k(A)$ to vertices in ∂A.

CLAIM. For any $B \in N_k(A)$, there are distinct vertices $v_-, v_+ \in \partial A$ such that for every path P connecting them, δB contains an edge lying on P.

PROOF OF CLAIM. For each $i \in \{-1, 1\}$, because B^i is connected and the corners $A \cap B^i$ and $A^c \cap B^i$ are nonempty, $\delta(A \cap B^i, A^c \cap B^i) \neq \emptyset$, so there is a vertex $v_i \in A \cap B^i$ that is incident to an edge in $(A \cap B^i, A^c \cap B^i)$. In particular, $v_i \in \partial(A) \cap B^i$.

Now, $v_1 \in B$ and $v_{-1} \in B^c$, so they are distinct. Moreover, any path connecting them has to intersect δB. ⊠

Thus, for each distinct pair $u, v \in \partial A$, we fix a path $P_{u,v}$ connecting them and we let F be the set of edges that lie on at least one of these paths. Because ∂A is finite and for each $u, v \in \partial A$, the number of edges lying on $P_{u,v}$ is finite, F is finite. By the claim, for each $B \in N_k(A)$, δB contains an edge from F. But δB is a minimal edge-cut of size k and, by Lemma 3.8, each edge is contained in only finitely many such edge-cuts, so the finiteness of F concludes the proof. □

NOTATION 3.12. For a set $A \subseteq X$, let ${d_k}^G(A)$ (or just $d_k(A)$) denote $|N_k(A)|$ and call it the *k-degree* of A.

3.B(ii). *Moving from $\mathcal{N}$-neighbors to their opposite corners.*

TERMINOLOGY 3.13. For sets A, B, corners $A^{i_0} \cap B^{j_0}$ and $A^{i_1} \cap B^{j_1}$ are said to be *opposite* if $i_0 i_1 = j_0 j_1 = -1$. For a corner C of (A, B), we denote its opposite corner by $-C$.

Our local goal is to show that moving from $\mathcal{N}$-neighbors to their opposite corners lowers the k-degree, provided all of the sets involved are in $\mathcal{C}_k$. This will imply that the thin cuts with minimum k_0-degree are $\mathcal{N}$-independent, i.e. nested.

LEMMA 3.14. *For sets A, B, C, if $A^c \cap B^c \neq \emptyset$ and C is not nested with $A \cap B$, then C is not nested with either A or B.*

PROOF. For some $i \in \{1, -1\}$, C^i intersects $A^c \cap B^c$. Also, C^{-i} intersects $A^c \cup B^c$. If C^{-i} intersects A^c, then C is not nested with A; otherwise, it is not nested with B. □

LEMMA 3.15. *For sets $A, B, C \subseteq X$, if there are opposite corners of (A, B) such that C is not nested with either of them, then C is not nested with either of A, B.*

PROOF. Because nestedness is immune to taking complements, we may assume that C is not nested with $A \cap B$ and $A^c \cap B^c$. In particular, the corners $C^i \cap A$ and $C^i \cap A^c$ are nonempty for each $i \in \{-1, 1\}$, so C is not nested with A. Similarly, C is not nested with B. □

LEMMA 3.16. *Let $A, B \in \mathcal{C}_k$. If $A \cap B$ and $A^c \cap B^c$ are also in $\mathcal{C}_k$, then*

$$d_k(A \cap B) + d_k(A^c \cap B^c) \leqslant d_k(A) + d_k(B),$$

and the inequality is strict if A and B are not nested; in fact,

$$d_k(A \cap B) + d_k(A^c \cap B^c) \leqslant d_k(A) + d_k(B) - 2.$$

PROOF. Lemma 3.14 implies that any cut in $\mathcal{C}_k$ that contributes exactly 1 to the left-hand side, contributes at least 1 to the right-hand side. Furthermore, by Lemma 3.15, any cut $C \in \mathcal{C}_k$ that contributes 2 to the left-hand side, also contributes 2 to right-hand side, so the first inequality follows.

If A and B are not nested, then $A \in N_k(B)$ and $B \in N_k(A)$, so they each contribute 1 to the right-hand side. On the other hand, both A and B are nested with any of their corners, so A, B contribute nothing to the left-hand side. □

Now we show that when $k = k_0$, the hypothesis of Lemma 3.16 is met, modulo taking complements.

LEMMA 3.17. *For any two thin A, B and any infinite corner C of (A, B), if the opposite corner $-C$ is also infinite, then both C and $-C$ are thin cuts.*

PROOF. By taking complements if necessary, we may assume that $C = A \cap B$ and we suppose that $-C = A^c \cap B^c$ is infinite and hence a cut. Putting

$$\begin{aligned} e_B &:= |\delta(A \cap B, A^c \cap B)| \\ e_A &:= |\delta(A \cap B, A \cap B^c)| \\ e_{B^c} &:= |\delta(A \cap B^c, A^c \cap B^c)| \\ e_{A^c} &:= |\delta(A^c \cap B, A^c \cap B^c)| \\ d_{AB} &:= |\delta(A \cap B, A^c \cap B^c)| \\ d_{AB^c} &:= |\delta(A \cap B^c, A^c \cap B)|, \end{aligned}$$

observe that

$$e_B + e_{B^c} + d_{AB} + d_{AB^c} = |\delta A| = |\delta B| = e_A + e_{A^c} + d_{AB} + d_{AB^c}. \tag{3.18}$$

Also, by the minimality of $|\delta A|$,

$$\begin{aligned} e_B + e_A + d_{AB} &= |\delta(A \cap B)| \geqslant e_B + e_{B^c} + d_{AB} + d_{AB^c} \\ e_{B^c} + e_{A^c} + d_{AB} &= |\delta(A^c \cap B^c)| \geqslant e_B + e_{B^c} + d_{AB} + d_{AB^c}, \end{aligned} \tag{3.19}$$

so $e_A \geqslant e_{B^c} + d_{AB^c}$ and $e_{A^c} \geqslant e_B + d_{AB^c}$. Plugging this into the right-hand side of (3.18) gives

$$\begin{aligned} e_B + e_{B^c} + d_{AB} + d_{AB^c} &= e_A + e_{A^c} + d_{AB} + d_{AB^c} \\ &\geqslant e_{B^c} + d_{AB^c} + e_B + d_{AB^c} + d_{AB} + d_{AB^c} \\ &= e_{B^c} + e_B + d_{AB} + 3d_{AB^c}, \end{aligned}$$

so $d_{AB^c} = 0$ and the inequality has to be equality, which, in the light of $e_A \geqslant e_{B^c}$ and $e_{A^c} \geqslant e_B$, implies $e_A = e_{B^c}$ and $e_{A^c} = e_B$. But now (3.18) and (3.19) together give $|\delta(A \cap B)| = |\delta A|$ and $|\delta(A^c \cap B^c)| = |\delta B|$, so $A \cap B$ and $A^c \cap B^c$ are also thin. □

Recalling that we denote by $\mathcal{C}'_k \subseteq \mathcal{C}_k$ the set of cuts of minimum $\mathcal{N}_k$-degree, we take $\hat{\mathcal{C}}' := \mathcal{C}'_{k_0}$, which turns out to be as desired.

PROPOSITION 3.20. *$\hat{\mathcal{C}}'$ is nested (i.e. $\mathcal{N}$-independent). It is also invariant, self-dual, and chain-vanishing.*

PROOF. The invariance self-duality are clear, and it is chain-vanishing because such is $\hat{\mathcal{C}}$ (Proposition 3.5). For nestedness, fix $A, B \in \hat{\mathcal{C}}'$. Because the sets A, A^c, B, B^c are infinite, the Pigeonhole Principle implies that there is a pair of oppositve corners C and $-C$ of (A, B) that are both infinite. By Lemma 3.17, both C and $-C$ are thin cuts, and by replacing one or both of A, B with their complements, we may assume without loss of generality that $C = A \cap B$ and $-C = A^c \cap B^c$. Thus, Lemma 3.15 applies, so if A and B were not nested, then the inequality in Lemma 3.15 would be strict, contradicting the minimality of $d_k(A)$ and $d_k(B)$. □

3.C. Stallings' theorem on ends of groups.

LEMMA 3.21. *Let $\Gamma \curvearrowright^\alpha X$ be an action of a group Γ on a set X with all point-stabilizers in the same commensurability class $\mathfrak{C}$ (e.g. $\mathfrak{C} :=$ finite subgroups of Γ). Then the set-stabilizers of finite subsets of X also belong to $\mathfrak{C}$.*

PROOF. Let $F \subseteq X$ be finite and let $H := \mathrm{Stab}_\alpha(F)$. The action of H on F gives a homomorphism $\varphi : H \to \Sigma(F)$. Then $[H : \ker(\varphi)] < \infty$, so H is in the same commensurability class as $\ker(\varphi)$. But $\ker(\varphi) = \bigcap_{x \in F} \mathrm{Stab}_\alpha(x)$, so $\ker(\varphi) \in \mathfrak{C}$ because F is finite. □

COROLLARY 3.22. *Let $\Gamma \curvearrowright^\alpha G$ be an action of a group Γ on a graph G with finite vertex-stabilizers. Then the set-stabilizers of cuts are also finite.*

PROOF. For any $\gamma \in \Gamma$ and a cut A of G, γA is also a cut and $\gamma(\partial A) = \partial(\gamma A)$, so $\mathrm{Stab}(A) \subseteq \mathrm{Stab}(\partial A)$ and Lemma 3.21 gives the conclusion. □

We are now ready to prove Theorem 1.2, which we state again here for the reader's convenience.

THEOREM 1.2. *If a group Γ admits a transitive action on a connected multi-ended graph G with all vertex-stabilizers being finite, then it also admits an action on a tree with finite edge-stabilizers and without fixed points.*

PROOF. By Proposition 3.20, we can apply Corollary 2.25 the to collection $\hat{\mathcal{C}}'$ as above and get an action on the tree $\mathcal{T}_{\hat{\mathcal{C}}'}$. The edges of this tree are exactly the cuts in $\hat{\mathcal{C}}'$, so their stabilizers are exactly the set-stabilizers of cuts and Corollary 3.22 concludes the proof. As for fixed points, let $[A]_{\hat{\mathcal{C}}'}$ be a vertex of $\mathcal{T}_{\hat{\mathcal{C}}'}$, so A is infinite but ∂A is finite. Take $x \in \partial A$ and $y \in A \setminus \partial A$ and let $\gamma \in \Gamma$ be such that $\gamma \cdot x = y$. Then $\partial(\gamma \cdot A) \neq \partial A$, so $\gamma \cdot A \neq A$, and also, $\gamma \cdot A \cap A \ni y$, so the hypothesis of the "moreover" part of Corollary 2.25 is satisfied, and thus, the action $\Gamma \curvearrowright \mathcal{T}_{\hat{\mathcal{C}}'}$ has no fixed points. □

Finally, combined with Bass–Serre theory [**Ser80**], the last theorem gives the following famous corollary, which states the nontrivial implication of Theorem 1.3.

COROLLARY 3.23 (Stallings theorem). *If a group* Γ *admits a transitive action* $\Gamma \curvearrowright^{\alpha} G$ *on a multi-ended graph* G *with all vertex-stabilizers being finite, then* Γ *splits over a finite subgroup* Δ.

PROOF. By Theorem 1.2, Γ admits an action on a tree with finite edge-stabilizers and without fixed points. By adding midpoints to the edges if necessary, we may assume that the action is without inversions in the sense of [**Ser80**, Section 3.1, first paragraph]. By the fundamental theorem of Bass–Serre theory [**Ser80**, Theorem 13], Γ is the fundamental group of a connected graph of groups $\mathcal{Y}$, where the groups are the vertex and edge stabilizers of the action α.

Now take an (undirected) edge e of $\mathcal{Y}$ and denote its stabilizer by Δ. Removal of e gives a new graph of groups $\mathcal{Z} := \mathcal{Y} - \{e\}$ and the associativity of the construction of the fundamental group gives the following: if $\mathcal{Z}$ is still connected, then $\Gamma = *_{\Delta} K$, where K is the fundamental group of $\mathcal{Z}$; otherwise, $G = K *_{\Delta} \Lambda$, where K, Λ are the fundamental groups of the two connected components of $\mathcal{Z}$. $\square$

4. A Stallings theorem for Borel equivalence relations

Throughout this section, let X be a standard Borel space (see [**Kec95**, 12.5] for the definition). We recall the necessary definitions here, referring to [**JKL02**] as a source. An equivalence relation E on X is said to be *Borel* if it is Borel as a subset of X^2. We say that E is *countable* if each E-class is countable. By a *Borel graph* G on X we mean an irreflexive and symmetric subset Borel of X^2. A *Borel graphing* of E is a Borel graph G on X whose connectedness relation E_G is exactly E. A graphing of E is called a *treeing* if it is acyclic. An equivalence relation on X is called *treeable* if it admits a Borel treeing.

All graph theoretic notions we have considered so far, such as multi-ended, cuts, basis, etc., were defined for a connected graph G. Given an arbitrary graph on X, we extend all these notions to G by restricting our attention to considering only E_G-related subsets of X, i.e. each set we consider will be contained in one G-connected component. More precisely, in the definitions of these notions, we replace

- $\mathscr{P}(X)$ with $\mathscr{P}_{E_G}(X)$—the collection of all nonempty E_G-related subsets of X,
- a complement of a set $A \in \mathscr{P}_{E_G}(X)$ with $A^* := [A]_{E_G} \setminus A$.

For example, we call G *multi-ended* if each of its connected components has more than one end. Furthermore, call $C \in \mathscr{P}_{E_G}(X)$ a *cut* of G if ∂C is finite and C, C^* are both infinite.

Next, we identify each cut C of G with

$$(\delta C, \delta C^*)_G := \left\{(x, y) \in X^2 : x \in C \text{ and } y \in C^*\right\},$$

so the set $\mathcal{C}_G$ is just a Borel subset of $[G]^{<\infty}$, hence a standard Borel space. The equivalence relation E_G naturally extends to that on $[G]^{<\infty}$ and we denote this extension by $\tilde{E}_G$. Note that if G is multi-ended, then $\mathcal{C}_G$ is a complete $\tilde{E}_G$-section.

We call a cut $C \in \mathcal{C}_G$ *thin* if it is thin for its G-connected component. We also define the non-nestedness graph $\mathcal{N}_G$ on $\mathcal{C}_G$ by putting an edge between $A, B \in \mathscr{P}_G(X)$ if A and B are in the same G-connected component and are non-nested.

Letting

- $\hat{\mathcal{C}}_G$ be the set of thin cuts of G,
- $\hat{\mathcal{C}}'_G \subseteq \hat{\mathcal{C}}_G$ be the set of those cuts C with minimum nestedness degree in the graph $\mathcal{N}$ restricted to the set of thin cuts in the G-connected component of C,

it easily follows by the Luzin–Novikov uniformization theorem that these subsets are Borel.

Furthermore, given a Borel set $\mathcal{C} \subseteq \mathcal{C}_G$, the binary relation $\sim_{\mathcal{C}}$ on $\mathcal{C}$, as in Definition 2.17, is also Borel, again by the Luzin–Novikov uniformization theorem. Recall that if $\mathcal{C}$ is nested and chain-vanishing, then Corollary 2.18 says that $\sim_{\mathcal{C}}$ is an equivalence relation on $\mathcal{C}$.

Theorem 1.4 is now easily derived from the constructions described above and is our main application. We restate it more precisely here, recalling that a set $\mathcal{C} \subseteq \mathcal{C}_G$ is called *G-complete* if it contains at least one cut from every connected component of G.

THEOREM 4.1 (Stallings for equivalence relations). *Let E be a countable Borel equivalence relation on a standard Borel space X and let G be a multi-ended Borel graphing of E. For any G-complete, nested, self-dual, chain-vanishing Borel collection $\mathcal{C} \subseteq \mathcal{C}_G$ (e.g. $\mathcal{C} := \hat{\mathcal{C}}'_G$), there is a treeable equivalence relation E_T and a Borel equivalence relation $E_{\mathcal{C}} \leqslant_B \sim_{\mathcal{C}}$ such that $E = E_T * E_{\mathcal{C}}$.*

Because free product of treeable equivalence relations is treeable, this immediately gives:

COROLLARY 4.2. *If a countable Borel equivalence relation E admits a multi-ended Borel graphing G such that $\sim_{\hat{\mathcal{C}}'_G}$ is treeable (e.g. finite, smooth, hyperfinite), then E is treeable.*

REMARK 4.3. Because the non-nestedness graph $\mathcal{N}$ on the set $\hat{\mathcal{C}}_G$ of thin cuts is locally finite by Proposition 3.11, there is a Borel maximal $\mathcal{N}$-independent set $\mathcal{C} \subseteq \hat{\mathcal{C}}_G$; indeed, by [**KST99**, Proposition 4.5] this graph admits a countable Borel coloring, using which one easily constructs a Borel maximal independent set. By maximality, $\mathcal{C}$ must be self-dual and it is chain-vanishing by Proposition 3.5, so Theorem 4.1 applies to $\mathcal{C}$.

PROOF OF *Theorem* 4.1. The Luzin–Novikov uniformization theorem, again, gives a Borel function $\pi : X \to \mathcal{C}$ that is a reduction of E to $\tilde{E}$, i.e. for each point x, $\pi(x)$ is a cut $C \in \mathcal{C}$ with $x \in [C]_E$. Let $E_{\mathcal{C}}$ be the π-pullback of the equivalence relation $\sim_{\mathcal{C}}$.

We now define the graph $T_{\mathcal{C}}$ similarly to Subsection 2.D, more precisely, for each $C \in \mathcal{C}$, we put the edge $(C, C^*) \in \mathcal{T}_{\mathcal{C}}$ as well as its inverse. The tree $\mathcal{T}_{\mathcal{C}}$ described in Subsection 2.D is actually defined on $\mathcal{C}/\sim_{\mathcal{C}}$, but it is easy to see that $T_{\mathcal{C}}$ is a lift of $\mathcal{T}_{\mathcal{C}}$ to $\mathcal{C}$. Moreover, the natural projection map $\mathcal{C} \twoheadrightarrow \mathcal{C}/\sim_{\mathcal{C}}$ induces a graph homomorphism $T_{\mathcal{C}} \to \mathcal{T}_{\mathcal{C}}$ that is injective on (the edges of) $T_{\mathcal{C}}$, so $T_{\mathcal{C}}$ is acyclic. Let $\mathcal{C}_0 := \pi(X)$.

Next, we apply the edge sliding argument as in either [**JKL02**, Proposition 3.3(i)] or [**Gab00**] and get an acyclic graph $T_{\mathcal{C}_0}$ whose projection to $\mathcal{C}_0/\sim_{\mathcal{C}}$ is a treeing of $\tilde{E}|_{\mathcal{C}_0}/\sim_{\mathcal{C}}$.

Let T be the π-pullback of $T_{\mathcal{C}_0}$, i.e. $xTy :\Leftrightarrow \pi(x)T_{\mathcal{C}_0}\pi(y)$. It follows from its definition that the projection $T_{E_{\mathcal{C}}}$ of T under quotient map $X \mapsto X/E_{\mathcal{C}}$ is a treeing of $E/E_{\mathcal{C}}$. This implies that $E = E_T * E_{\mathcal{C}}$, finishing the proof of the theorem. $\square$

Acknowledgments

I thank Institut Mittag-Leffler (Sweden) and the organizers of the program "Classification of Operator Algebras: Complexity, Rigidity, and Dynamics" in Spring of 2016 as the present research was done within this program at the institute. I am most grateful to Damien Gaboriau for encouraging this line of thought and for his feedback. I thank Martin Dunwoody and Dawid Kielak for pointing out various closely related papers and for helpful discussions. I also thank Clinton Conley, Andrew Marks, and Robin Tucker-Drob for going through my tree construction with me. Last but not least, many thanks to Vibeke Quorning for going through the paper with me and helping me discover some missing nontriviality assumptions in an earlier version of the current paper. Last but not least, I thank the anonymous referee for many useful suggestions and corrections, in particular, for pointing out the missing hypothesis of self-duality from Proposition 2.15 and onward.

References

[DD89] Warren Dicks and M. J. Dunwoody, *Groups acting on graphs*, Cambridge Studies in Advanced Mathematics, vol. 17, Cambridge University Press, Cambridge, 1989. MR1001965

[Dun79] M. J. Dunwoody, *Accessibility and groups of cohomological dimension one*, Proc. London Math. Soc. (3) **38** (1979), no. 2, 193–215, DOI 10.1112/plms/s3-38.2.193. MR531159

[Dun82] M. J. Dunwoody, *Cutting up graphs*, Combinatorica **2** (1982), no. 1, 15–23, DOI 10.1007/BF02579278. MR671142

[Dun17] Martin J. Dunwoody, *Structure trees, networks and almost invariant sets*, Groups, graphs and random walks, London Math. Soc. Lecture Note Ser., vol. 436, Cambridge Univ. Press, Cambridge, 2017, pp. 137–175. MR3644007

[DK15] M. J. Dunwoody and B. Krön, *Vertex cuts*, J. Graph Theory **80** (2015), no. 2, 136–171, DOI 10.1002/jgt.21844. MR3385727

[EP14] Anastasia Evangelidou and Panos Papasoglu, *A cactus theorem for end cuts*, Internat. J. Algebra Comput. **24** (2014), no. 1, 95–112, DOI 10.1142/S0218196714500076. MR3189668

[Gab00] Damien Gaboriau, *Coût des relations d'équivalence et des groupes* (French, with English summary), Invent. Math. **139** (2000), no. 1, 41–98, DOI 10.1007/s002229900019. MR1728876

[Ghy95] Étienne Ghys, *Topologie des feuilles génériques* (French), Ann. of Math. (2) **141** (1995), no. 2, 387–422, DOI 10.2307/2118526. MR1324140

[HK18] Sebastian Hensel and Dawid Kielak, *Nielsen realization by gluing: limit groups and free products*, Michigan Math. J. **67** (2018), no. 1, 199–223, DOI 10.1307/mmj/1519095620. MR3770860

[HM09] Greg Hjorth and Benjamin D. Miller, *Ends of graphed equivalence relations. II*, Israel J. Math. **169** (2009), 393–415, DOI 10.1007/s11856-009-0016-y. MR2460911

[JKL02] S. Jackson, A. S. Kechris, and A. Louveau, *Countable Borel equivalence relations*, J. Math. Log. **2** (2002), no. 1, 1–80, DOI 10.1142/S0219061302000138. MR1900547

[Kec95] Alexander S. Kechris, *Classical descriptive set theory*, Graduate Texts in Mathematics, vol. 156, Springer-Verlag, New York, 1995. MR1321597

[KST99] A. S. Kechris, S. Solecki, and S. Todorcevic, *Borel chromatic numbers*, Adv. Math. **141** (1999), no. 1, 1–44, DOI 10.1006/aima.1998.1771. MR1667145

[Krö10] Bernhard Krön, *Cutting up graphs revisited—a short proof of Stallings' structure theorem*, Groups Complex. Cryptol. **2** (2010), no. 2, 213–221, DOI 10.1515/GCC.2010.013. MR2747150

[Krö01] B. Krön, *Quasi-isometries between non-locally-finite graphs and structure trees*, Abh. Math. Sem. Univ. Hamburg **71** (2001), 161–180, DOI 10.1007/BF02941469. MR1872721

[Mil04] Benjamin David Miller, *Full groups, classification, and equivalence relations*, ProQuest LLC, Ann Arbor, MI, 2004. Thesis (Ph.D.)–University of California, Berkeley. MR2706636

[Mil09] Benjamin D. Miller, *Ends of graphed equivalence relations. I*, Israel J. Math. **169** (2009), 375–392, DOI 10.1007/s11856-009-0015-z. MR2460910

[Möl92a] Rögnvaldur G. Möller, *Ends of graphs*, Math. Proc. Cambridge Philos. Soc. **111** (1992), no. 2, 255–266, DOI 10.1017/S0305004100075344. MR1142744

[Möl92b] Rögnvaldur G. Möller, *Ends of graphs. II*, Math. Proc. Cambridge Philos. Soc. **111** (1992), no. 3, 455–460, DOI 10.1017/S0305004100075551. MR1151324

[Ser80] Jean-Pierre Serre, *Trees*, Springer-Verlag, Berlin-New York, 1980. Translated from the French by John Stillwell. MR607504

[Sta68] John R. Stallings, *On torsion-free groups with infinitely many ends*, Ann. of Math. (2) **88** (1968), 312–334, DOI 10.2307/1970577. MR228573

[Sta71] John Stallings, *Group theory and three-dimensional manifolds*, Yale University Press, New Haven, Conn.-London, 1971. A James K. Whittemore Lecture in Mathematics given at Yale University, 1969; Yale Mathematical Monographs, 4. MR0415622

Department of Mathematics, University of Illinois at Urbana-Champaign, Illinois 61801

Email address: anush@illinois.edu

Selected Published Titles in This Series

752 **Samuel Coskey and Grigor Sargsyan, Editors,** Trends in Set Theory, 2020

750 **A. Bourhim, J. Mashreghi, L. Oubbi, and Z. Abdelali, Editors,** Linear and Multilinear Algebra and Function Spaces, 2020

749 **Guillermo Cortiñas and Charles A. Weibel, Editors,** K-theory in Algebra, Analysis and Topology, 2020

748 **Donatella Danielli and Irina Mitrea, Editors,** Advances in Harmonic Analysis and Partial Differential Equations, 2020

747 **Paul Bruillard, Carlos Ortiz Marrero, and Julia Plavnik, Editors,** Topological Phases of Matter and Quantum Computation, 2020

746 **Erica Flapan and Helen Wong, Editors,** Topology and Geometry of Biopolymers, 2020

745 **Federico Binda, Marc Levine, Manh Toan Nguyen, and Oliver Röndigs, Editors,** Motivic Homotopy Theory and Refined Enumerative Geometry, 2020

744 **Pieter Moree, Anke Pohl, L'ubomír Snoha, and Tom Ward, Editors,** Dynamics: Topology and Numbers, 2020

743 **H. Garth Dales, Dmitry Khavinson, and Javad Mashreghi, Editors,** Complex Analysis and Spectral Theory, 2020

742 **Francisco-Jesús Castro-Jiménez, David Bradley Massey, Bernard Teissier, and Meral Tosun, Editors,** A Panorama of Singularities, 2020

741 **Houssam Abdul-Rahman, Robert Sims, and Amanda Young, Editors,** Analytic Trends in Mathematical Physics, 2020

740 **Alina Bucur and David Zureick-Brown, Editors,** Analytic Methods in Arithmetic Geometry, 2019

739 **Yaiza Canzani, Linan Chen, and Dmitry Jakobson, Editors,** Probabilistic Methods in Geometry, Topology and Spectral Theory, 2019

738 **Shrikrishna G. Dani, Surender K. Jain, Jugal K. Verma, and Meenakshi P. Wasadikar, Editors,** Contributions in Algebra and Algebraic Geometry, 2019

737 **Fernanda Botelho, Editor,** Recent Trends in Operator Theory and Applications, 2019

736 **Jane Hawkins, Rachel L. Rossetti, and Jim Wiseman, Editors,** Dynamical Systems and Random Processes, 2019

735 **Yanir A. Rubinstein and Bernard Shiffman, Editors,** Advances in Complex Geometry, 2019

734 **Peter Kuchment and Evgeny Semenov, Editors,** Differential Equations, Mathematical Physics, and Applications, 2019

733 **Peter Kuchment and Evgeny Semenov, Editors,** Functional Analysis and Geometry, 2019

732 **Samuele Anni, Jay Jorgenson, Lejla Smajlović, and Lynne Walling, Editors,** Automorphic Forms and Related Topics, 2019

731 **Robert G. Niemeyer, Erin P. J. Pearse, John A. Rock, and Tony Samuel, Editors,** Horizons of Fractal Geometry and Complex Dimensions, 2019

730 **Alberto Facchini, Lorna Gregory, Sonia L'Innocente, and Marcus Tressl, Editors,** Model Theory of Modules, Algebras and Categories, 2019

729 **Daniel G. Davis, Hans-Werner Henn, J. F. Jardine, Mark W. Johnson, and Charles Rezk, Editors,** Homotopy Theory: Tools and Applications, 2019

728 **Nicolás Andruskiewitsch and Dmitri Nikshych, Editors,** Tensor Categories and Hopf Algebras, 2019

727 **André Leroy, Christian Lomp, Sergio López-Permouth, and Frédérique Oggier, Editors,** Rings, Modules and Codes, 2019

For a complete list of titles in this series, visit the AMS Bookstore at **www.ams.org/bookstore/conmseries/**.